SECRETS OF METHAMPHETAMINE MANUFACTURE

Including Recipes for MDA, Ecstasy, and Other Psychedelic Amphetamines
Revised and Expanded

7TH EDITION

by Uncle Fester

Festering Publications
Green Bay, Wisconsin
www.unclefesterbooks.com

Secrets of Methamphetamine Manufacture
Including Recipes for MDA, Ecstasy, and Other Psychedelic Amphetamines
Revised and Expanded, Seventh Edition

Published by:
Festering Publications
826 S. Baird St.
Green Bay, WI. 54301

Drawings by Donald B. Parker and Ray Bosworth

ISBN: 0-9701485-6-9
Library of Congress Card Catalog Number 205900622

Contents

Introduction

Welcome to the Seventh Edition of *Secrets of Methamphetamine Manufacture*. Beyond any doubt, this is the best book ever written on the subject of clandestine chemistry, by anyone, anywhere, anytime, period! Your humble and gracious Uncle has been training champions for almost 20 years now, and this tour de force of clandestine ingenuity is living testament to the fact that the game hasn't passed me by.

What this work reveals, even more than my other books, is the utter *futility* of the so-called "War on Drugs." Of course, there can be no such thing as a "war" on inanimate objects — there can only be a war on *people*. Endlessly adding more common chemicals to lists to be watched by America's secret police has done nothing to stem this nation's voracious appetite for illegal drugs. Any laws against victimless crimes can be easily evaded — "criminals" are just plain *smarter* than the Drug Clowns. Even the most cursory reading of this text shows that most of my references are from common standard chemical literature — that's right, folks, "drugs" are merely *chemicals*, and knowledge of how they are produced can *never* be removed from the body of civilized knowledge. So grow up, "Drug Warriors," and get a life! Try to do something useful for the society you feed on instead of destroying our freedoms.

So what new treats do we have in the Seventh Edition? For starters, I have developed a method for defeating the new generation of polymer-filled Sudafed and ephedrine pills. The method is so laughably simple that it was inspired by watching Granny Clampett cooking lye soap by the Cement Pond. By using easily available off-the-shelf materials, I have once again rendered their Patents worthless so meth production can resume in all its glory.

Tired of scrounging around for pills a few at a time, and sick of those attempts to make them unextractable? How does brewing your own ephedrine sound, using yeast and brewing equipment and brew supplies? I thought you'd like that! How about forgetting all about those pills and cooking crank from cinnamon oil? I know you'll like that! Or maybe setting up shop using common flavoring ingredients? Or any number of most common and easily available industrial chemicals that can never be put under any sort of effective sales scrutiny?

The police state goon squads and their lowest common denominator, pandering-politician masters, have once again been exposed for what they are with this Seventh Edition of *Secrets of Methamphetamine Manufacture*. They have bitten off more than they can chew, and it's going to be my pleasure to jam it right to them! Public ridicule and a practical demonstration of their impotence are the only things which politicians and police-staters dread: This Seventh Edition will heap both upon their heads in liberal portions. If they thought they were getting their butts kicked before, well, they haven't seen anything yet! Let us pray they take their newfound humility well.

Enjoy this latest installment of the Journal for Clandestine Cookers. It will educate, entertain, and shake pillars all at the same time. You'll be quite pleased, I'm sure!

Uncle Fester

Postscript

The freedom to read whatever one chooses and to be unencumbered in the access to those books has been central to the Western concept of freedom for the past several hundred years. This cornerstone of our liberty is now under heavy assault.

You may be familiar with The Patriot Act, and the provisions it contains allowing the Federal Police access to which books you check out at the library, which websites you visit, and which books you buy from booksellers with anal retentive meticulous bookkeeping on their customers. Be assured that this is only the opening salvo upon your freedom to read.

An agenda has been devised by our central scrutinizers to sweep all books they consider to be troublesome off the marketplace of ideas. It's being conducted as another battle in the endless "War on Terror," but it is so shameful that nobody will admit they are taking part in it, nor that it is going on.

Bookselling has undergone a remarkable transformation and centralization over the past decade. Distribution of books is now largely controlled by two entities: amazon.com and Barnes & Noble. Access to information at large is dominated by a few search engines, in particular Google and Yahoo. Pressure applied to these few outlets allows effective censorship of what all of us get a chance to read. This has not gone unnoticed by those who consider you to be enemies in "the War on Terror," or at least troublesome.

I have compiled a library of correspondence with these central sources, and rest assured it will show up on a website advertised widely on their sites until they catch on, but the point is who they consider to be "enemies in the War on Terror." Yes, look in the mirror and you will see their perceived enemy. Toss out the 19 mugshots of the 9-11 hijackers, the similarly illegal alien Washington Sniper duo, The Shoe Bomber from Jamaica, dirty bomber Juan Padilla, and the Buffalo Sleeper Cell. It is you book readers who pose a threat, and you will be dealt with in spite of the fact that the previously mentioned characters couldn't read any book except maybe the Koran if it was read to them.

Since the Feds perceive you as the enemy, your reading choices will be curtailed. The only way I could properly guide you through what they have done is to invite you over to my kitchen table to wade through my stacks of printed out correspondence, or to assemble it all into a website. Just go to barnes&noble.com and note that all of my books have had their back covers, the reader reviews and the promo material stripped from their sites.

Similarly at Amazon, all my books have had their reader reviews removed, been listed as "out of print"...but you can get a 4th edition for $99, and the search functions have been rendered nonfunctional. My appeal to Jeff Bezos, the owner of Amazon went unanswered.

The computer search engines have behaved in a similar way. I guarantee that all these operations didn't suddenly get the same idea themselves, but they all cover for the perps with the same nonresponsive arrogant manner that somebody hiding behind a computer screen and following orders will do. The low level grunt doing as she is told is not to blame.

It is a shameful act to burn books, just as it is a shameful act to suppress them. I understand the shame they feel as they follow the orders received from higher up. That's why they don't come clean and just say that they are suppressing chemistry books at the "suggestion" of the central scrutinizers. To come clean might lead to adverse publicity for their operations. It looks to me like we are now in the midst of a very "dirty war," and the worst may be yet to come.

•

Chapter One
Chemicals and Equipment

The heart of the chemical laboratory is the set of glassware collectively called "the kit." It consists of several round bottom flasks, a claisen adapter, a still head with thermometer holder, a thermometer, a condenser, a vacuum adapter and a separatory funnel (sep funnel, for short). These pieces each have ground glass joints of the same size, so that the set can be put together in a variety of ways, depending on the process being done. For the production of a quarter to a third of a pound batches, $^{24}/40$ size ground glass joints are used. Also necessary are one each of the following sizes of round bottom flasks: 3000 ml, 2000 ml and 500 ml; and two each of 1000 ml and 250 ml. Two condensers are also required, both of the straight central tube variety, one about 35 cm in length, the other about 50 cm in length.

The standard taper glassware kit suitable for use in doing distillations and reactions has become very risky and difficult to obtain by any method other than theft or diversion from a friendly source. We can take this as further evidence of the creeping police state we are caught in. The aim of the system is to produce a disarmed and dumbed-down populace, a goal which is being achieved to an alarming degree. The very thought that home chemical experimentation might have some purpose other than drug manufacture has been banished from the general consciousness. In keeping with this, the suppliers of glassware keep their purchase records open for regular inspection by agents of the police state. Further, most of the suppliers will not do business on a "cash-and-carry" basis. Rather, they insist upon setting up an account, whereby they compile a dossier on their customer before doing business.

Fortunately, this is not a real obstacle to production. The chemical manufacturing industry gets along just fine without using standard taper glassware in which to cook their chemicals. They rightly view it as expensive, and very prone to breakage. Instead, they construct their reaction vessels and distillation apparatus from materials like stainless steel and Teflon. The only thing lost from the use of these materials versus glass is that you can't watch the batch cook or distill, and magnetic stirring is prevented.

The inability to stir magnetically is quite easily circumvented by use of a mechanical stirrer. These are available commercially, or may be constructed at home. The preferred construction materials are a stainless-steel shaft, and a Teflon paddle. Teflon-coated steel is also acceptable. A typical stirrer is shown in Figure 1.

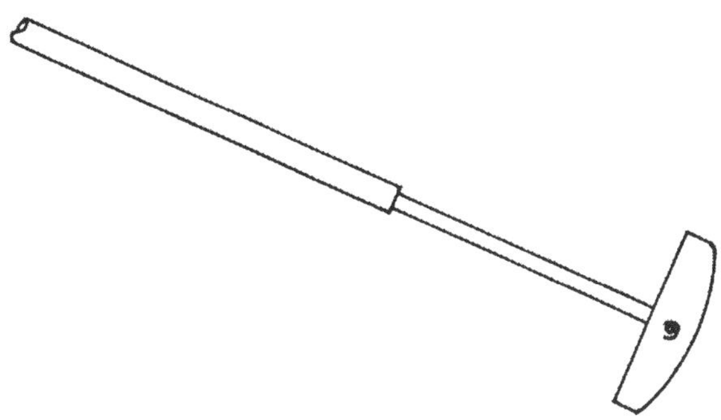

Figure 1

Construction of a stainless-steel cooking apparatus is simplified if the necks of cooking vessels and condensers are made wide enough for the shaft of the stirrer to fit down into them, and yet have enough free space for the condensation of vapors and their easy return to the cooking pot.

The inability to see what is going on inside the apparatus is more troublesome. An oil bath must be used to heat the vessel, so that by tracking the bath temperature, one can guard against overheating the contents beyond the ability of a condenser to return the reflux. During distillations, both the oil bath temperature and the thermometer reading at the stillhead must be used to keep track of what is going on inside the vessel. For those with lots of practice doing distillations, this is an easy transition. For those who are less experienced, some practice filled with trial-and-error will be required before completely getting the hang of it.

The actual construction of a stainless-steel "kit" is a fairly easy and low-profile operation, thanks to the widespread knowledge and use of the metal working arts. A good glassblower is almost impossible to find, but good welders and metal workers are a dime a dozen. Let's start with the items, which comprise the majority of parts in the kit: round bottom flasks of various sizes. Figure 2 is the typical round bottom flask. Figure 3 is its stainless-steel counterpart.

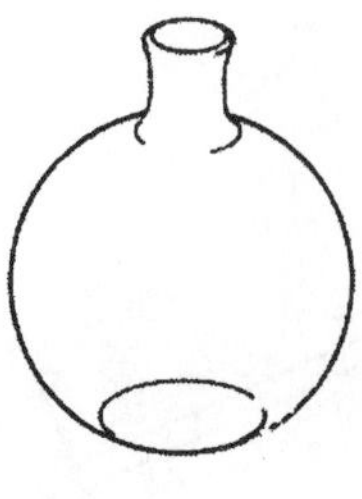

Figure 2

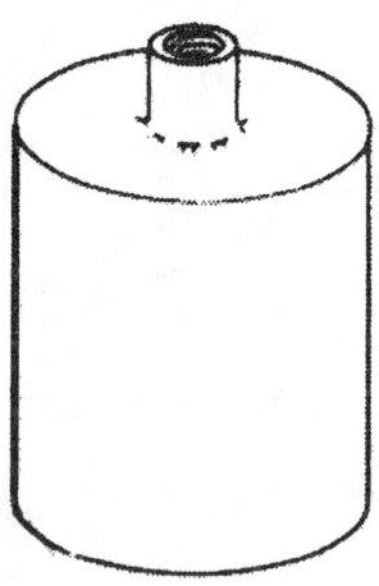

Figure 3

This is merely a stainless-steel (preferably 304 or 316 alloy) cylindrical canister with a round hole cut in the top. To that round hole is then welded a ¾-inch-long section of stainless-steel pipe with fine threads on the inner surface. A series of pots with different volumes are constructed, each with the same size stainless-steel threaded-pipe neck welded to the top.

Construction of a condenser is similarly straightforward. Note the standard glass condenser in Figure 4.

Figure 4

A stainless-steel condenser is made by taking a section of stainless-steel pipe 1½ to 2 feet long, and cutting fine threads on the outside at the "male" end and inside at the "female" end. The pipe's diameter should be chosen so that it will screw into the top of one's stainless-steel cooking vessels. Then a pipe of larger diameter is overlaid around the central section of the pipe to form the water jacket. It should cover at least ¾ of the length of the inner pipe. Welding the two together, and then drilling in a couple of nipple adapters for water flow in and out of the jacket, completes the construction.

Making a claisen adapter, as shown in Figure 5, is self-explanatory.

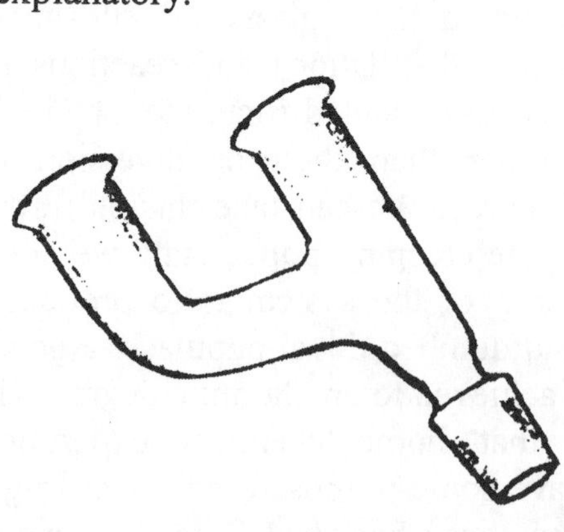

Figure 5

For distillations, two additional pieces are required:

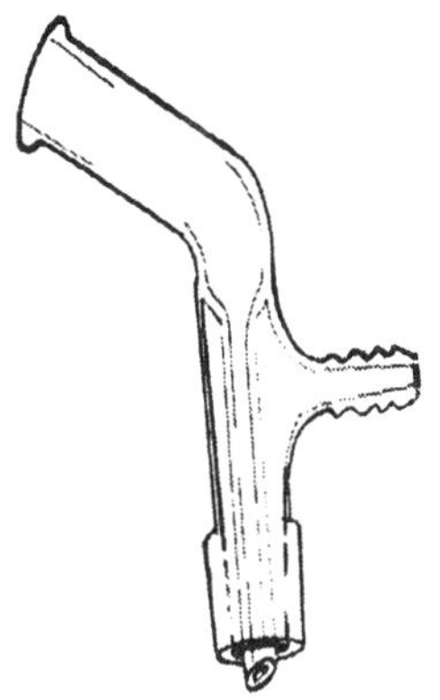

Figure 6

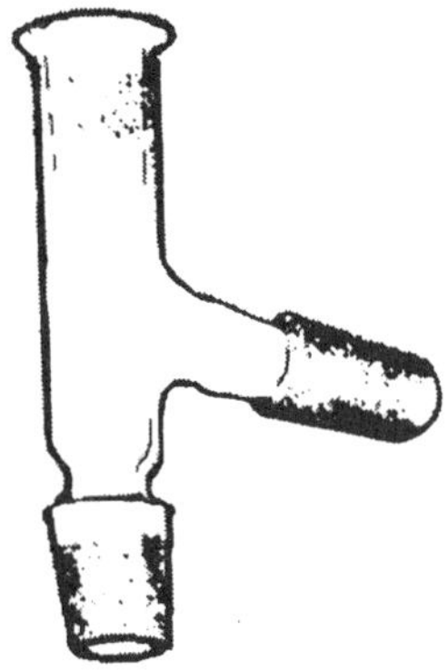

Figure 7

Note the drip tip inside the delivery end of the vacuum adapter (Figure 6). This is very important, as the drip tip prevents the product from being sucked down the vacuum line during vacuum distillation. The stillhead (Figure 7) can be used for vacuum distillations as shown by plugging the top with a one-hole stopper into which is inserted a thermometer. The thermometer should be pushed through to the usual depth for monitoring the temperature of the distilled vapors.

There are some applications where stainless steel isn't enough to stand up to the chemicals being used. For instance, strong hot acids will dissolve even 316 stainless. This means that a stainless pot and reflux column isn't suitable for use in the HI and red P reduction of ephedrine, pseudoephedrine or phenylpropanolamine to meth or dexedrine. In other reactions in this book, boiling hot hydrochloric acid is used. A stainless-steel reaction vessel and reflux condenser will be eaten away here as well.

For these conditions, another alternative should be the first choice. That alternative is to construct the reaction pot and reflux condenser from copper, and then coat the inside surfaces of these two pieces with either Teflon or a Teflon-loaded paint such as Xylan 1006. Of these two, the latter is preferable because the Teflon paint doesn't require that a primer first be applied to the metal surface. The basic procedure is to thoroughly clean the metal surface to remove all dirt and grease. Then the surface of the metal is roughened, either by sandblasting the metal, or by scouring it with a wire brush. This roughening makes footholds for the Teflon paint to grip onto. Then a thin coat of Teflon paint is applied to the inner surface, and after allowing any excess to drip back into the paint container, the coated piece is bake-cured at about 300° F. This is followed by a second coat, applied the same way, and bake-cured at 300° F. If one can follow this second bake with a third bake at around 500° F, even better results are obtained, as the Teflon in the paint will melt throughout the coating. See *Vestbusters* for complete details on how to order and apply either Teflon or a Teflon-based paint such as Xylan 1006.

Your Uncle works as a chemist in an electroplating and metal finishing shop. I have personally seen Teflon and Teflon-based paints in action. Once applied properly, nothing is going to touch these coatings. The only way to remove them is to either burn them off, or mechanically scrape them off. A regular check of the inside of the coated reaction vessels with a flashlight and dentist's mirror is all the maintenance they need.

That is all there is to making one's own organic glassware "kit." With it, another layer of snoopervision is rendered useless, and the shop can be set up with a far greater degree of the safety to be found in anonymity.

For some simple meth production methods, clandestine operators like to turn everyday household items into reaction vessels. For example, the HI and red P reduction of ephedrine to meth requires that the mixture be boiled, and the vapors which rise up off the boiling mixture be cooled, condensed back to liquid, and returned to the reaction vessel. To accomplish this, some people put the reaction mixture into a Pyrex casserole dish, then put the lid on upside down, and put ice cubes into the lid. As the ice melts, they will remove the meltwater with a turkey baster and add more ice.

For the lithium metal in liquid ammonia reduction of ephedrine to meth, a plastic pail works just fine as a reaction vessel. A countryside location where the fumes of ammonia won't be noticed is required with this crude setup, but it works nonetheless. Similarly, a pressure cooker can be used to steam the active ingredients out of the new formulation ephedrine pills.

Using household items as "cookware" when possible has the advantages of easy acquisition and super stealthiness. Friends, snoopy landlords and even police can look at a Mr. Coffee pot, and see just that rather than the boiling reaction vessel it really is.

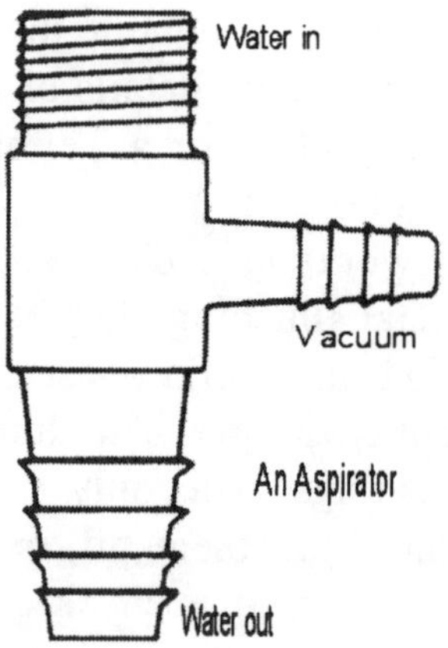

Figure 8

Another necessary piece of equipment is a source of vacuum for vacuum distillation and filtering the crystal product. Here there are two choices, each with its advantages and disadvantages.

One choice is the aspirator, also called a water pump. It works by running tap water through it under good pressure, producing a vacuum in the side arm theoretically equal to the vapor pressure of the water being run through it (see Figure 8). For this reason, the best vacuum is obtained with cold water, since it has a lower vapor pressure. The vacuum is brought from the side arm to the glassware by an automotive-type vacuum hose such as can be purchased at an auto parts store. The vacuum adapter and filtering flask each have nipples to which the other end of the hose is attached, making it possible to produce a vacuum inside the glassware. The top end of the aspirator is threaded so it can be threaded into the water source. Alternatively, the threaded head can be pushed inside a section of garden hose and secured by a pipe clamp. The hose can then be attached to a cold water faucet. The bottom end of the aspirator, where the water comes out, is rippled and can also be pushed and clamped inside a section of garden hose leading to the drain. The aspirator is kept in an upright position and at a lower level than the glassware it serves. This is because water has a habit of finding its way into the vacuum hose and running into the batch. Keeping the aspirator lower forces the water to run uphill to get into the glassware. The aspirator has the disadvantage that it requires constant water pressure flowing through it, or the vacuum inside the glassware draws water from it inside to make a mess of the batch. For this reason, only city water is used. And, unless the vacuum line is disconnected from the glassware before the water flow through the aspirator is turned off, the same thing will happen. The aspirator has these advantages: it flushes fumes from the chemicals down with the water flow, costs only about $10, and produces no sparks. A well-working aspirator produces a vacuum of 10 to 20 torr (2 to 3% of normal air pressure). (The unit "torr" means one millimeter of Mercury pressure. Normal air pressure is 760 torr.) A good aspirator is getting hard to find these days.

The other choice for a source of vacuum is an electric vacuum pump, which costs about $200, not including the electric motor, purchased separately. To avoid the danger of sparks, the motor must be properly grounded. The pump has the advantage that it can be used in the country, where steady water pressure is not available. It also produces a better vacuum than the aspirator, about 5 torr, for faster and lower temperature distillation. It has the disadvantage of exhausting the chemical fumes it pumps into the room air, unless provision is made to pump them outside. The oil inside the pump also tends to absorb the vapors of ether or toluene it is pumping, thereby ruining the vacuum it can produce and making it necessary to change the oil.

Another necessary piece of equipment is a single-burner element buffet range with infinite temperature control. It is perfect for every heating operation and only costs about $20 at a department store. Finally, a couple of ringstands with a few Fisher clamps are used to hold the glassware in position.

Since I wrote the first edition of this book 15 years ago, a whole slew of new restrictive laws have been enacted in a futile attempt to prevent clandestine cookers from practicing their craft. Restrictions on the sale and possession of glassware have already been mentioned. Many chemicals are also subject to the reporting of their sale as a result of the Chemical Diversion Act. The chemicals subject to reporting are given later in this chapter. Some communities in California are limiting sales of cold pills to a few packages at a time. Similarly, Wal-Mart is limiting sales of cold medicines to three packages at one time. All of this is just a waste of time on the part of those posing, pandering politicians. It is the purpose of this book to expose them, and hold them up to the ridicule they deserve. If there is one thing a politician can't stand, it is ridicule and a practical demonstration of their impotence.

An even more noxious, yet similarly futile law has been enacted in California. Since this is bound to be the model for similar laws enacted throughout the country, let's examine it more closely.

The most easily defeated part of the law concerns the sale of chem lab equipment and chemicals. The law states that purchasers of equipment and/or chemicals in excess of $100 must present proper ID, and that the seller must save the bill of sale for inspection by officers of the law. Since most individual pieces of chem lab equipment go for less than $100, this law is gotten around by keeping one's equipment purchases under $100, and splitting up one's business between various suppliers. The five-finger discount method while attending college chem lab courses is another option. Similarly, transfers between friends, and the old-fashioned heist from well-stocked labs are other ways around this law.

The most stringent section of the law is aimed primarily at production of meth, LSD, MDA and MDMA, PCP, and the barbiturates. Of those chemicals relevant to this book, it lists: phenylacetone, methylamine, phenylacetic acid, ephedrine, pseudoephedrine, norpseudoephedrine, phenylpropanolamine, isosafrole, safrole, piperonal, benzyl cyanide, chlorephedrine, thionyl chloride, and N-methyl derivatives of ephedrine.

This section of the law states that anyone wishing to purchase these chemicals must obtain a permit. Anyone wishing to obtain such a permit must submit two sets of his ten fingerprints to the authorities. It is interesting to note here that the over-the-counter stimulants which contain ephedrine or pseudoephedrine are exempt from these restrictions. Sudafed and those mail-order white crosses, have not been made illegal. The determined experimenter can extract the needed starting material out of the legal "stimulant" pills.

A third, and less restricted, class of chemicals deals mainly with meth and PCP. The chemicals of interest here are: sodium and potassium cyanide, bromobenzene, magnesium turnings (the last two also have PCP implications), mercuric chloride, sodium metal, palladium black, and acetic anhydride. For this class of chemicals, the law requires presentation of proper ID (i.e., state-issued photo ID) and calls for the seller to record said ID. The obvious ways around this section of the law are to do business in less nosy states, or to obtain false identification.

Clandestine operators also keep in mind that the law allows the central scrutinizers to add chemicals to the lists without warning or approval. So the new precursors mentioned in this book could go on the lists of restricted chemicals at any time.

Waste Exchanges

A really great source of chemicals which has appeared in recent years is the surplus market. This market has arisen because of increasingly stringent environmental laws which prohibit the haphazard dumping of surplus chemicals. To avoid having to pay exorbitant fees to hazardous waste disposal companies, universities and firms list their surpluses with waste exchanges who act as matchmakers to pair one firm's waste with another firm's need. This second-hand market seems to be, at the time of this writing, completely unpoliced and full of eager sellers. It looks like all one needs to do is get some fake company letterhead printed up, send it to the waste exchanges and ask to get on their mailing list. The desired chemicals listed can then be obtained, usually at no charge other than shipping.

The United States Environmental Protection Agency has released a new publication entitled, *Review of Industrial Waste Exchanges*. Copies are available, free of charge, from the RCRA Hotline. Call 800-424-9346 or TDD (hearing impaired) 800-412-7672 and request publication number EPA-53-K-94-003.

There is some information which is contained in the *Chemical Diversion and Trafficking Act (21 CFR* part 1300 onward) that is essential for those who are interested in the chemicals which are required for many of the processes outlined in this book. The government has listed many precursor and essential chemicals, and provided threshold amounts for them. Any transactions which equal or exceed these threshold amounts must, by law, be reported. This list was established 4/1/98, and will never get smaller. It can only grow larger, as the feds become aware of additional chemicals that can be used for drug syntheses. A careful reading of the *Chemical Diversion and Trafficking Act* would be a wise course of action for those who wish to acquaint themselves with the subject matter at hand.

Listed Precursor Chemicals

1. Anthranilic acid and its salts (30 kg)
2. Benzyl cyanide (1 kg)
3. Ergonovine and its salts (10 gr)
4. Ergotamine and its salts (20 gr)
5. N-Acetylanthranilic acid and its salts (40 kg)
6. Norpseudoephedrine, its salts, optical isomers, and salts of optical isomers (2.5 kg)
7. Phenylacetic acid, its esters (like ethylphenylacetate!) and its salts (1 kilo)
8. Phenylpropanolamine, its salts, optical isomers, and salts of optical isomers (2.5 kg)
9. Piperidine and its salts (500 gr)
10. Pseudoephedrine, its salts, optical isomers, and salts of optical isomers (1 kg)
11. 3, 4-Methylenedioxyphenyl-2-propanone (4 kg)
12. Methylamine and its salts (1 kg)
13. Ethylamine and its salts (1 kg)
14. D-lysergic acid, its salts, optical isomers, and salts of optical isomers (10 gr)
15. Propionic anhydride (1 gr)
16. Isosafrole (4 kg)
17. Safrole (4 kg)
18. Piperonal (4 kg)
19. N-Methylephedrine, its salts, optical isomers, and salts of optical isomers (1 kg)
20. N-Ethylephedrine, its salts, optical isomers, and salts of optical isomers (1 kg)
21. N-methylpseudoephedrine, its salts, optical isomers, and salts of optical isomers (1 kg)
22. N-ethylpseudoephedrine, its salts, optical isomers, and salts of optical isomers (1 kg)
23. Hydriotic acid (57%) (1.7 kg or 1 liter by volume)
24. Ephedrine; with no threshold, i.e., all sales are reported
25. Benzaldehyde (4 kg)
26. Nitroethane (2.5 kg)
27. GBH
28. Red Phosphorus
29. White or Yellow Phosphorus
30. Hypophosphoric Acid, and its salts (like sodium hypophosphite!)

Listed Essential Chemicals

1. Imports and Exports:
   A. Acetic anhydride (250 gal or 1,023 kg)
   B. Acetone (500 gal or 1,500 kg)
   C. Benzyl chloride (4 kg)

   D. Ethyl ether (500 gal or 1,364 kg)
   E. Potassium permanganate (500 kg)
   F. 2-Butanone (MEK) (500 gal or 1,455 kg)
   G. Toluene (500 gal or 1,591 kg)

2. Domestic sales:
   A. Acetic anhydride (250 gal or 1,023 kg)
   B. Acetone (50 gal or 150 kg)
   C. Benzyl chloride (1 kg)
   D. Ethyl ether (50 gal or 135.8 kg)
   E. Potassium permanganate (55 kg)
   F. 2-Butatone (MEK) (50 gal or 145 kg)
   G. Toluene (50 gal or 159 kg)
   H. Iodine

The most low-profile method of obtaining chemicals is to make a run to the hardware store or grocery store, and subvert the common place items found there to our needs. Solvents such as toluene, xylene, acetone or naphtha, or even Coleman camper fuel are easily obtained and work very well for making meth. Toluene is getting harder to find on hardware store shelves because it is so useful, but xylene is still universally available and it substitutes well for toluene. Similarly, Liquid Fire drain opener substitutes well for concentrated sulfuric acid, and hardware store muriatic acid is almost as good as lab grade hydrochloric acid. Lithium batteries are a good source of lithium metal. The ephedrine or pseudoephedrine so widely used for making meth is easy to get in the form of Sudafed pills or the ephedrine "bronchodilator" pills sold in gas stations.

Internet sites catering to hobbyists are another wonderful source of useful chemicals. I'm not talking about lab supply companies here, as even the smallest of them may have been bludgeoned into "cooperation". Rather, I'm referring to sites such as soap making hobby shops which can be a good source of potassium hydroxide. This cousin of lye is used to make soft soaps, and is VERY useful in breaking up the polymers which infest the present day Sudafed and ephedrine pills.

The more one can stock and operate a meth lab with ordinary materials, the more clandestine it will be. Care, however, should be taken in disposing the empty containers of these household items. Search warrants can be obtained if the heat examines the trash of a suspect and finds empty Sudafed boxes, the remains of disassembled lithium batteries, or solvent cans. Materials discreetly tossed into a dumpster are not traceable to clandestine chemists.

Chapter Two
The Leuckardt-Wallach Reaction: An Overview

A good way to produce batches of up to ½ pound in size is by the Leuckardt-Wallach reaction. Back when I was a "producer," this was the method I liked to use, so it has a strong sentimental attachment to me. It tends to be a touchy reaction that if not done correctly will instead give you a bunch of useless red tar. I still love it nonetheless, as it brings back fond memories. Other methods for converting phenylacetone to meth are covered in Chapters Eleven and Twelve.

A variation on the standard Leuckardt procedure is very popular in Europe, and is used to make benzedrine from phenylacetone reliably. Their batch sizes using this European variation can work up to over a pound. We'll talk more about this European variation later. A Russian Advance, which uses nickel in the mixture as catalyst, is also worth checking out.

The Leuckardt-Wallach reaction involves reacting a ketone with two molecules of a formamide to produce the formyl derivative of an amine, which is then hydrolyzed with hydrochloric acid to produce the desired amine. In this case, the reaction shown in the next column.

There are several reviews of this reaction in the scientific literature, the best of them by Crossley and Moore in the *Journal of Organic Chemistry* (1949).

The conditions which favor the production of high yields of fine-quality products are as follows. There should be a small amount of formic acid in the reaction mixture, because it acts as a catalyst. It should be buffered by the presence of some free methylamine, to prevent the pH of the reaction mixture from falling too low (becoming too acidic). The presence of water in the reaction mixture is to be avoided at all costs, because this really messes up the reaction. It prevents the phenylacetone from dissolving in the N-methylformamide, leading to low yields of purple-colored crystal. The recipe I give in a later chapter for making N-methylformamide makes a product which is perfect for this reaction.

It is also important that the reaction be done at the lowest temperature at which it will proceed smoothly, and that the heating be continued for as long as the reaction is still going. In this way nearly all the phenylacetone is converted to methamphetamine.

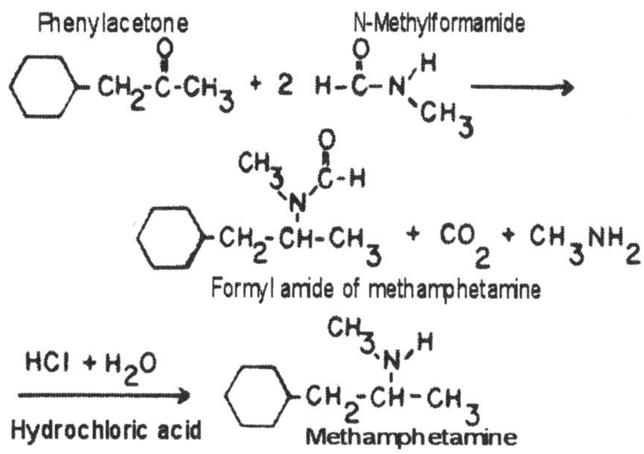

There is one stumbling block in the path of underground chemists: in 1979, the DEA made phenylacetone illegal to purchase or possess. N-methylformamide is also risky to obtain, although it is not illegal and is used in industries as a solvent.

However, they are both easy to make. And, because of these restrictions, the street price of methamphetamine has gone above $100 per gram, while it costs only $1 or $2 per gram to make.

Chapter Three
Preparation of Phenylacetone

Phenylacetone, also known as methyl benzyl ketone, or 1-phenyl-2-propanone, is easy to make if one can get the chemicals. In this reaction, phenylacetic acid reacts with acetic anhydride with pyridine catalysts to produce phenylacetone plus carbon dioxide and water. In chemical writing:

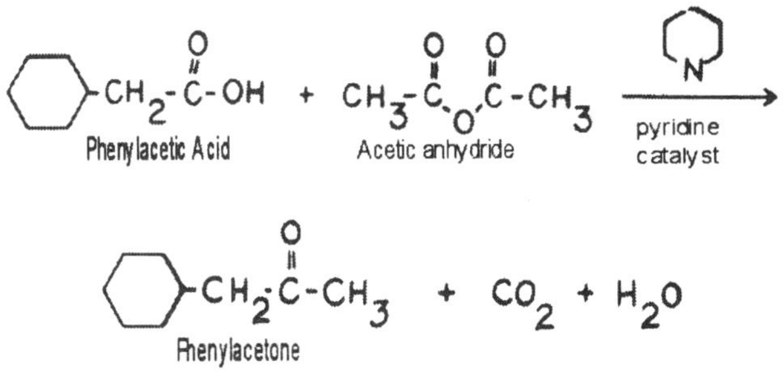

A Russian journal tells of using sodium acetate instead of pyridine. For nearly two decades now, I have let one sour experience with this reaction convince me it is useless, an example of lying commie science. I have been convinced that my judgment on this Russian variation was premature, and it really does work. We'll talk more about that method at the end of this chapter, but first let's cover the version using pyridine. I have done this reaction many times.

The reaction is done as follows: Into a clean, dry 3000 ml round bottom flask is placed 200 grams of phenylacetic acid, 740 ml of acetic anhydride and 740 ml of pyridine. This is done on a table covered with a sheet of newspaper, because phenylacetic acid, once it is exposed to the air, smells like cat urine, and the smell is next to impossible to get rid of. Pyridine also smells awful. The pyridine and acetic anhydride are measured out in a large glass measuring cup.

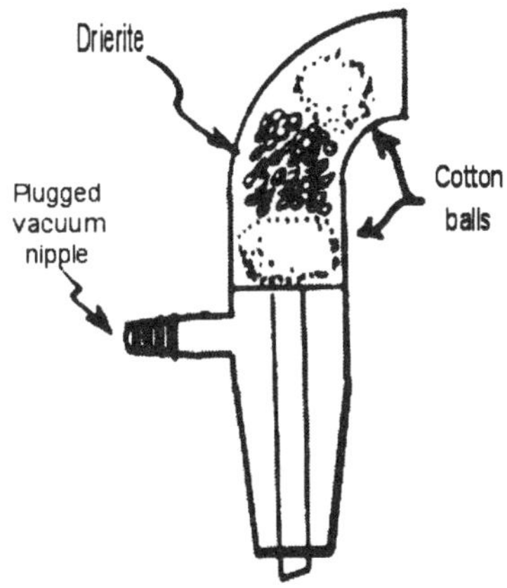

Figure 9

The flask is then gently swirled until the phenylacetic acid is dissolved. The flask is then assembled with the 50 cm condenser and the vacuum adapter, as shown in Figure 9. Before assembly, the joints are lightly greased with silicone-based stop cock grease. This prevents the pieces from getting stuck together. All pieces should be clean and dry. The vacuum nipple of the vacuum adapter is plugged with a piece of tape. In the rounded section of the vacuum adapter is a plug of cotton, then about two teaspoons of Drierite (anhydrous calcium sulfate), then another plug of cotton. This makes a bed of Drierite, which is prevented from falling into the flask by a ball of cotton. The purpose of this is to keep moisture from the air away from the reaction.

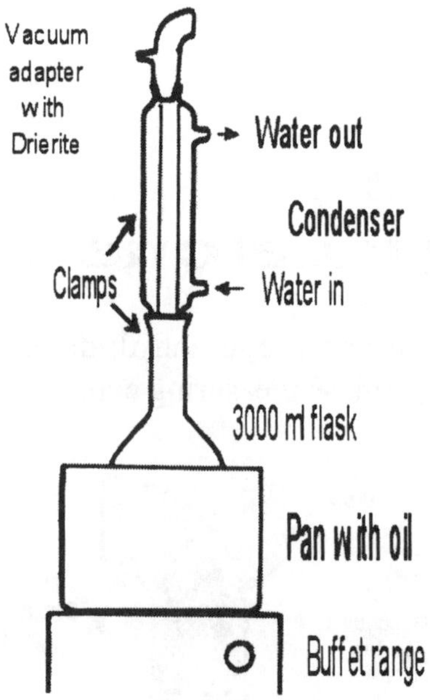

Figure 10

Now the underground chemist is ready to begin heating the flask. Notice that in Figure 10, the flask is in a large pan, which sits on the buffet range. The pan is filled about half-full of cooking oil (Wesson works fine). This is so that the flask is heated evenly. The heat is turned about half way to maximum, and the flow of cold water through the condenser is begun. A length of plastic or rubber tubing runs from the cold-water faucet to the lower water inlet of the condenser. The cold water runs through the condenser and out of the top water exit, through another length of tubing to the drain. In this way, the rising vapors from the boiling pyridine are condensed and returned to the flask. A rate of water flow of about one gallon per minute is good.

Within a half hour, the flask is hot enough to begin boiling. The heat is then turned down to stabilize the flask at a gentle rate of boiling. This is called a reflux. The boiling is allowed to continue for seven hours. During this time, the reaction mixture turns from clear to brownish-red in color. Periodically, the rate of water flow coming out of the condenser is checked, because faucet washers tend to swell after a while and slow down the rate of water flow.

After seven hours, the heat is turned off. Twenty minutes after the boiling stops, the glassware is set up as shown in Figure 11. The cotton and Drierite are removed from the vacuum adapter. Then four pea-sized pieces are broken off a pumice foot stone (purchased at the local pharmacy). These are called boiling chips, because they cause liquids to boil faster and more evenly. They are added to the flask with the reaction mixture in it. But they are not added until 20 minutes after the boiling stops; otherwise they could produce a geyser of hot chemicals.

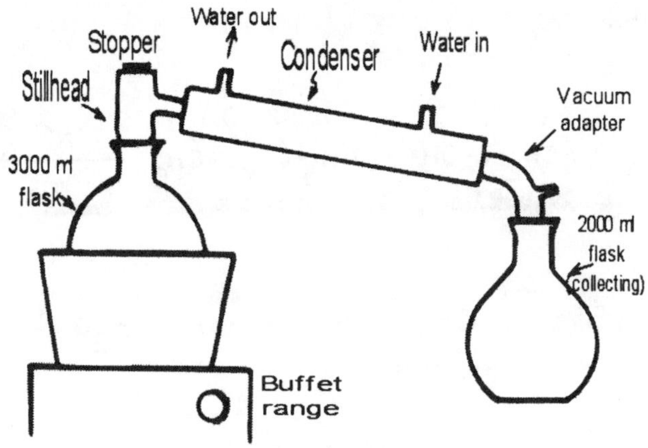

Figure 11

Now the heat is turned back on, a little hotter than when refluxing the reaction mixture. Water flow to the condenser is resumed. The mixture soon begins boiling again and the vapors condense in the condenser and flow to the collecting round bottom flask. What is being boiled off is a mixture of pyridine and acetic anhydride. The phenylacetone remains behind in the distilling round bottom flask, because its boiling point is about 100 degrees Celsius higher than the pyridine and acetic anhydride. This process is called simple distillation. Distillation continues until 1300 ml has been collected in the collecting round bottom flask, then the heat is turned off. The 1300 ml is poured into a clean dry glass jug about one gallon in size, which is then stoppered with a cork. Later in this chapter, I will describe a process by which this pyridine is recycled for future use. Since pyridine is so expensive, this cuts production costs considerably.

What is left in the distilling round bottom flask is a mixture of phenylacetone, some acetic anhydride and pyridine, and a high-molecular-weight tarry polymer, which is reddish-brown in color. The next step is to isolate and purify the phenylacetone.

The flask is taken out of the hot oil and allowed to cool down. Three-quarters of a gallon of 10% sodium hydroxide solution (NaOH) is needed. So a gallon-size glass jug is filled three-quarters full of cold water and about 10 ounces of sodium hydroxide pellets are added to it. A good quality lye, such as Red Devil or Hi-Test, is a substitute that saves a good deal of money and works fine. Eye protection is always worn when mixing this up. It is mixed thoroughly by swirling, or by stirring with a clean, wooden stick. The dissolution of NaOH in water produces a great deal of heat. It is allowed to cool off before the chemist proceeds.

About 500 ml of the 10% NaOH is put in a 1000 ml sep funnel, then the crude phenylacetone mixture from the round bottom flask is poured in the sep funnel also. The top of the sep funnel is stoppered and mixed by swirling. When the funnel gets hot, it is allowed to set for a while. Then the mixing is continued, with the underground chemist working his way up to shaking the sep funnel, with his finger holding in the stopper. What he is doing is removing and destroying the acetic anhydride. Acetic anhydride reacts with the sodium hydroxide solution to produce sodium acetate, which stays dissolved in the water, never to be seen again. Some of the pyridine and red-colored tar also goes into the water. The destruction of the acetic anhydride is what produces the heat.

After it has cooled down, about 100 ml of toluene is added to the sep funnel and shaken vigorously for about 15 seconds. The sep funnel is unstoppered and allowed to sit in an upright position for about one minute. The liquid in the funnel will now have separated into two layers. On top is a mixture of toluene, phenylacetone, and red tar. On the bottom is the water layer, which has some phenylacetone in it. Pyridine is in both layers.

Two 500 ml Erlenmeyer flasks are placed on the table, one marked "A," the other marked "B." The stop cock on the sep funnel is opened, and the water layer is drained into B. The top layer is poured into A. B is poured back into the sep funnel, and 50 ml of benzene is added. The funnel is shaken for 15 seconds, then the water layer is drained back into B. The top layer is poured into A. The purpose of this is to get the phenylacetone out of the water. Once again the water in B is put in the sep funnel. Fifty ml of toluene is added, and shaken. Xylene is for almost all purposes substitutable for toluene, and is at present easier to get at the hardware store. The water is drained into B and the toluene layer poured into A. The water in B is poured down the drain and the contents of A put into the sep funnel along with 400 ml of 10% NaOH solution from the jug. After shaking, the water layer is drained into B and the toluene layer poured into A. The contents of B are put back in the sep funnel and 50 ml of toluene added. After shaking, the chemist drains the water layer into B and pours it down the drain. The contents of A are added to the funnel again, along with 400 ml of 10% NaOH solution; the funnel is shaken again. The water layer is drained into B and the toluene layer poured into A. The contents of B are returned to the sep funnel, along with 50 ml toluene, and shaken again. The water layer is poured into B and poured down the drain. The toluene layer is poured into A. The sep funnel is washed out with hot water.

Now the last traces of pyridine are removed from the phenylacetone. For this purpose, some hydrochloric acid is needed. Hardware stores usually have the 28% strength sometimes called muriatic acid. A bottle in which the acid seems clear-colored is used; the ones with a green tint have been sitting around too long.

The contents of A are returned to the clean sep funnel. Then 10 ml of hydrochloric acid, mixed with 10 ml of water, is added to the sep funnel and shaken for 30 seconds. The stopper is pulled out to check whether or not the odor of pyridine has disappeared. If not, another 20 ml of the acid-water mix is added and shaken. The odor should now be gone, but if it is not, some more of the mix is added and shaken. Now 200 ml of water is added and shaken. Flask A is rinsed out with hot water; the water layer is drained into B and poured down the drain. The toluene layer is poured into A. What has just been done is to convert the pyridine into pyridine hydrochloride, which dissolves in water, but not in toluene. It is now down the drain.

Finally, for one last time, the contents of A are returned to the sep funnel, along with 200 ml of the 10% NaOH solution. This is shaken and the water layer drained into B. The toluene layer is allowed to stay in the sep funnel for the time being; more water will slowly fall out to the area of the stop cock, where it can be drained out. It is now ready to be distilled, and stray water must be removed beforehand.

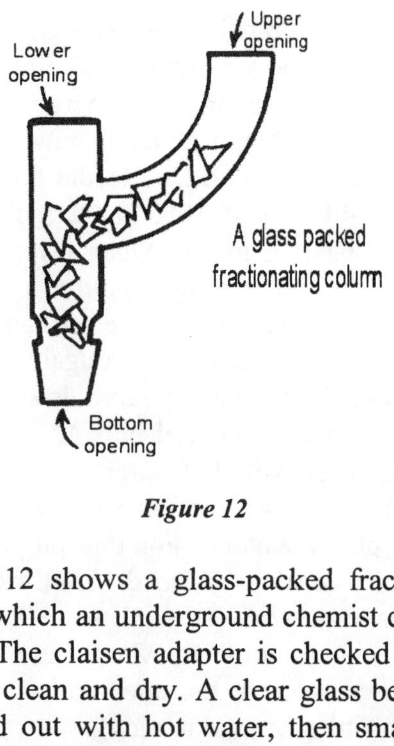

Figure 12

Figure 12 shows a glass-packed fractionating column, which an underground chemist can make himself. The claisen adapter is checked to make sure it is clean and dry. A clear glass beer bottle is washed out with hot water, then smashed on the cement floor. A few pieces are picked out that are small enough to fit in the lower opening of the claisen, yet big enough that they will not fall out of the bottom opening of the claisen adapter. Pieces of the broken bottle are dropped in the lower opening until that section of the claisen adapter is filled to about the level shown in the drawing. The chemist tries to get it to land in a jumbled pattern, as shown in the drawing. Then more similarly sized pieces of glass are dropped in the upper opening of the claisen adapter until it is filled to the level shown. Again a jumbled pattern is striven for. The lower opening is then stoppered with the proper size of glass or rubber stopper. Finally, the outside is wrapped with a layer or two of aluminum foil, except for the ground glass joint.

The underground chemist will now distill the phenylacetone. First, here is some information on the process to be performed. The crude phenylacetone the underground chemist has is a mixture of toluene, phenylacetone, red tarry polymer, some water and maybe some dibenzyl ketone. These substances all have very different boiling temperatures. By distilling this mixture through a fractionating column, the chemist can separate them very effectively and get a high-purity product. The way it works is easy to understand. The vapors from the boiling mixture in the distilling flask rise up into the fractionating column and come into contact with the pieces of glass inside. Here the vapors are separated according to boiling point. The substance in the mixture with the lowest boiling point is able to pass on through, while the other substances are condensed and flow back into the distilling flask. This is why the pieces of glass in the column can't be tightly packed, as that would interfere with the return flow, leading to a condition called flooding. Once all of the lowest-boiling substance has been distilled, the substance with the next higher boiling point can come through the fractionating column. In the distillation process to be described, the order is as follows: toluene-water azeotrope, 85° C; toluene, 110° C; phenylacetone, 120-130° C (under a vacuum of about 20 torr).

Why must the phenylacetone be distilled under a vacuum? Because its boiling point at normal pressure is 216° C, which is much too hot. Distilling it at that temperature would ruin the product. By distilling it under a vacuum, it boils at a much lower temperature. The exact temperature depends on how strong the vacuum is; the stronger the vacuum, the lower the temperature. For example, at a vacuum of 13 torr, the boiling point goes down to about 105° C.

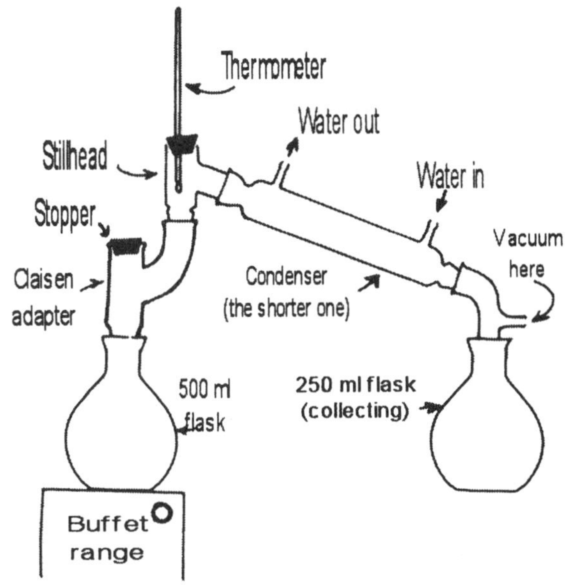

Figure 13

The glassware is set up as shown in Figure 13. The distilling flask is no more than 2/3 full. If the underground chemist has more crude phenylacetone than that, he has to wait until some of the toluene has distilled off, then turns off the heat, waits until the boiling stops and adds the rest of it to the distilling flask.

The glassware should be clean and dry. A fast way to dry glassware after washing is to put it in the oven at 400° F for 20 minutes. Rubber stoppers do not go in the oven. Water tends to stay inside round bottom flasks dried in this way. So, while they are still hot, the chemist takes a piece of glass tubing and puts it inside the flask. He sucks the moist air out of the flask with the glass tubing before it has a chance to cool down and condense. For the distillation, two 250 ml round bottom flasks are needed, one to collect the toluene in, the other to collect the phenylacetone in. Five boiling chips are put in the distilling flask.

The heat source is turned on, to the low range, about ¼ maximum. Water must be flowing through the shorter condenser at about one gallon per minute. When the mixture has begun boiling, the heat is adjusted so that about one or two drops per second drip into the collecting flask. The temperature on the thermometer should say about 68° C. For accurate temperature readings, the tip of the thermometer extends into the stillhead to the depth shown in Figure 14.

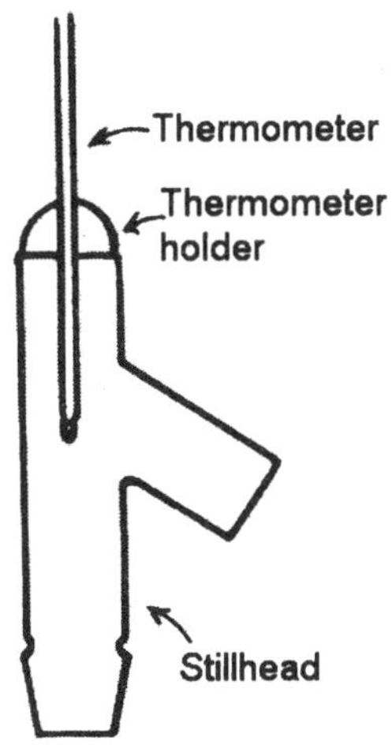

Figure 14

The material distilling at 85° C is the toluene-water azeotrope. It is about 80% toluene and 20% water. It is milky white from suspended droplets of water. Once the water is all gone, pure toluene is distilled at about 80° C. It is clear in color. If the liquid in the collecting flask is not clear or white in color, then undistilled material is being carried over from the distilling flask. This is caused either by having the distilling flask too full or by having the heat turned too high. In either case, the chemist must correct accordingly and redistill it. Once the temperature reaches 115° C on the thermometer, or the rate of toluene appearing in the collecting flask slows to a crawl, the heat is turned off because the chemist is ready to vacuum distill the phenylacetone.

There is a problem that is sometimes encountered while distilling off the toluene. Sometimes the toluene in the distilling flask will foam up in the distilling flask instead of boiling nicely. These bubbles refuse to break and they carry undistilled material along with them to the collecting flask, leaving a red liquid over there. This cannot be allowed to happen. One effective method of dealing with this is to turn on the water supply to the aspirator at a slow rate so that a weak vacuum is produced. Then the vacuum hose is attached to the vacuum adapter and a weak vacuum produced inside the glassware. This causes the bubbles to break. Every few seconds, the vacuum hose is removed, then reattached. In a while, the toluene begins to boil normally and the vacuum can be left off.

After it has cooled off, the collected distilled toluene is poured into a labeled glass bottle. It can be used again in later batches of phenylacetone. The same 250 ml round bottom flask is reattached to the collecting side, and the vacuum hose attached to the vacuum adapter. The vacuum source is turned on. If an aspirator is being used, the water is turned all the way on. All the pieces of glassware must be fitted snugly together. A strong vacuum quickly develops inside the glassware. The heat is turned on to about $^{1}/_{3}$ maximum. The boiling begins again. At first, what distills over are the last remnants of toluene and water left in the distilling flask. Then the temperature shown on the thermometer begins to climb. The phenylacetone begins to distill. When the thermometer reaches 100° C, the vacuum hose is removed and the collecting 250 ml flask is replaced with the clean, dry 250 ml flask, then the vacuum hose is reattached. If a good vacuum pump is being used, the flasks are changed at about 80° C. This flask changing is done as fast as possible to prevent the material in the distilling flask from getting too hot during the change over. If it gets too hot, it distills too rapidly when the vacuum is reapplied, resulting in some red tar being carried over along with it.

The vacuum is reapplied, and the phenylacetone is collected. With a properly working aspirator, the phenylacetone will all be collected once the temperature on the thermometer reaches 140-150° C. With a good vacuum pump, it will all come over by the time the temperature reaches 110-115° C. Once it is all collected, the heat is turned off, the vacuum hose is removed from the vacuum adapter and the vacuum source is turned off.

The yield is about 100 ml of phenylacetone. It should be clear to pale yellow in color. It has a unique but not unpleasant smell. The flask holding this product is stoppered and stored upright in a safe place. Although phenylacetone can be stored in a freezer to keep it fresh, the chemist now proceeds to making N-methylformamide, as described in the next chapter.

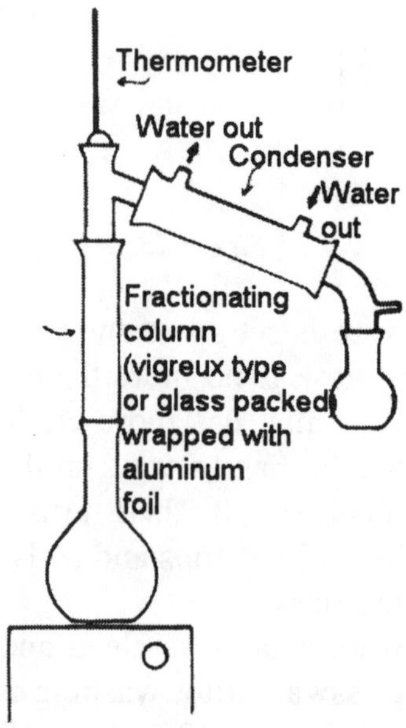

Figure 15

Once the distilling flask has cooled down, the glassware is taken apart and cleaned. The red tar left in the distilling flask and the fractionating column is rinsed out with rubbing alcohol. Then hot soapy water is used on all pieces. A long, narrow brush comes in handy for this.

One last word about vacuum distillation. To keep the vacuum strong, the vacuum hose is no more than three feet long. This forces the chemist to do the distilling close to the source of the vacuum.

Now for that pyridine recycling process I mentioned earlier in this chapter. After the underground chemist has made a few batches of phenylacetone, he will have accumulated a fair amount of pyridine-acetic anhydride mixture in the gallon-sized glass jug. He will now fractionally distill it to recover the pyridine from it. The clean dry glassware is set up as shown in Figure 15. It has a long fractionating column instead of the short type just used. This is because pyridine and acetic anhydride are harder to separate, so a longer column is needed to do the job.

The distilling flask is a 3000 ml round bottom flask with 5 boiling chips in it. The chemist pours 2000 ml of the acetic anhydride-pyridine mixture into it. The heat is turned on to about $^1/_3$ maximum and the cold water is started flowing slowly through the condenser. Within a half hour, the mixture will begin to boil. A couple of minutes later, the vapors will have worked their way through the fractionating column and begin appearing in the 2000 ml collecting flask. The heat source is adjusted so that it is collecting at the rate of one or two drops per second. Distilling is continued until 1000 ml have accumulated in the collecting flask. If the temperature reading on the thermometer goes above 135° C, the heat is turned down a little to slow the rate of distillation.

Once 1000 ml has been collected, the heat is turned off and it is allowed to cool down. After it is cool, the distilling flask is removed and its contents (mainly acetic anhydride) poured down the drain. The contents of the collecting flask (mainly pyridine) are poured into a clean, dry 2000 ml round bottom flask with 5 boiling chips, or 5 boiling chips are simply added to the 2000 ml round bottom flask that the pyridine collected in and that flask is put on the distilling side in place of the 3000 ml flask. A clean, dry 1000 ml round bottom flask is put on the collecting side. The heat is turned back on and in a while the distilling begins again. As before, the rate of distillation is adjusted to one or two drops per second. The distillation is continued until 750 ml of pyridine has been collected. Sometimes it does not keep well, but so long as it is used to make another batch of phenylacetone within a few hours after it is made, this pyridine works just as well as new pyridine.

Now let's talk about the Russian recipe, and how I messed it up when I was a neophyte cooker 20 years ago, and the right way to do it. The Russian recipe, which dates to about 1940 during the height of Stalin's wackiness when all sorts of politically motivated "science" was turned out by people fearing losing their lives if they didn't get the "politically correct" results, calls for mixing 420 grams of phenylacetic acid with 700 ml of acetic anhydride and 210 grams of sodium acetate in a 2000 ml flask. Two advantages are obvious here. The expensive reagent pyridine has been replaced with the cheap chemical sodium acetate. Also, the reaction is being done considerably more concentrated than with the pyridine recipe, i.e., more phenylacetic acid is getting poured into the 2000 ml flask, so more phenylacetone will be produced at a single cooking session.

Back in, I think it was 1979, I tried this recipe using a standard round bottom flask, which I just set upon a magnetic stirrer hot plate, and began to cook. Just setting a round bottom flask on a hot plate surface is a poor way to heat this flask. I had to turn the heat up to maximum just to get it to start to boil. Shortly thereafter, the magnetic stirrer motor burned out from all the heat from the hot plate, and the sodium acetate just sat at the bottom of the flask with the weak boiling I was making. At the end of the 18 hours of prescribed cooking, I got maybe a 20% yield of product, and was soured on this method for 20 years.

Now for the right way to do this reaction. First of all, an oil bath or heating mantle should be used to heat the reaction flask, because a weak and puny boil isn't sufficient. The reaction mixture must reach 145-150° C. Acetic anhydride boils at 139° C, but with the higher boiling phenylacetic acid mixed into the solution, and similarly high boiling point phenylacetone being produced, it's not too hard to make the solution reach this temperature if it is being boiled good and hard. With a good hard boil like that, just one condenser isn't enough. One must use a three-necked flask, and attach a condenser on two of the necks, and plug the third neck with a glass stopper, as shown in Figure 16:

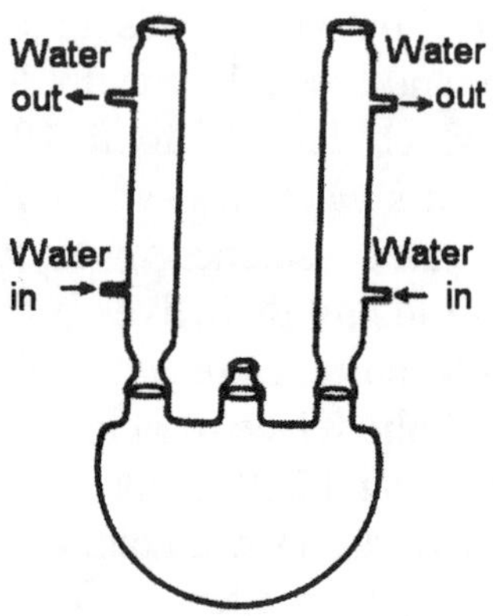

Figure 16

No stirrer is really necessary with this reaction, as the strong boil will lift sodium acetate off the bottom of the flask, and mix it with the solution. Three-necked flasks are a bitch to get these days, but stainless steel is fine for this reaction, as is a Teflon-coated metal replica. As an added bonus, using metal replicas makes it impossible to bust one for having glassware for the purpose of making meth, a 10-year felony under the Meth Act of 1996. Drying tubes should be attached to the tops of the condensers, as in the example using pyridine.

Heat the oil to about 160° C or so to get a good boiling inside the flask, and let cook for 18 hours. At the end of the cook, allow the flask to cool, then pour the reaction mixture into a gallon of cold water. Phenylacetone will float on top of the water, and the acetic anhydride and sodium acetate will stay in the water.

Using a sep funnel, separate off the phenylacetone from the water, and pour it into a convenient container. Now extract the water layer with two 200 ml portions of toluene, and add these extracts to the phenylacetone. The water can now be thrown away. The combined phenylacetone and toluene extracts should next be poured into a large sep funnel, and washed two times with 500 ml portions of 5% sodium hydroxide or lye solution in water. This will destroy any acetic anhydride left floating around.

The toluene solution containing the phenylacetone is poured into a beaker, and allowed to sit for a few hours. Some water will fall out of the solution, and sit on the bottom of the beaker.

Next, the toluene solution containing the phenylacetone is poured into a distilling flask, and a distillation is done, just like in the recipe using pyridine. First, the toluene-water azeotrope distills, drying the mixture, then pure toluene distills. The toluene can be reused. When the toluene has distilled, the mixture is allowed to cool some before commencing vacuum distillation of the phenylacetone. The Russians would have you believe that about 400 ml of phenylacetone will result from the cook. I have been told by people in a position to know that 300-350 ml is more likely. Not a bad day's work, by any means.

An alternative procedure for making phenylacetone from phenylacetic acid can be found in *The Journal of the Society of Chemical Industry*, Volume 44, pages 109-112 (1925) and in the *Journal of the Chemical Society*, Volume 59, pages 621-629 (1891). In this method, the calcium salts of phenylacetic acid and acetic acid are mixed together and then heated. The product we want, phenylacetone, distills out of the reaction mixture. The advantage to this method is that acetic anhydride and pyridine don't need to be obtained. They are replaced with the very cheap and easily available chemical calcium hydroxide. Phenylacetic acid still is required in this method, and as a List One chemical, it should never be purchased. See *Advanced Techniques of Clandestine Psychedelic & Amphetamine Manufacture* for a good phenylacetic acid recipe starting with the common industrial solvent ethylbenzene.

To do the reaction, first one must make the calcium salt of phenylacetic acid. To do this, put 500 grams of phenylacetic acid into a glass container. Add about two quarts of a 50-50 mixture of denatured alcohol and water. This is the reaction solvent. Stir and warm the mixture using a hot water batch until all the phenylacetic acid has dissolved producing a clear water-like solution. Now to this solution, add 135 grams of finely powdered calcium hydroxide. It should be added slowly with strong stirring of the solution. Heat will be produced by the reaction to make calcium phenylacetate, so take care that the mixture doesn't boil during the addition.

Once all the calcium hydroxide has been added, continue stirring for an additional hour, and allow the mixture to cool. A white-colored precipitate of calcium phenylacetate will have been formed. This product should be filtered out, rinsed with some water, then spread out to dry on wax paper or clean dishes.

The next thing one needs is a reaction vessel to produce phenylacetone from the calcium phenylacetate. Luck is on our side here. It has been found that iron and steel reaction vessels are superior to glass when doing this reaction. They don't need to be coated on the inside with any protective paint. I would avoid the use of galvanized steel because the zinc metal coating may interfere with the reaction. The reason why an iron reaction pot is superior, is because it conducts heat so well. The top of the reaction vessel gets almost as hot as the bottom. In this way, the phenylacetone formed and boiled out of the reaction mixture doesn't condense on a cold top of the pot and drip back into the reaction mixture. At that point a lot of it would be destroyed. Now go to your metal workshop and construct a reaction pot that looks pretty much like the one shown in Figure 17.

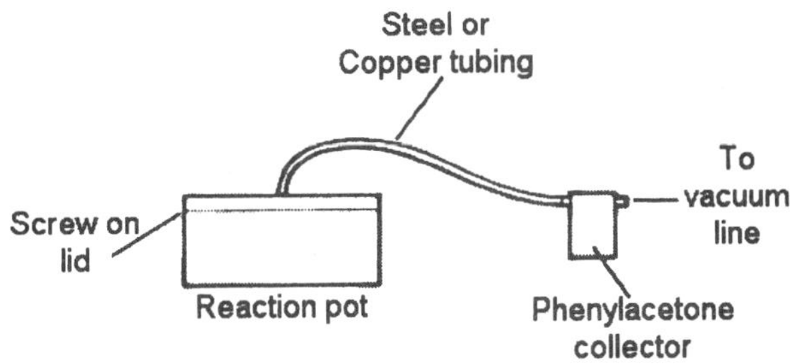

Figure 17

One starts with a steel or iron pot about as big around as a large frying pan. It should measure in height less than half the diameter of the pot. Threads should be cut around the top of the pot, and a flat iron or steel lid should be obtained with threads that match the ones on the reaction pot. Into the center of the lid, a hole should be drilled. The hole should have at least a half-an-inch diameter. Larger would be better to allow for easier escape of the phenylacetone. A section of steel or copper tubing should next be mated with that central hole and welded into place. The tubing should be bent so that it first rises a couple of inches off the top of the pot, and then begins sloping downward so that the cooled condensed phenylacetone liquid will run towards the phenylacetone collector.

The phenylacetone collector is a metal cup of about one quart capacity. Towards the top on one side, a threaded hole should be drilled in it which matches the tubing used. In this way, the collector can be screwed onto and off of the tubing. On the opposite side of the collector, a smaller hole should be drilled, and a nipple to which a vacuum line can be attached should be screwed in. One may want to put a section of water jacket around the lower portion of the tubing to assure easy condensing of the phenylacetone product. One could get around this by wrapping some cloth around the lower section of tubing, and keeping it wet with cold water during the reaction. One could also chill the collector vessel with an ice bath.

Now the reaction is ready to be run. Take the calcium phenylacetate, which was made earlier (roughly 650 grams product), and stir it up together with about 900 grams of calcium acetate. This will give about three moles of calcium acetate for each mole of calcium phenylacetate. Now take this rough mixture, and put about a cup full at a time into a blender. Grind the mixture together while shaking the blender for about a minute. When the dust has settled in the blender, pour it into the reaction pot, and repeat the process with another cup of the rough mixture. Continue until the reaction pot is no more than a little over half full. Fill no further because the mixture froths at first while it is being heated.

Now screw the lid onto the reaction pot, attach the collector vessel to the tubing, and begin heating the reaction pot. A natural-gas-fired ring heater which will blow flames part way up the sides of the reaction vessel is what was used in the scientific papers. I suppose one could also stick the reaction vessel into a pile of charcoal. One wants to heat the pot to a temperature of around 350° to 400° C. This is in the neighborhood of 700° F or so. When the pot starts heating up, attach the vacuum line to the vacuum nipple, and apply a vacuum to the system. An aspirator will give enough vacuum, and also flush the fumes from the reaction down the drain. One of the fumes is acetone. This flammable material is not a good mixture with open flames. If one uses a vacuum pump, make sure that its discharge port vents fumes away from flame through some tubing. The vacuum serves to help pull the product out of the reaction pot and to the collector vessel. It helps to eliminate the problem of product condensing and dripping back into the pot.

After about two hours of heating, the reaction will be done. The heating can be stopped, the vacuum removed, and then the collector vessel unscrewed from the tubing. It should be roughly half full of a brownish-colored liquid that is mostly phenylacetone. A decent vacuum will prevent much of any acetone from being collected. If there isn't much of any product in the collector, the heating wasn't strong enough. In this case, reheat the reaction pot with higher flames. In a well-done reaction, there will be nothing left in the reaction vessel except a solid residue of calcium carbonate colored funky with some dark tar. This can be cleaned out, and another production is run.

The crude dark-colored phenylacetone product obtained must be purified before it is used to make meth. The simplest way to do this is to vacuum fractionally distill the mixture, just as described earlier in this chapter. The alternative method for getting pure phenylacetone from the mixture is to steam distill out the phenylacetone. The procedure for doing this is described in Chapter Nine of this book.

One of the biggest hassles in doing this reaction is the need to first convert the phenylacetic acid to the calcium salt. I read a similar recipe in *Organic Syntheses* where they omitted that step. To follow their procedure here, one would mix 500 grams of phenylacetic acid with 660 ml of glacial acetic acid. When the two substances have mixed together with stirring to form a clear solution, then with stirring, slowly add 550 grams calcium hydroxide to the mixture. In their procedure, they then just added this reaction mush to the reaction pot and cooked away. Does this work as well in the case of making phenylacetone? Damned if I know. I doubt I'll be trying it out anytime soon. It's worth a try, though.

A somewhat similar procedure can be found in the *Journal of Organic Chemistry*, Volume 28, pages 880-882. In this method, phenylacetic acid, acetic acid, and iron are heated together in chemical glassware to around 300° C. Phenylacetone distills out and can be collected. Check out the reference for more details.

Chapter Four
Preparation of N-Methylformamide

N-methylformamide is best made by the reaction of methylamine with formic acid. The reaction proceeds like this:

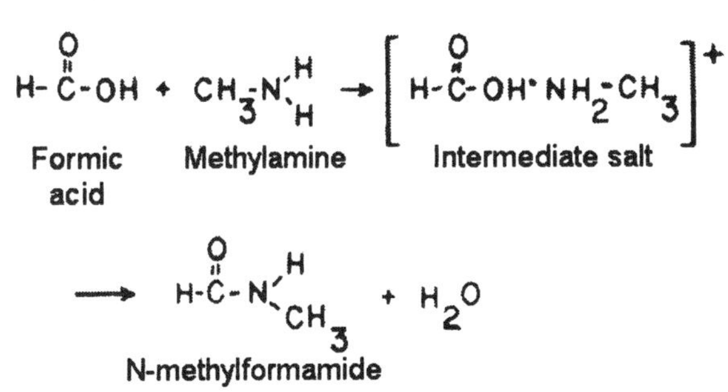

The methylamine (a base) reacts with formic acid to form the methylamine salt of formic acid. The heat that this reaction builds up then causes this intermediate salt to lose a molecule of water and form N-methylformamide. Since water is a product of this reaction, the underground chemist wants to keep water out of his starting materials as far as is possible. That is because having less water in them will shift the equilibrium of the reaction in favor of producing more N-methylformamide.

Both of the starting materials have water in them. The usual grade of formic acid is 88% pure and 12% water. It cannot be made any purer by distilling. The chemist can put up with the 12% water, but if a higher purity formic acid is available, it is worth the extra cost. The usual grade of methylamine is 40% by weight in water. The majority of this water can be removed by using the apparatus shown in Figure 18. Methylamine may also be obtained as a gas in a cylinder. In that case, the methylamine can be piped directly into the formic acid.

The glassware is set up as shown in Figure 18. The 40% methylamine is in a 1000 ml round bottom flask attached to a long condenser. In the top of this condenser is a one-hole stopper. A bent piece of glass tubing is pushed all the way through this stopper so that the end of the piece of tubing extends about one or two millimeters below the bottom of the stopper. This bent piece of tubing then extends down through the center of the other condenser into the flask containing the formic acid. It should extend below the surface of the formic acid and end about one centimeter above the bottom of the flask containing the formic acid. The idea here is simple. The 40% methylamine is heated, causing methylamine gas to be boiled out along with some water vapor. These gases then travel up the condenser, where the water is condensed out, allowing nearly pure methylamine gas to be forced by pressure through the glass tubing into the formic acid.

The bent tubing has to be bent by the chemist himself from a 3-foot-long piece of glass tubing. Its outer diameter should be about ¼ inch. The glassware is set up as shown in Figure 18 and he decides about where the tubing should be bent. If necessary, he will consult the chapter on bending glass in an *Organic Chemistry* lab manual. With a little practice, it is easy. A good source of flame to soften up the glass is a propane torch with the flame spreader attachment. After it is bent, he will blow through the tubing to make sure he did not melt it shut.

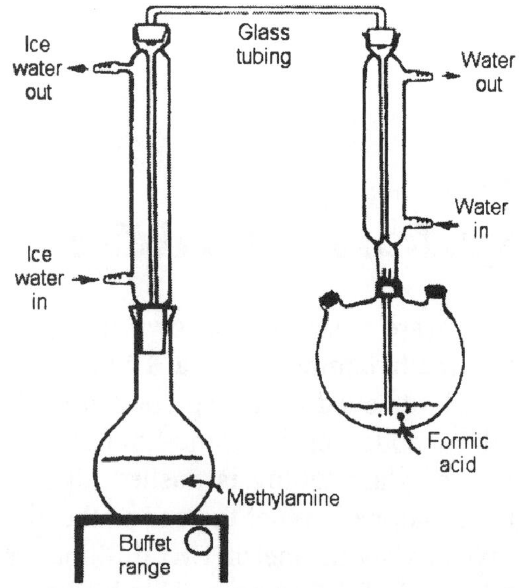

Figure 18

He is now ready to proceed. All pieces of glassware are clean and dry. Into the round bottom 1000 ml flask sitting on the heat source is placed 500 grams (about 500 ml) of 40% methylamine in water, along with 3 or 4 boiling chips. Into the other 1000 ml round bottom flask is placed 250 ml of 88% formic acid. Water flow is begun through the longer condenser. It is advantageous to use ice cold water in this condenser, because it will then do a better job of removing water vapor from the methylamine. A good way to get ice cold water for the condenser is to get a couple of 5-gallon pails. One of them is filled with ice cubes no bigger than a fist and topped off with water. Then the section of plastic tubing that runs to the lower water inlet of the condenser is placed in the pail. Its end is weighted to keep it on the bottom of the pail. This pail is placed on the table along with the glassware. The other pail is placed on the floor and the plastic tubing from the upper water exit of the condenser is run to this pail. By sucking on the end of the water exit tubing, the ice cold water can be siphoned from the pail on the table, through the condenser, to the pail on the floor. The rate of water flow can be regulated to about one gallon per minute by putting a clamp on the tubing to slow its flow. When the pail on the table is about empty, the water that has flowed to the pail on the floor is returned to the table.

The heat on the methylamine is turned on to about ¼ maximum. Soon the methylamine begins boiling out and moving through the tubing into the formic acid. The underground chemist checks for gas leaks in the system by sniffing for the smell of escaping methylamine. If such a leak is detected, the joint it is escaping from is tightened up.

The methylamine bubbling into the formic acid produces a cloud of white gas inside the flask containing the formic acid. It makes its way up to the condenser, then returns to the flask as a liquid. For this condenser, tap water flow is fine. The rate of methylamine boiling is adjusted so that the white gas does not escape out the top of the condenser. As more methylamine is boiled out, a higher heat setting is required to maintain the same rate of methylamine flow.

In this process, the formic acid gets very hot. It must get hot to produce good yields of N-methylformamide. It sometimes gets hot enough to boil a little bit (105° C), but this is no problem. As the chemist continues bubbling methylamine into the formic acid, its volume increases until it is double its starting volume, about 500 ml. At about this time, the cloud of white gas thins and then disappears. This white gas is formed by the fumes of formic acid reacting with methylamine above the surface of the liquid formic acid. It disappears because there is no longer much formic acid left. The chemist now begins checking to see if the reaction is complete. He pulls out one of the stoppers from the 3-necked flask that contains the N-methylformamide and sniffs the escaping fumes for the odor of methylamine. He does this periodically until he smells methylamine. Once he smells it, he turns off the heat on the methylamine. When the methylamine stops bubbling into the N-methylformamide, he immediately lowers the level of the 3-necked flask so that the bent glass tubing is above the surface of the N-methylformamide. This is done because, as the methylamine cools, it will contract and create a vacuum which would suck the N-methylformamide over into the other flask in a flash, ruining his work.

Both flasks are allowed to cool down. The methylamine is almost gone, so it can be poured down the drain. The next step is to fractionally distill the N-methylformamide. The glass-packed claisen adapter is used as the fractionating column. The glassware is set up as shown in Figure 13, back in Chapter Three. The distilling flask is a 1000 ml round bottom flask with 5 boiling chips in it. The collecting flask is a 250 ml round bottom flask. Unlike the distillation of phenylacetone, in this case the distillation is done under a vacuum from the beginning. The ice water siphoning system is used for the condenser, because N-methylformamide has a very high latent heat of vaporization, and, without this precaution, it may collect very hot in the collecting flask.

The underground chemist is now ready to distill the N-methylformamide. All of the crude product is put in the 1000 ml round bottom flask. It will fill it about half full. The vacuum is applied at full strength, and the heat source is turned on to 1/3 to ½ maximum. The water in the mixture begins distilling. The temperature shown on the thermometer will show a steady climb during the process.

In a while, the temperature rises high enough that the chemist can begin collecting the distilled liquid as suspected N-methylformamide. If he is using an aspirator, he begins collecting in a clean, dry 250 ml round bottom flask when the temperature reaches 95-100° C. If he is using a good vacuum pump, he begins saving the distilled material at about 85° C. As the N-methylformamide distills, the temperature rises a little bit above the temperature at which he first began collecting the N-methylformamide, then holds steady. This temperature is noted. Distilling is continued until he has collected 100 ml. Then the heat is turned off. When the boiling stops, the vacuum hose is disconnected from the glassware.

During the distillation process, a fair amount of methylamine was lost, leaving the N-methylformamide with too much formic acid. The next step is to correct this problem.

The 100 ml of N-methylformamide that has been distilled is poured back in the distilling flask with the undistilled material. The distilled material is clear, while the undistilled material has turned yellow from the heat of distilling. The glassware is set up again as shown in Figure 18. This time, the round bottom flask holding the methylamine is a 500 ml flask. It has 100 ml of fresh 40% methylamine in water in it. The bent glass tubing leads into the flask containing the N-methylformamide. This flask does not need to have a condenser on it.

The heat is turned on the methylamine and the flow of ice water through its condenser is begun. Soon the methylamine gas is bubbling into the N-methylformamide, reacting with the excess formic acid in it. Within about 10 seconds, the odor of methylamine can be detected above the N-methylformamide. The heat is turned off, and when the bubbling stops, the level of the N-methylformamide is lowered so that it is not sucked into the other flask. Once the methylamine has cooled off, it can be poured back in with the good methylamine, because it is not exhausted. Once a bottle of methylamine has been opened, it should be reclosed tightly and the cap sealed with vinyl electrical tape in order to hold in the methylamine gas.

Now the N-methylformamide is to be distilled again. The glassware is set up again for fractional distillation as shown in Figure 13. The distilling flask is a 500 ml round bottom flask, while the collecting flask is 250 ml. All pieces are clean and dry.

The N-methylformamide is placed inside the distilling flask with 5 boiling chips. (Fresh chips are used every time.) The vacuum is reapplied and the heat is turned on again to 1/3 to ½ maximum. A little bit of water is again distilled. The temperature shown on the thermometer climbs as before. When it reaches a temperature 7° C below the temperature at which it leveled off the first time around, the chemist begins collecting in a clean dry 250 ml flask. The distilling continues until it has almost all distilled over. About 10 or 15 ml is left in the distilling flask. If he is using an aspirator, the chemist makes sure that no water is backing into the product from the vacuum line. The yield is about 250 ml N-methylformamide. If he gets a little more, it won't all fit in the 250 ml collecting flask. If that happens, he pours what has collected into a clean dry Erlenmeyer flask and continues distilling. N-methylformamide is a clear liquid with no odor.

The N-methylformamide the underground chemist has just made is perfect for the Leuckardt-Wallach reaction. Because he began collecting it 7° below the leveling off temperature, it contains a mixture of N-methylformamide, formic acid and methylamine. To get good results, he uses it within a few hours after distilling it.

References

Journal of the American Chemical Society, Volume 53, page 1879 (1931).

Chapter Five
Making Methamphetamine

I explained the general theory behind this reaction in Chapter Two. Now, after doing the reactions described in the previous two chapters, the underground chemist has phenylacetone and N-methylformamide suitable for making methamphetamine. He will want to get going before the chemicals get stale.

The first thing he does is test the chemicals. He puts 5 ml of phenylacetone and 10 ml of N-methylformamide in a clean dry test tube or similar small glass container. Within a few seconds they should mix together entirely. At this point, he may offer a prayer to the chemical god, praising his limitless chemical power and asking that some of this power be allowed to flow through him, the god's High Priest. He may also ask to be delivered from the red tar that can be the result of this reaction. If they do not mix, there is water in the N-methylformamide. In this case, he must distill it again, being more careful this time.

Having tested the chemicals, he is ready to proceed with the batch. (However, if the underground chemist was reckless enough to obtain N-methylformamide ready made, he will have to distill it under a vacuum before it can be used in this reaction.) The phenylacetone he made (about 100 ml) is mixed with the N-methylformamide. The best amount of N-methylformamide to use is about 250 ml, but any amount from 200 to 300 ml will work fine. With 200 ml of N-methylformamide, there are about four molecules of N-methylformamide to one of phenylacetone. This is the bare minimum. With 300 ml, the ratio is nearly six to one. Any more than this is a waste of N-methylformamide. The best flask for mixing them is a 500 ml round bottom flask. After they are mixed, this flask is set up as shown in Figure 19. The flask is sitting in an oil bath, to supply even heating to the flask. The oil (once again, Wesson is a good choice) should extend about $^{2}/_{3}$ of the way up the side of the flask. A metal bowl makes a good container for this oil bath. This is better than a pan, because it will be important to see into the flask. The fact that the oil will expand when heated is kept in mind when filling the bowl with oil. A thermometer is also needed in the oil bath to follow its temperature.

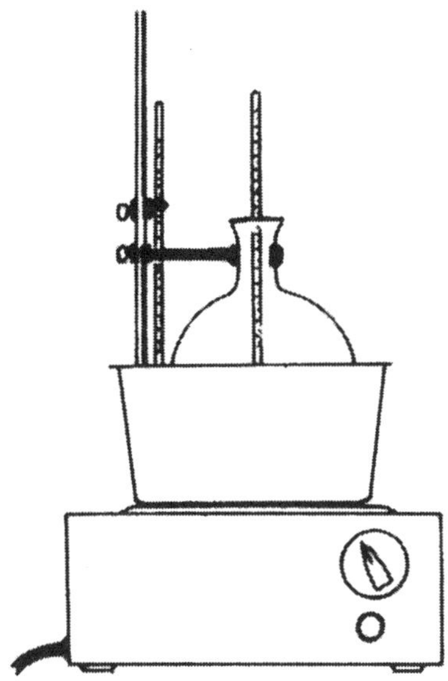

Figure 19

The test material is added to the flask. The heat source to the flask is turned on. A low heat setting is used so that the rise in temperature can be closely controlled. The thermometer used in the distillations is placed (clean and dry) inside the flask.

The rise in temperature of both the oil bath and the flask is monitored. The contents of the flask are stirred regularly with the thermometer. The temperature of the oil bath is brought to 100° C over the course of about 45 minutes. Once it reaches this level, the heat is turned back down a little bit to stabilize it in this area. The chemist must closely control every degree of temperature increase from here on. The temperature of the contents of the flask is worked up to 105° C. The contents of the flask are stirred every 15 minutes. At about 105° C, the reaction kicks in, although sometimes the heat must go as high as 110° C before it starts. When the reaction starts, the contents of the flask begin to bubble, sort of like beer, except that a head does not develop. A trick to get this reaction going at a nice low temperature is to gently scrape the thermometer along the bottom of the flask. Although I have never had the sophisticated equipment to prove it, it is a pet theory of mine that this is because ultrasonic waves are generated, producing a condition of resonance with the reactants that causes the reaction to start.

The chemist wants to keep the temperature down at the same level at which the reaction first kicked in for as long as the reaction will continue at that level. Generally, it can go for a couple of hours at this level before the reaction dies down and an increase in temperature is necessary. The reaction mixture has the same color as beer and gently bubbles. The bubbles rise up from the bottom of the flask, come to the surface, and then head for where the thermometer breaks the surface. Here they collect to form bubbles about 1 centimeter in size before they break. This may look like boiling, but it is not. Everything inside the flask has a much higher boiling point than the temperatures being used. These are actually bubbles of carbon dioxide gas being formed as byproducts of the reaction. The chemist can tell how well the reaction is going by the amount of bubbling going on.

When the rate of bubbling slows down to almost stopping, it is time to raise the temperature. It should only be raised about 3° C. This requires turning up the heat only slightly. The highest yield of product is obtained when the lowest possible temperature is used. For the duration of the reaction, the contents of the flask are stirred with the thermometer every half hour.

And so the reaction is continued. As the reaction dies down at one temperature setting, the temperature is raised a few degrees to get it going again. It will be able to stay in the 120° to 130° C range for a long time. The reaction has a lot of staying power in this range. Finally, after 24 to 36 hours, 140° or 145° C is reached. The reaction stops. The chemist takes his time working up to this temperature because the amount and quality of the product depends on it.

Once 140° to 145° C is reached and the reaction stops, the heat is turned off and the contents allowed to cool down. It should still look like beer. A reddish tint means that his prayer failed and he was not delivered from the tar. Even so, there is still lots of good product in it.

While it is cooling down, the underground chemist gets ready for the next step in the process. He is going to recover the unused methylamine for use in the next batch. This cuts his consumption of methylamine to about half of what it would be without this technique. What he is going to do is react the unused N-methylformamide with a strong solution of sodium hydroxide. The N-methylformamide is hydrolyzed to form methylamine gas and the sodium salt of formic acid (sodium formate). In chemical writing, this reaction is as follows:

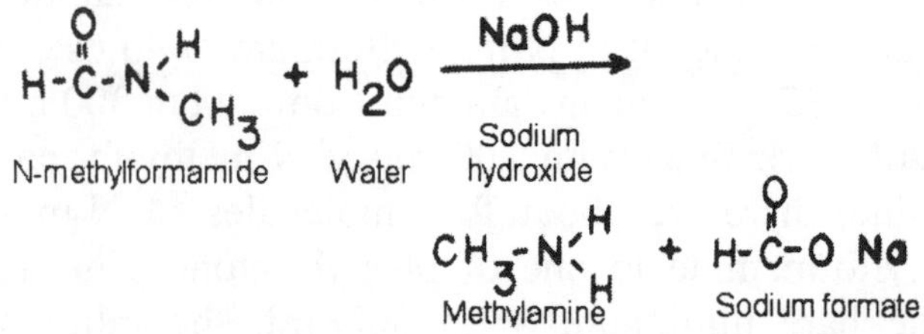

The methylamine gas produced is piped into formic acid to make N-methylformamide for use in the next batch.

First, 6 ounces (about 180 grams) of sodium hydroxide pellets are added to 450 ml of water. A good quality lye is an acceptable substitute. Eye protection is worn. Once the solution has cooled down, it is poured in a 2000 ml round bottom flask with 5 boiling chips. Then all of the methamphetamine reaction mixture is poured into the flask along with it. It is swirled around a little bit to try to get some of the N-methylformamide dissolved into the water. This does not accomplish much, however, as the reaction mixture floats on the sodium hydroxide solution. The glassware is set up as shown in Figure 18 in Chapter Four. The 2000 ml flask containing the NaOH solution and the methamphetamine reaction mixture sits on the heat source. The bent piece of glass tubing once again leads to a 1000 ml round bottom flask equipped with a condenser. The 1000 ml flask once again contains 250 ml of 88% formic acid.

The heat source is turned on to about $^{1}/_{3}$ maximum. The flow of ice water through the long condenser is begun. In a while, the boiling chips float up to the interface of the sodium hydroxide solution and the reaction mixture, and some bubbling and frothing of the reaction mixture begins. The heat is turned down some, since the temperature of the mixture should rise slowly from now on. That is because the hydrolysis reaction forming methylamine tends to kick in all at once, if this precaution is not taken, leaving the chemist in a dangerous situation with a runaway reaction.

After the first rush of the reaction has subsided and the bubbling of the methylamine into the formic acid has slowed down, the heat applied to the 2000 ml flask is increased to maintain a good rate of methylamine flow to the formic acid. Eventually, all the methylamine will be boiled out. This will be when methylamine no longer flows evenly into the formic acid. The flask must not be heated so strongly that water is forced through the bent glass tubing.

The heat is turned off and the level of the flask containing formic acid is lowered so that the acid is not sucked back into the other flask. This formic acid is about half reacted with methylamine. When it has cooled down, it is poured in a tall glass bottle and kept in the freezer until the next batch is made, when it is used for the production of N-methylformamide. Since it is already half reacted, the amount of methylamine used is reduced accordingly.

Meanwhile, back in the 2000 ml flask, the methamphetamine reaction mixture is about 100 ml in volume and has a red color. It floats above the sodium hydroxide solution. Once it has cooled down, the contents of this flask are poured into a 1000 ml sep funnel. The sodium hydroxide solution is drained out and thrown away. The red methamphetamine formyl amide is poured into a 500 ml round bottom flask with 3 boiling chips. Two hundred ml of hydrochloric acid is measured out. (The 28% hardware store variety is fine for this purpose.) It is poured into the sep funnel and swirled around to dissolve any product left behind in the sep funnel. Then it is poured into the 500 ml flask with the product. When swirled around, they mix easily.

The glassware is set up as shown in Figure 10 in Chapter Three. Tap water flow is proper for use in the condenser. The heat is turned on to the 500 ml flask, and a gentle rate of boiling is maintained for 2 hours. The mixture quickly turns black. The reaction going on here is methamphetamine formyl amide reacting with hydrochloric acid to produce methamphetamine hydrochloride and formic acid. This is a hydrolysis reaction.

After the two hours have passed, the heat to the flask is turned off. While the flask is cooling down, 80 grams of sodium hydroxide and 250 ml of water are mixed in a 1000 ml round bottom flask. Once again, a good quality lye is acceptable. If the 35% laboratory grade of hydrochloric acid was used in the last step, then 100 grams of sodium hydroxide is mixed with 300 ml of water.

When both flasks have cooled down, the black reaction mixture is cautiously added to the sodium hydroxide solution. It is added in small portions, then swirled around to mix it. They react together quite violently. The reaction here is sodium hydroxide reacting with hydrochloric acid to produce table salt, with formic acid to produce sodium formate, and with methamphetamine hydrochloride to produce methamphetamine free base. When the sodium hydroxide solution gets very hot, the chemist stops adding the reaction mixture to it until it cools down again.

After all the black reaction mixture has been added to the sodium hydroxide solution, there is a brown liquid layer floating above the sodium hydroxide solution. This brown layer is methamphetamine free base. It also has a good deal of unreacted methamphetamine hydrochloride dissolved in it. This latter has to be neutralized because it will not distill in its present form. The 1000 ml flask is stoppered and shaken vigorously for 5 minutes. This gets the methamphetamine hydrochloride into contact with the sodium hydroxide so it can react.

The bottom of the flask is full of salt crystals that cannot dissolve in the water because the water is already holding all the salt it can. The chemist adds 100 ml of water to the flask and swirls it around for a few minutes. If that does not dissolve it all, he adds another 100 ml of water.

After the flask has cooled down, it is poured into a 1000 ml sep funnel, and 100 ml of toluene is added. The sep funnel is stoppered and shaken for 15 seconds. It is allowed to stand for a couple of minutes, then the lower water layer is drained into a glass container. The brown methamphetamine-toluene layer is poured into a clean, dry 500 ml round bottom flask. The water layer is extracted once more with 100 ml toluene, then thrown away. The toluene layer is poured into the 500 ml flask along with the rest of the methamphetamine.

The chemist is now ready to distill the methamphetamine. He adds three boiling chips to the 500 ml round bottom flask and sets up the glassware for fractional distillation as shown in Figure 13. The 500 ml flask sits directly on the heat source. The glass-packed claisen adapter is the proper fractionating column. The collecting flask is a 250 ml round bottom flask. Tap water is used in the condenser.

The heat source is turned on to ¼ to $^{1}/_{3}$ maximum. Soon the mixture begins boiling. The first thing that distills is toluene-water azeotrope at 85° C. Then pure toluene comes over at 110° C. Once again, as in the distillation of phenylacetone, foaming can sometimes be a problem. In that case, it is dealt with in the same way as described in Chapter Three.

When the temperature reaches 115° C, or the rate of toluene collecting slows to a crawl, the heat is turned off and the flask allowed to cool down. The collected toluene is poured into a bottle. It can be used again the next time this process is done. The same 250 ml flask is put on the collecting side.

The distilling flask is now cool, so vacuum is applied to the glassware at full strength. The last remnants of toluene begin to boil, and the heat is turned back on to $^{1}/_{3}$ maximum. The temperature begins to climb. If an aspirator is being used, when the temperature reaches 80° C, the chemist quickly removes the vacuum hose and replaces the 250 ml flask with a clean dry one. If he is using a good vacuum pump, he makes this change at about 70° C. The flask change is done quickly to avoid overheating in the distilling flask.

The methamphetamine distills over. With an aspirator, the chemist collects from 80° C to about 140° or 150° C, depending on how strong the vacuum is. With a vacuum pump, he collects to about 120° or 130° C. Once it has distilled, the heat is turned off and the vacuum hose disconnected.

The product is about 90 ml of clear to pale yellow methamphetamine. If the chemist is feeling tired now, he may take out a drop on a glass rod and lick it off. It tastes truly awful and has a distinctive odor, somewhat biting to the nostrils.

He is now ready to make his liquid methamphetamine free base into crystalline methamphetamine hydrochloride. Half of the product is put into each of two clean, dry 500 ml Erlenmeyer flasks.

The chemist now has a choice to make. He can use either toluene or ethyl ether as the solvent to make the crystals in. Toluene is cheaper, and less of it is needed because it evaporates more slowly during the filtering process. Ether is more expensive, and flammable. But since it evaporates more quickly, the crystals are easier to dry off. If ether is used, it is anhydrous (contains no water).

A third choice is also possible for use as a crystallization solvent. This is mineral spirits available from hardware stores in the paint department. Mineral spirits are roughly equivalent to the petroleum ether or ligroin commonly seen in chem labs. Those brands which boast of low odor are the best choice, such as Coleman camper fuel. Before using this material it is best to fractionally distill it, and collect the lowest boiling point half of the product. This speeds crystal drying. Since the choice of mineral spirits or naphtha eliminates ether from the supply loop, the clandestine operator may well go this route. Toluene is also a very acceptable solvent.

With the solvent of his choice, the chemist rinses the insides of the condenser, vacuum adapter and 250 ml flask to get out the methamphetamine clinging to the glass. This rinse is poured in with the product. Solvent is added to each of the Erlenmeyer flasks until the volume of liquid is 300 ml. They are mixed by swirling.

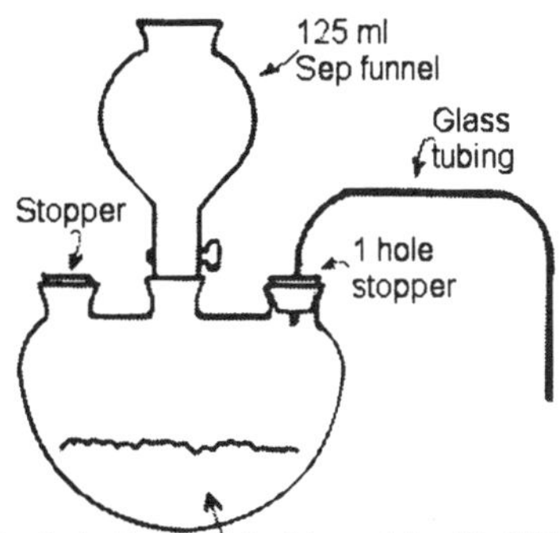

Figure 20

A source of anhydrous hydrogen chloride gas is now needed. The chemist will generate his own. The glassware is set up as in Figure 20. He will have to bend another piece of glass tubing to the shape shown. It should start out about 18-inches long. One end of it should be pushed through a one-hole stopper. A 125 ml sep funnel is the best size. The stoppers and joints must be tight, since pressure must develop inside this flask to force the hydrogen chloride gas out through the tubing as it is generated.

Into the 1000 ml three-necked flask is placed 200 grams of table salt. Then 35% concentrated hydrochloric acid is added to this flask until it reaches the level shown in the figure. The hydrochloric acid must be of laboratory grade.

Some concentrated sulfuric acid (96-98%) is put into the sep funnel and the spigot turned so that 1 ml of concentrated sulfuric acid flows into the flask. It dehydrates the hydrochloric acid and produces hydrogen chloride gas, and also goes on to make more HCl from reaction with table salt. This gas is then forced by pressure through the glass tubing.

One of the Erlenmeyer flasks containing methamphetamine in solvent is placed so that the glass tubing extends into the methamphetamine, almost reaching the bottom of the flask. Dripping in more sulfuric acid as needed keeps the flow of gas going to the methamphetamine. If the flow of gas is not maintained, the methamphetamine may solidify inside the glass tubing, plugging it up.

An even more distressing phenomena can occur when using hardware store sulfuric acid. This material is not quite as concentrated as lab grade sulfuric acid, and the initial "push" of HCl gas in the flask can turn into a "pull." This vacuum will suck your product up the tubing and into the HCl generating flask. You can imagine how that really sucks! Hah, Hah!

When using lower grade sulfuric acid, one needs to watch carefully, or use a closely related procedure. In this procedure, the HCl generating flask is filled ¼ full of hardware store sulfuric acid and the sep funnel is filled with muratic acid. By dripping the muriatic acid into the sulfuric acid, the HCl solution (muriatic) is dehydrated to dry HCl gas, and the gas is pushed out the flask and tubing. It is advisable to do a dry run first without product at risk when using hardware store acids in this procedure to avoid heart breaking losses.

Another popular clandestine method using lower grade sulfuric acid is to put the salt in a Mason jar, and punch two holes in the lid. Plastic tubing is attached to both holes. One line of tubing goes into the solution of free base meth in solvent. The other line of tubing is run to an aquarium pump or other source of air. The low grade sulfuric acid is then added to the mason jar, and the air blower is started to slowly keep the fumes of dry HCl flowing into the meth free base rather than ever being sucked backwards.

Within a minute of bubbling, white crystals begin to appear in the solution. More and more of them appear as the process continues. It is an awe-inspiring sight. In a few minutes, the solution becomes as thick as watery oatmeal.

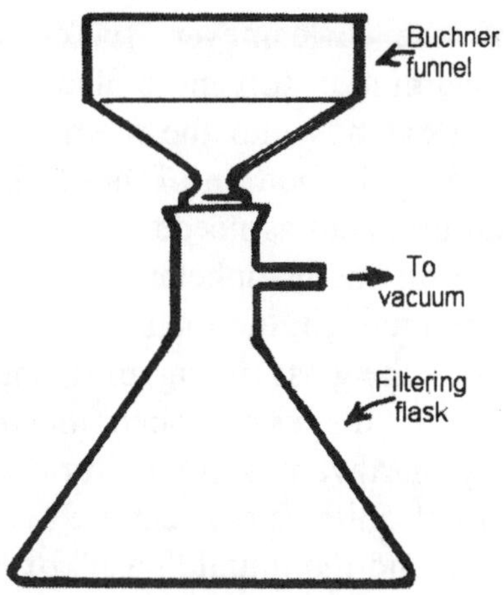

Figure 21

It is now time to filter out the crystals, which is a two-man job. The flask with the crystals in it is removed from the HCl source and temporarily set aside. The three-necked flask is swirled a little to spread around the sulfuric acid and then the other Erlenmeyer flask is subjected to a bubbling with HCl. While this flask is being bubbled, the crystals already in the other flask are filtered out.

The filtering flask and Buchner funnel are set up as shown in Figure 21. The drain stem of the Buchner funnel extends all the way through the rubber stopper, because methamphetamine has a nasty tendency to dissolve rubber stoppers. This would color the product black. A piece of filter paper covers the flat bottom of the Buchner funnel. The vacuum is turned on and the hose attached to the vacuum nipple. Then the crystals are poured into the Buchner funnel. The solvent and the uncrystallized methamphetamine pass through the filter paper and the crystals stay in the Buchner funnel as a solid cake. About 15 ml of solvent is poured into the Erlenmeyer flask. The top of the flask is covered with a palm and it is shaken to suspend the crystals left clinging to the sides. This is also poured into the Buchner funnel. Finally, another 15 ml of solvent is poured over the top of the filter cake.

Now the vacuum hose is disconnected and the Buchner funnel, stopper and all, is pulled from the filtering flask. All of the filtered solvent is poured back into the Erlenmeyer flask it came from. It is returned to the HCl source for more bubbling. The Buchner funnel is put back into the top of the filtering flask. It still contains the filter cake of methamphetamine crystals. It will now be dried out a little bit. The vacuum is turned back on, the vacuum hose is attached to the filtering flask, and the top of the Buchner funnel is covered with the palm or a section of latex rubber glove. The vacuum builds and removes most of the solvent from the filter cake. This takes about 60 seconds. The filter cake can now be dumped out onto a glass or China plate (not plastic) by tipping the Buchner funnel upside-down and tapping it gently on the plate.

And so the filtering process continues, one flask being filtered while the other one is being bubbled with HCl. Solvent is added to the Erlenmeyer flask to keep their volumes at 300 ml. Eventually, after each flask has been bubbled about seven times, no more crystal will come out and the underground chemist is finished.

If ether was used as the solvent, the filter cakes on the plates will be nearly dry now. With a knife from the silverware drawer, the cakes are cut into eighths. They are allowed to dry out some more then chopped up into powder. If toluene was used, this process takes longer. Heat lamps may be used to speed up this drying, but no stronger heat source.

The yield of product is about 100 grams of nearly pure product. It should be white and should not get wet, except in the most humid weather. It is suitable for any purpose. It can be cut in half and the underground chemists will still have a better product than their competition. But they will not cut it until a few days have passed, so that their options are not limited should one of the problems described in the next few paragraphs arise.

Here are some of the common problems that arise with the crystals, and how they are dealt with. To spot these possible problems, the crystals are first left on the plate to dry out, and then transferred to glass jars or plastic baggies.

Yellow Crystals. This is caused by not properly rinsing off the crystals while in the Buchner funnel, or not using enough solvent to dissolve the methamphetamine in the Erlenmeyer flasks. To whiten them up, they are allowed to soak in some ether or toluene in a glass jar, then filtered again.

Yellow Stinky Crystals. The smell takes a few days to develop fully. They are left alone for 5 days, then soaked in ether and filtered again. The smell should not return. (The problem is caused by heating the reaction mixture above the 145° C upper limit.)

Crystals Refuse to Dry. This can especially be a problem using toluene as a solvent. It can also be a problem on very humid days. The crystals are placed in the clean, dry filtering flask, the top is stoppered and vacuum applied at full strength for 15 minutes. Warming the outside of the filtering flask with hot water while it is under vacuum speeds the process.

Crystals Melt. Here the crystals soak up water from the air and melt. This is usually caused by raising the temperature of the reaction too rapidly, or by collecting too much high boiling material during the distillation. First, they are put into the filtering flask and a vacuum applied to dry them out. They are soaked in ether or toluene and filtered. If this doesn't cure the problem, cutting the material to 50% purity should take care of it.

Crystals Are Sticky. Here the crystals seem covered by a thin layer of oily material, causing them to stick to razor blades, etc. The problem is dealt with in the same way as melting crystals.

Crystals Fail to Form. This problem occurs during the process of bubbling HCl into the methamphetamine. Instead of forming crystals, an oil settles to the bottom of the flask. This is generally caused by incomplete hydrolysis of the formyl amide. Perhaps it didn't mix with the hydrochloric acid. It is put in a flask and the solvent boiled off under a vacuum. Then 200 ml of hydrochloric acid is added and the process is repeated, starting from the hydrolysis of the formyl amide of methamphetamine. The 35% laboratory grade of hydrochloric acid is used.

In the event of melting or sticky crystals, cutting is first tried on a small sample of the crystals to see if that will solve the problem. If it does not, then a recrystallization must be resorted to. This is done by dissolving the crystals in the smallest amount of warm alcohol that will dissolve them. 190 proof grain alcohol, 95% denatured alcohol, or absolute alcohol may be used. Then 20 times that volume of ether is added. After vigorous shaking for three minutes, the crystals reappear. If not, more ether is added, followed by more shaking. After being filtered, the crystals should be in good shape.

A technique which may be used in especially stubborn cases is to dissolve the crystals in dilute hydrochloric acid solution, extract out the oily impurities with toluene, and then isolate the methamphetamine. This is done as follows:

For every 100 grams of crystal, 200 ml of 10% hydrochloric acid is prepared by mixing 60 ml of 35% hydrochloric acid with 140 ml of water. The crystals are dissolved in the acid solution by stirring or shaking in the sep funnel. 100 ml of toluene is added to the solution in the sep funnel, which is then shaken vigorously for about 2 minutes. The lower layer is drained out into a clean beaker. It contains the methamphetamine. The toluene layer is thrown out. It contains the oil grunge which was polluting the crystals.

The acid solution is returned to the sep funnel and the acid neutralized by pouring in a solution of 70 grams of sodium hydroxide in 250 ml of water. After it has cooled down, the mixture is shaken for 3 minutes to make sure that all the methamphetamine hydrochloride has been converted to free base. Then 100 ml of toluene is added and the mixture shaken again. The lower water layer is drained out and thrown away. The toluene-methamphetamine solution is distilled as described earlier in this chapter. Then, as described earlier in this chapter, dry hydrogen chloride gas is bubbled through it to obtain clean crystals. (Hydrogen chloride gas must be made in a well-ventilated area; otherwise, it will get into the chemist's lungs and do real damage.)

There is an alternative method for converting amphetamine free base into the crystalline hydrochloride. It is based on the method that South American cocaine manufacturers use to turn coca paste into cocaine hydrochloride. This method does not give the really high-quality crystals that the bubble-through method gives, but its use is justified when really big batches are being handled.

In this alternative procedure, the free base is dissolved in two or three volumes of acetone. Concentrated hydrochloric acid (37%) is then added to the acetone while stirring until the mixture becomes acid to litmus paper. Indicating pH paper should show a pH of 4 or lower. The hydrochloride is then precipitated from solution by slowly adding ether with stirring. It will take the addition of 10 to 20 volumes of ether to fully precipitate the hydrochloride. Toluene or mineral spirits may be substituted for the ether. Then the crystals are filtered out using a Buchner funnel as described before, and set aside to dry. The filtrate should be tested for completeness of precipitation by adding some more ether to it.

For really big batches, it would seem that using sulfuric acid to make the crystalline sulfate salt of the amphetamine has its advantages. This is a transcript of a conversation I had with another cooker on the Internet:

Someone: Methamphetamine sulphate. Amphetamine sulphate can be made very easily mixing freebase and H_2SO_4 in alcohol solvent. Is there any difference between hydrochloride and sulphate end product? Can one make methamphetamine sulphate the same way?

Fester: My experience with the sulfate salt is limited to what I heard from somebody who did it. With the sulfate salt, recrystallizing the product is required to get out excess sulfuric acid, which turns the product to mush over time. Must hurt like hell snorting it too. HCl is perfect, the excess gas of HCl just evaporates away.

Someone: The following is tested with amphetamine freebase only: Usually one adds solution of freebase to the solution of H_2SO_4 until pH about 8.0 or so. But when one does the procedure in different order (that means adding diluted acid to freebase) there cannot be any excess acid, which must be washed out. By the way, using much ether-alcohol as freebase solution, very dry and extremely white powder without any smell will be the result. HCl-salt via gas bubbling can be used only in very small amounts.

Fester: I don't know what you consider to be small amounts. I used to make quarter pound (100 gr) batches regularly, and we used dry HCl to crystallize. Doing half or even full pound batches by the same method would have been no problem. Ten pounds at a time, maybe we're talking about some problems. The HCl is great because you can't overdo it, and no recrystallization is needed.

Someone: Big batch starts from 1 kg. With HCl on normal pressure it is VERY time consuming. We tried crystallization under pressure (HCl from steel-made gas-generator — max pressure 5-7 bar) piped into ether solution of freebase. Maximum batch was 100g in time (needs 1000 ml ether). Process generates blue-colored product and solvent heats up to 40° C. It was very difficult to measure needed amount of gas. Excess HCl results in yellow smoking solution — product cannot be filtered out.

Maybe problem was in incomplete dehydration of HCl.

By the way, 5 bars of hydrogen chloride routed through H_2SO_4 dehydrator was real hell when it went out of control... One of these laboratory gas-wash-bottles exploded under 1m distance from my face. After that, we used only steel. With sulphate 1 Kg batch needs 3L of spirit-ether, cooling, constant stirring (no lab stirplates) and pH-monitoring. NO WASH is needed if one does not add excess acid to free-base solution.

Fester: Thanks for the info on big batches.

The European Variation

A few years ago, I received this letter from Geert Hendricks. I've known Geert for several years now, and I know him to be a veteran cooker, and a reliable source of inside information. This is what he has to say about how the Leuckardt reaction is done in Europe. With this variation, the 98% formamide commonly available as an industrial solvent can be used to make amphetamine from phenylacetone. The key to this variation is to mix the phenylacetone and formamide, and allow the mixture to sit for several days at least prior to commencing heating of the mixture. Without this preliminary "aging" of the mixture, only 99% formamide could be used in the reaction and still avoid the dreaded red tar. Read on:

Dear Uncle Fester,

Thanks for the information.

The magic in the reaction is to put the phenylacetone and the formamide together a couple of days before doing the reaction.

1 part phenylacetone and 2 parts formamide must stand at room temperature at least 24 hours or more. I think one week is best. It should be shaken twice a day or more. I believe the formamide is slowly reacting with the phenylacetone, so the longer it stands, the better!

Maybe if the temperature is higher, it reacts faster. Some people leave it for only a day, but if you have the time, why not let it stand for one or two weeks?

Just before I got arrested, I was measuring the pH of the mixture, and I believe that it changes every day. With a little patience, you should be able to figure out the best length of time to let it stand.

Warning: NEVER put the formic acid with the mixture. It will ruin everything, and you will have to start all over.

So now you have waited for a week, and you're ready to go for it. Put your mixture in a glass flask with boiling chips. You should use as many as possible, because you need small bubbles. This is very important!! With a condenser on top for refluxing, the water inside the condenser must not be very cold, because the ammonia will block the hole in the condenser, so 30° C is a good temperature for it. You can even let all the water out, if you don't mind the smell of ammonia.

Just before you start, you take a little bit of formic acid and add it to the mixture. You only need a little bit; for instance, for a 3-liter mixture, add 5cc. More won't do any harm. If you put in too much, just heat the flask without the condenser until the temperature has gone up to 160-165° C, and put the condenser back on.

Now you must raise the temperature to 180° C, and sometimes a little bit higher, depending on the quality of your formamide. Here they use industrial grade. I've never had any problems, as long as it's clean.

Let the mixture reflux for one hour after it has reached the desired temperature. The mixture will change from light yellow to dark yellow. If it starts to darken, then your temperature is too high.

After one hour, let it cool down, or when it is a small batch, take a sep funnel filled with water and mix it with your batch. Don't use any lye at this stage.

Now take out the oil and mix it with twice the amount of hydrochloric acid. Let it reflux for one hour.

Now you have to separate this with lye. The best thing to do is to first let it cool down, but if it's a small batch and a big sep funnel, you can take the risk. Also, very good shaking is needed here.

Take the oily layer and start to distill it under maximum vacuum.

Take the distilled product and mix this with twice the amount of alcohol, methanol, acetone, or whatever you think is best. Now take 20% H_2SO_4 and add this slowly until the pH is 7.

If you added too much H_2SO_4, just bring the pH back with some lye.

Now filter out the product and let it dry. Big batches take a long time to dry, so people here put them in a centrifuge. Take a big bedsheet, put your batch in there, and let it spin. (Be careful with sparks and such!)

Instead of formamide, you can also use ammonium formate. You don't need formic acid here.

I'm now trying to do the reaction with only ammonia, but I'm still working on that one.

If you have any questions, just write me and ask. Keep up the good work and have a good time.

Geert Hendricks
Eindhoven, Holland
PS. This reaction also works with methylenedioxyphenyl acetone.

That is an interesting recipe, isn't it? I wish I'd have known about it back when I was cooking. Can this European variation using formamide be adapted to use with N-methyl formamide to make meth instead of benzedrine? Your Uncle would say that the answer is "Yes!"

Your Uncle suggests this: Go to Chapter Four, describing how to make N-methyl formamide. Methylamine is bubbled into formic acid to make N-methyl formamide. Then it is distilled, and some more methylamine is pumped into the mixture. I would stop right there. Instead of distilling it again, I would just use that mixture. If the mixture smells mildly of methylamine, all the formic acid has been neutralized. I would just mix the phenylacetone with this mixture, and let it sit for a few days as in the European variation. Then I'd put a hand full of boiling chips in the reaction flask, and heat the mixture just as described in the main body of this chapter, stopping when 145° C is reached. The NaOH hydrolysis should be used to recover the unused methylamine, then HCl hydrolysis of the formyl amide of meth, just as described in this chapter. Yields should go up, and there should be no upper limit on possible batch sizes.

The Russian Advance

A variation upon the standard Leuckardt Reaction method is given in *Zhurnal Obshchei Khimii* Volume 25, pages 1432-7 (1955). This variation claims very high yields and is well worth checking out by anyone pursuing this reaction method.

In their method, the general reaction procedure is the same as given earlier in this chapter. They simply add on two little details. The first change they make to the standard reaction method is that they add a little bit of nickel hydrogenation catalyst to the reaction mixture. In their procedure, they add about 1% by weight of Raney nickel to the N-methylformamide. This form of nickel is discussed in Chapter Eleven, and I'm pretty sure that other forms of nickel catalyst could be substituted for it.

The simplest nickel catalyst to make is prepared by dissolving a few grams of nickel sulfate or nickel chloride in water. Then to that water solution of the nickel one adds sodium borohydride. The sodium borohydride is first dissolved in some water, and slowly, with stirring, the sodium borohydride solution is added to the nickel solution. The nickel solution will fizz during this addition and turn black. The black particles are the nickel catalyst, and when the stirring is stopped they will settle to the bottom of the beaker.

One can check to see if the reaction is complete by stopping the stirring and letting the black particles settle. If the water solution is still blue (for nickel sulfate) or green (for nickel chloride), then not enough borohydride has yet been added to convert all of the nickel into the active catalyst. The reaction is complete when the black powder settles leaving a clear water solution.

Next, one cleans this catalyst by decanting off the water from the settled black powder. Then add some alcohol to the beaker and stir. Then let the black powder settle again, and decant off the alcohol. Finally add just enough fresh alcohol to the beaker so that the black powder can be suspended when the mixture is stirred. This is your catalyst, and it contains an amount of nickel which is roughly half the weight of the nickel sulfate or nickel chloride originally used. It should be used immediately, or bottled up and stored in a refrigerator.

The next point on which these Russians vary from the standard procedure is that they don't just mix the phenylacetone together with the N-methyl formamide and begin heating. What they like to do is to put the N-methyl formamide into the flask along with that little bit of nickel catalyst. Then they start to heat this mixture in the oil bath. When it warms up to roughly 90° C, they begin to drip the phenylacetone into the N-methyl formamide. They take about an hour to add all the phenylacetone to the N-methyl formamide, and while they are adding the phenylacetone they allow the temperature to slowly rise to about 120° C in the reaction mixture.

From there on out, the reaction is done exactly the same as the standard method given in this chapter. Heating is continued as the reaction bubbles, and the temperature is slowly raised to keep it going. Then the product is obtained from the reaction mixture in exactly the same way as described earlier in this chapter.

References

Journal of Organic Chemistry, Volume 14, page 559 (1949).

Journal of the American Chemical Society, Volume 58, page 1808 (1936); Volume 61, page 520 (1939); Volume 63, page 3132 (1941).

Organic Syntheses, Collective Volume II, page 503.

Chapter Six
Industrial-Scale Production

In the previous five chapters, I described a process by which underground chemists make smaller amounts of methamphetamine, up to about one-half pound of pure methamphetamine. The process takes about three days with two people working in shifts around the clock. Thus, the maximum production level is stuck at one pound per week.

There is a way to break through this production limit, which is to produce phenylacetone and turn it into methamphetamine by different methods. These methods produce more in less time, and they are cheaper. Two of them, the tube furnace and the hydrogenation bomb, are major engineering projects. But they are no problem for those with a Mr. Handyman streak. If constructing a factory for meth production wasn't what you had in mind, you will also find in the next few chapters many methods which use standard labware and which also can potentially produce large amounts of meth fairly quickly.

However, underground chemists will not move up to industrial-scale production until they are sure that they are going to be able to sell it without having to deal with strangers — unless, of course, they want to get busted.

One major difference in the logistics of a large-scale operation versus a smaller one is that a different source of chemicals is required. An outlet that specializes in pints and quarts of chemicals is not going to be much help when multi-gallons are needed. Here a factor comes into play which cannot be taken advantage of at lower levels of production. Most chemical suppliers will not deal with individuals, only with corporations and companies.

Now the underground chemist can turn this situation to his advantage by means of subterfuge. First he develops a false identity. He gets some of the books on false ID and — *Abracadabra!* — he's Joe Schmoe. He uses this identity to form several companies. If he wants to be official, he consults the book, *How to Form Your Own Corporation for Under 50 Dollars,* available in most libraries. Otherwise, he just has some invoice order forms printed up for his company. He may also open a checking account for his company to pay for chemicals. He uses checks with high numbers on them so that they don't think that he just appeared out of thin air. As an alternative, he may pay with certified checks from the bank.

The next step is to rent some space as his company headquarters and chemical depot. Indeed, he'll probably rent a couple such depots to house his various companies.

Now he starts contacting chemical dealers, ordering enough of one or two chemicals to last for a couple of years. Then he contacts another dealer and orders a similar quantity of one or two other chemicals under a different company name. He continues this process until he has everything he needs. He offers to pick them up so that they do not see the dump he's rented as his headquarters. As a precaution, he equips these dumps with a phone and answering machine so that they can call him back. If he doesn't live in a large city, he does business out of town. That way they won't be surprised that they never heard of him. But he does not do business too far away from home base, so they won't wonder why he came so far.

There is a better strategy to follow in getting the equipment and chemicals needed for clandestine meth production. The best method to use is to first order the equipment and a couple of the most suspicion-arousing chemicals. Then the underground operator lays low for a while. The narco swine have a habit of going off half-cocked on their search warrants. If the initial purchases caught their eyes, they will likely swoop right in, planning on finding an operating lab, or at least enough to make a conspiracy charge stick. If they move now, the meth meister will not be prosecutable, so long as he does not admit guilt. An alternative narco swine strategy would be for them to initiate intense surveillance upon Joe Schmoe. So long as Joe is not brain dead, this will be pretty obvious after awhile. If surveillance is noticed, it is time to put the plan into a deep freeze, and consider the initial purchases a long-term investment rather than a quick payoff. If Joe is able to get the most sensitive materials unnoticed, it is then time to quickly get the more mundane items needed and immediately turn to the production end of the operation.

When it is time for the underground chemist to pick up the chemicals, he uses a pickup or van registered in Joe Schmoe's name. As a precaution, he equips his vehicle with a radio scanner. He buys the book, *U.S. Government Radio Frequencies,* and tunes the scanner to pick up the FBI, the DEA, the state and local police. He picks up the chemicals and returns with them to his headquarters and depot. He takes a roundabout route to make sure he isn't being followed. Two tricks he may use to detect a tail are to turn into a dead-end street and to drive either too fast or too slow. He leaves Joe's vehicle at the depot and takes a roundabout route home. He stops at a few bars and leaves by the back exit.

A very common and quite stale trick is for the narco swine to place a radio-tracking device in the packing materials surrounding jugs of chemicals purchased by suspected drug manufacturers. All items purchased should be carefully inspected during the drive away from the point of purchase. If such a device is found, it is cause for clear thinking action, rather than panic. While using such a device, the heat will usually lay quite far back in their pursuit to avoid being noticed. They will rely on the transmitter to tell them where you are going. It is best not to smash such a transmitter, but rather keep it in hand, and toss it into the back of another pickup truck at a stoplight. This is then followed by putting the plan into a deep freeze until the heat grows bored with you.

The next thing the underground chemist needs is a laboratory location. A country location makes any surveillance very obvious and keeps chemical smells out of the way of nosy neighbors. Electricity and running water are absolutely necessary. Now he loads the chemicals onto Joe's wheels and heads for the laboratory in a very roundabout manner, keeping an eye open for any tail and paying close attention to the scanner. He leaves the scanner at the lab for entertainment in the long hours ahead.

A nice addition to any underground laboratory is a self-destruct device. This consists of a few sticks of dynamite armed with a blasting cap, held inside an easily opened metal can. The purpose of the metal can is to prevent small accidental fires from initiating the self-destruct sequence. If Johnny Law pays an uninvited visit to his lab, the underground chemist lights the fuse and dives out the window. The resulting blast will shatter all the glass chemical containers and set the chemicals on fire. This fire will destroy all the evidence. He keeps his mouth shut and lets his lying lawyer explain why the blast happened to come at the same time as the raid. He has no reason to fear the state crime lab putting the pieces of his lab back together. These guys learned their chemistry in school and are truly ignorant when it comes to the particulars of a well-designed lab.

The feds, on the other hand, have a higher grade of chemist working for them, but they are tiny individuals who are haunted by nagging self-doubt, wondering why after obtaining a Ph.D., they are just faceless cogs in a machine. To compensate for this, they will claim to make great discoveries of the obvious. Case in point is an article published in the *Journal of Forensic Sciences.* This is a petty journal published by Johnny

Law where the aforementioned tiny individuals can stroke their egos by getting published. In an article covering the lithium in ammonia reduction of ephedrine to meth production method featured in Chapter Fifteen of this book, the unnamed tiny, frustrated chemists trumpeted "we found that a nitrogen atmosphere to protect the reaction was unnecessary, contrary to the claims of the authors who said it was essential."

The authors to which they refer here are Gary Small and Arlene Minnella, legitimate scientists who were published in a legitimate scientific journal, the *Journal of Organic Chemistry*. In their article covering the lithium in ammonia reduction of benzyl alcohols, they used really tiny batches that might actually need a nitrogen atmosphere to protect them, and in no place claimed that it was essential. See the *Journal of Organic Chemistry* article cited in Chapter Fifteen of this book. It was obvious that the steady boiling away of liquid ammonia would form its own protective gas blanket when done on a scale corresponding to real meth production.

They further went on to nitpick the purification procedure used by the real scientists claiming it was unnecessary. Everyone who reads the journals knows that it is unnecessary. This is just the protocol that has been followed by research scientists for the past god-knows-how-many ages. They just do this so that if they get unexpected results in their research, they will know that it is not due to impurities in the reaction mix. To make a great discovery out of finding that these rigorous purification schemes are not needed for practical production methods just shows how shallow these people are.

Chapter Seven
Phenylacetone From B-Keto Esters

In this chapter, I will cover another method of making phenylacetone. The chemicals must be weighed and measured fairly exactly. This is unlike the method described in Chapter Three, where anything within a ballpark range will work. This method requires a reliable scale.

This reaction uses sodium metal, which is some nasty stuff. It reacts violently with water to produce sodium hydroxide and hydrogen. It will also react with air. The chemist never touches it intentionally; if he does touch it, he washes it off with warm water. Sodium metal comes in a can, covered with a bath of petroleum distillate. This is to protect it from water and air. As long as it stays covered, it causes the chemist no problems.

In this reaction, sodium metal is reacted with absolute alcohol to make sodium ethoxide ($NaOCH_2CH_3$). Ethyl acetate and benzyl cyanide are then added to this to produce a beta keto ester. Reaction with acid then produces phenylacetone.

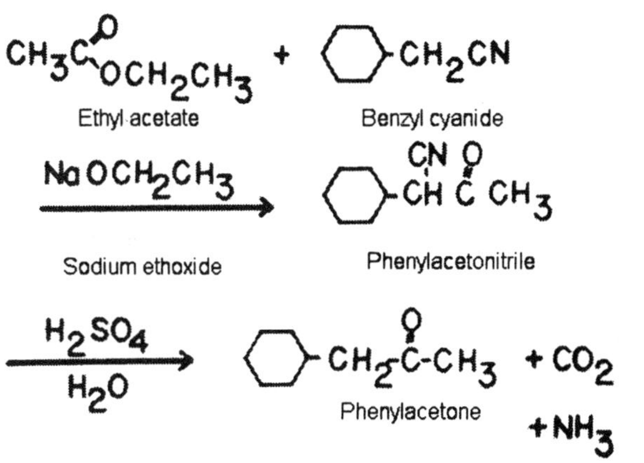

Figure 22 shows the glassware used. The glassware must be very dry, so it is dried out in the oven for an hour or so. If the sep funnel has a plastic valve, the valve is taken out before the sep funnel is put in the oven. The magnetic stirring bar does not go in the oven either. It is coated with Teflon, so it does not have any water on it. A magnetic stirrer is necessary to do this reaction, because good stirring is very important. An extra claisen adapter is needed for this reaction; one is filled with broken pieces of glass for use as a fractionating column, the other is kept as is for use in the Figure 22 apparatus.

To begin, the underground chemist puts a bed of Drierite in the vacuum adapter as shown in Figure 9, being sure to plug up the vacuum nipple. The water lines are attached to the condenser and cold water started flowing through it. But if it is humid, the water flow is not started until the glassware is assembled.

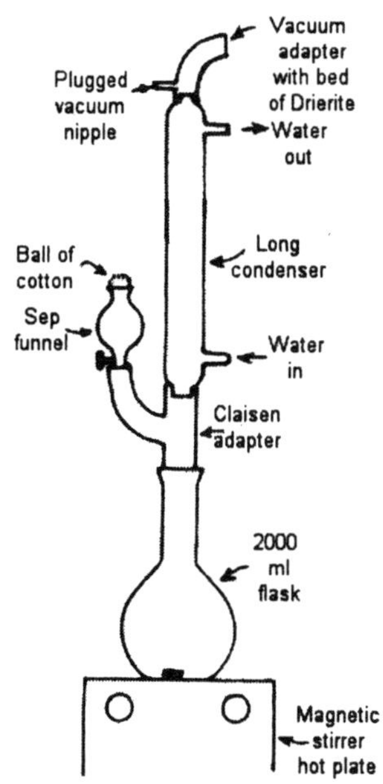

Figure 22

In this case, the main reactant is benzyl cyanide, also called phenylacetonitrile or alpha-tolunitrile. Benzyl cyanide is now a controlled substance precursor, and so must be made.

Benzyl cyanide is not outrageously poisonous like sodium cyanide. It is an organic cyanide, called a nitrile. As long as the chemist doesn't drink the stuff, he's OK. It is a somewhat smelly liquid, clear in color. For a recipe to make it from benzyl chloride and cyanide, see *Organic Syntheses,* Collective Volume I.

The chemist starts with a chunk of clean sodium metal that weighs 128 grams. It is weighed out in a 300 ml beaker half-filled with petroleum distillate or xylene. Then the sodium metal is transferred to another beaker half-filled with anhydrous ether and chopped into small pieces with a clean knife. Then it is scooped out with a spoon and put in the 3000 ml flask. The glassware is quickly assembled as shown in Figure 22, with the 3000 ml flask sitting in a pan. Water flow through the condenser is begun, and 300 ml of absolute ethyl alcohol is put in the sep funnel. As the alcohol is allowed to flow in onto the sodium, the reaction is kept under control by putting cold water in the pan and wrapping the flask in a wet towel.

When the reaction is under control, more alcohol is added until a total of 1500 ml has been added. The alcohol is gently boiled until the sodium metal is dissolved.

Now the chemist mixes 500 grams (490 ml) of benzyl cyanide with 575 grams (637 ml) of ethyl acetate and stops the heating of the ethanol solution. Just as it stops boiling, the mixture of ethyl acetate and benzyl cyanide is added with good magnetic stirring. This addition takes about 15 minutes. The stirring is continued for about 10 minutes after the addition is complete, then the mixture is heated in a steam bath or in a pan of boiling water for about 2 hours. Then it is taken out of the heat and allowed to sit overnight, or at least for a few hours.

The underground chemist has just made the sodium salt of phenylacetacetonitrile. To collect it, he cools the flask in a mixture of salt and ice. With a clean wooden stick, he breaks up the chunks of crystals that have formed, as the flask is cooling down. When it reaches -10° C, he keeps it at this temperature for a couple of hours, then filters out the crystals. They are rinsed a couple of times with ether or another solvent such as hexane, then, while still wet with ether, added to a large flask or beaker containing 2000 ml of water. They are dissolved by stirring, then the flask or beaker is cooled down to 0° C by packing it in ice mixed with salt. When it reaches this temperature, 200 ml of glacial acetic acid is added to it with vigorous stirring. The chemist must make sure that the temperature does not go up more than a few degrees while he is adding it.

He has now made phenylacetacetonitrile. He filters the crystals off it and rinses them a few times with water. The crystals must now be kept moist in order for them to be turned into phenylacetone.

All is now ready for producing phenylacetone from these crystals. In a 2000 ml flask, he puts 700 ml of concentrated sulfuric acid. It is cooled down to -10° C by packing the flask in a mixture of salt and ice, then magnetic stirring is begun. The crystals are slowly added to the sulfuric acid. They must be moist, or he will get a mess. It takes about an hour to add the crystals to the sulfuric acid. Once they are added, the flask is heated in a pan of boiling water and swirled around to dissolve the crystals. After they have dissolved, the flask is heated for a couple more minutes, then removed from the pan of boiling water. It is cooled down slowly to 0° C by first letting it cool down, then packing it in ice.

The underground chemist puts 1700 ml of water in a 3000 ml flask. Half of the sulfuric acid solution is added to it. It is heated in a pan of boiling water for a couple of hours. It is given a couple of good shakes every 15 minutes. A layer of phenylacetone forms in the mixture.

After 2 hours of heating, the mixture is poured into a gallon-size glass jug to cool off. Another 1700 ml of water is put in the flask and the rest of the chilled sulfuric acid solution is poured into it. It is also heated for 2 hours in a pan of boiling water, then poured into another glass jug.

The chemist is ready to separate the phenylacetone from the water and distill it. The liquid in the first jug is slowly poured into a 1000 ml sep funnel until the sep funnel is full. Most of the phenylacetone layer will be in the sep funnel, because it is floating on top of the water. The water layer is drained back into the jug, and the phenylacetone layer is poured into a large beaker. He adds 300 ml of toluene to the jug, stoppers it and shakes it for 15 seconds. Then he stops and lets the layer of toluene containing phenylacetone float up to the surface. It is slowly poured into the sep funnel, and the water layer is drained back into the jug. The water is thrown away. This process is repeated with the other jug.

This phenylacetone has some sulfuric acid in it. The chemist puts 150 ml of a 5% solution of lye in water in the 1000 ml sep funnel. He also pours half of the phenylacetone and toluene mixture he got from the two jugs into the sep funnel. He shakes it with the water to remove the sulfuric acid. The water is drained out, and the phenylacetone-toluene layer is poured into a 1000 ml round bottom flask. Another 150 ml of water is put into the sep funnel. It is shaken also, and the water layer is drained off. He pours as much of this toluene-phenylacetone mixture into the 1000 ml round bottom flask as he can until it reaches $^{2}/_{3}$ full.

The glassware is set up as shown in Figure 13 in Chapter Three, with a few boiling chips in the 1000 ml flask. The collecting flask is 250 ml. He distills off a couple of hundred ml of toluene to make room for the rest of the product. When there is some room, he turns off the heat and waits for the boiling to stop. Then the rest of the toluene-phenylacetone mixture in the sep funnel is added to the 1000 ml flask. The distillation is continued until the toluene stops coming over. About 500 to 600 ml of toluene will be collected.

When the rate of toluene distillation slows down to just about stopping, the heat is turned off and it is allowed to cool down. Then the last of the toluene is removed under a vacuum. When the toluene is gone, the collecting flask is changed to a 500 ml flask and the phenylacetone is distilled under a vacuum at the usual temperature range. The yield is about 300 ml of phenylacetone. Once the toluene is gone, virtually all of the material left in the flask is phenylacetone. If there is a high boiling residue, it is unchanged phenylacetacetonitrile.

A claimed improvement upon this recipe is found in *Chem Abstracts,* Volume 36, column 2531 (1942). Here once again we come to the question of Russian science, and do we believe them, or are they lying through their rotten, stinking teeth? Your Uncle is tending towards believing them, because the variation they use is a general one that has been found to give improved yields in related reactions. Let the reader know that I haven't tried this improved procedure, so I'm just passing along the recipe from the literature for what it's worth.

In the above example, absolute alcohol is used both to react with the sodium metal to produce sodium ethoxide, and as the solvent for the reaction. It has been found in general with other reactions of this type that the presence of all that extra unreacted alcohol in the solution tends to cause side reactions which lower the yield of the desired product. In the Russian recipe, the extra alcohol beyond that amount needed to react with the sodium metal is left out, and replaced with anhydrous ether. The yield of phenylacetacetonitrile as a result goes up to 86% from the roughly 50% or so obtained when all the extra alcohol is in the reaction mixture.

For clandestine purposes, anhydrous ether is a very undesirable material. It's very flammable, and evaporates so quickly that an explosive cloud of vapors can easily fill a room. All it's lacking is a spark to set it off. It also smells to high heaven, and can be detected far downwind. This could easily attract unwanted attention. For strike number three, one just doesn't run down to the hardware store and pick up a jug or pail of anhydrous ether. It is associated strongly with drug manufacture.

One need not be discouraged on this point. Anhydrous ether isn't the only possible replacement solvent. No doubt it was just lying around the commie lab, so they grabbed it and used it. Alternative solvents would include anhydrous acetone, hexane, toluene, xylene, or even that Coleman camper fuel. The latter three can be picked up at the hardware store. None of them produce the smell or fire hazards that ether does.

Now to go back to the recipe. The 3000 ml flask is set up as in the previous example, and 1000-1200 ml of the alternative solvent is put into the flask. The sodium metal is cut up and added to the flask as in the example given. Then just enough absolute alcohol to react with the sodium metal is added to the solution. For 128 grams of sodium metal, this works out to 325-350 ml of absolute alcohol. 350 ml would give a little excess alcohol to speed up the reaction of the last bits of sodium metal. It will take considerably longer for the sodium metal to react using a dilute solution of absolute alcohol as compared to just dissolving it in straight absolute alcohol. A mixture of sodium ethoxide with the solvent will be produced. Methanol can be substituted for the absolute alcohol if this material is easier for you to obtain. In this case, roughly 215 ml of methanol would be used for this example reaction.

Next, the benzyl cyanide and ethyl acetate mixture is slowly added just as in the standard example given, and the mixture boiled for a couple of hours to produce the sodium salt of phenylacetacetonitrile. It is filtered out and then reacted with acetic acid to produce phenylacetacetronitrile, just as in the standard recipe.

This is another claimed improvement in this process given in the Russian article. To produce phenylacetone from the phenylacetacetonitrile, they claim that phosphoric acid is superior to sulfuric acid. One would just take the standard procedure given here, which by the way can be found in *Organic Syntheses,* Collective Volume Two, and substitute phosphoric for the sulfuric in that step. Rather than heat the sulfuric acid and phenylacetacetonitrile mixture on a steam bath, they prefer to heat the mixture to 150° C. An oil bath heated on a hot plate would work well for this variation. From there the procedure works exactly the same way as in the standard procedure. The claimed yield should increase to over 400 ml of phenylacetone, as compared to maybe 300 ml with the standard procedure. It's worth checking out.

References

Journal of the American Chemical Society, Volume 60, page 914 (1938).

Chapter Eight
Phenylacetone Via the Tube Furnace

The best way to produce phenylacetone on a large scale and continuous basis is by a catalyst bed inside a tube furnace. This has several advantages over the other methods described in this book. Cheap and very common acetic acid is used to react with phenylacetic acid instead of the expensive and more exotic acetic anhydride and pyridine. Use of the tube furnace frees up the glassware for use in other operations. The furnace requires very little attention while it is in operation, which allows the underground chemist to spend his time turning the phenylacetone into methamphetamine. There is no reason why this process cannot be used in small-scale production. It is just that its advantages really come out when large amounts of phenylacetone must be produced.

In this process, a mixture of phenylacetic acid and glacial acetic acid is slowly dripped into a Pyrex combustion tube which is filled with pea-sized pumice stones covered with a coating of either thorium oxide or manganous oxide catalyst. This bed of catalyst is heated to a high temperature with a tube furnace and the vapors of phenylacetic acid and acetic acid react on the surface of the catalyst to produce ketones. Three reactions result.

The acid mixture is prepared so that there are three molecules of acetic acid for every molecule of phenylacetic acid. This makes it much more likely that the valuable phenylacetic acid will react with acetic acid to produce phenylacetone rather than with another molecule of phenylacetic to produce the useless dibenzyl ketone.

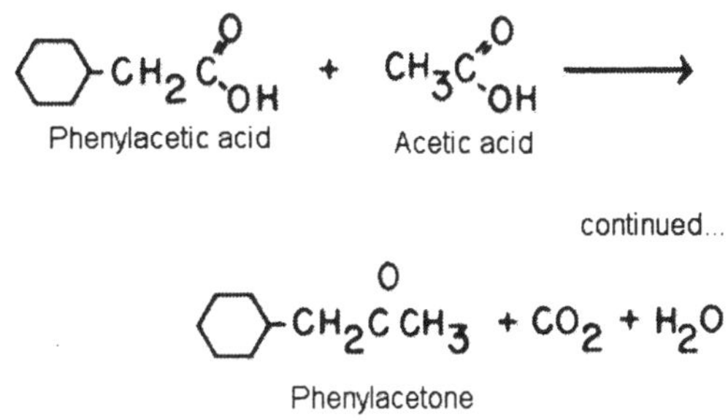

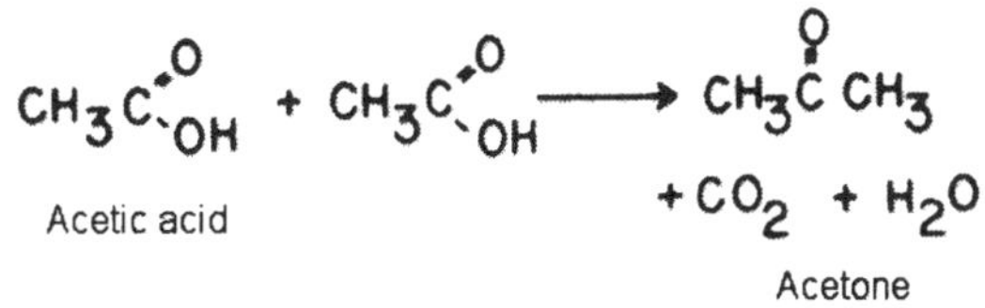

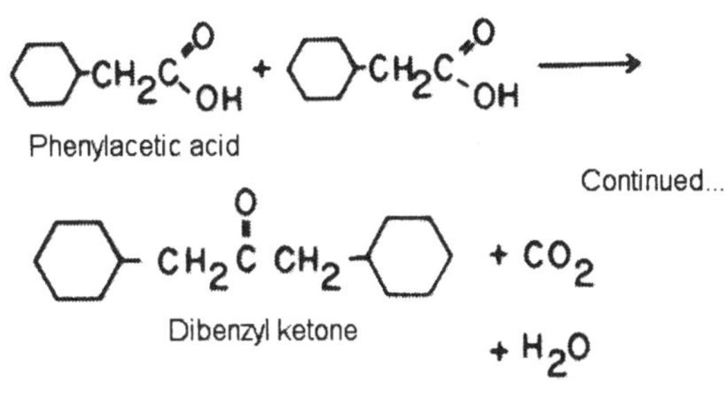

The vapors are kept moving in the catalyst tube by a slow stream of nitrogen and eventually the product comes out the far end of the catalyst tube. The vapors are then condensed and collected in a flask.

The complete apparatus for doing this reaction is shown in Figure 23. The combustion tube is made of Pyrex and is about one meter long. It is about 2 centimeters in internal diameter, with a male $^{24}/40$ ground glass joint on one end and a female $^{24}/40$ ground glass joint on the other end. If the underground chemist cannot buy the tube with the glass joints already on it, there are many places which will weld these glass joints onto the tube. He can find such a place by asking around and checking the Yellow Pages.

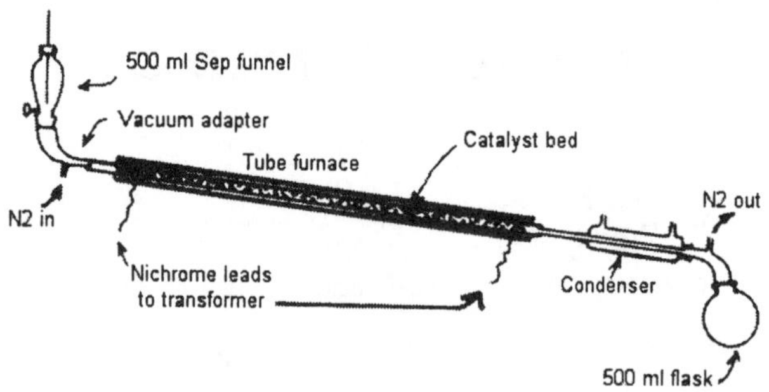

Figure 23

The tube furnace must be 70 centimeters in length. The only commercially available tube furnace that I know of is the Hoskins tube furnace. It is a fine furnace, but only 35 cm in length. Two of these would have to be run end-to-end to get the required 70 cm length. The cost, including a transformer for each of the furnaces, would be over $700. It is better and cheaper for the chemist to build his own tube furnace.

The tube furnace starts with a section of thinwall iron tubing about 75 cm long and 3 to 3.2 cm in internal diameter. Thinwall iron tubing has a metal thickness of .024 inch. The outside of the tubing is wrapped with asbestos cloth or asbestos paper to a thickness of about 2 millimeters. Asbestos cloth or paper is available at hardware stores.

Fifty feet of 28-gauge AWG nichrome wire is wrapped around the central 70 cm of the tube. The windings are made fairly taut so that the wire sinks slightly into the asbestos paper. Each winding is evenly spaced from the previous one, about ½ cm apart. One winding must not be allowed to come into contact with another, or there will be a short circuit.

The outside of the tubing is insulated with 6 or 7 layers of asbestos paper or cloth. This insulation is held in place by using copper wire ligatures about 6 inches long, wrapped around the outside of the insulation, and tied at the ends to make it tight.

The two ends of the nichrome wire are attached to insulated connectors (two of them) and then to a transformer. The Variac autotransformer is perfect for this job. It can adjust 115-volt house current anywhere from 140 volts down to zero. The transformer can handle 5 amps of current.

The chemist picks up a couple of pumice foot stones (Dr. Scholl's are suitable) at the pharmacy. With a hammer and screwdriver, he breaks them into round pieces somewhat smaller than a pea. Any sharp or protruding edges are knocked off. He makes enough of these pumice pebbles to fill the combustion tube for a length of 70 cm.

The pumice must now be purified to remove traces of metals and other garbage. This prevents the catalyst from being poisoned. The pumice pebbles are put into a 1000 ml beaker along with a wad of glass wool (Angel Hair) somewhat larger than a fist. The glass wool will be going into the combustion tube, so it must be cleaned off along with the pebbles. The glass wool is packed down. Then nitric acid is added until both the pumice and glass wool are covered. The beaker is put on an electric hot plate and the nitric acid boiled for half an hour. This converts metal impurities into soluble nitrates, and oxidizes other garbage. The nitric acid is all poured off and down the drain. The pumice and glass wool are then covered with distilled water and soaked for 5 minutes. This water is then drained off and replaced with more water. The water is boiled for 10 minutes, then drained off. This boiling water rinse is repeated two more times using distilled water. Finally, the water is drained out and the beaker laid on its side to drip out the last drops of water.

The pumice pebbles are now ready to be coated with catalyst. About 450 ml of distilled water is put into a clean 1000 ml beaker. The chemist dissolves 276 grams of thorium nitrate into this water. In another clean beaker, he dissolves 106 grams of anhydrous sodium carbonate into 400 ml of distilled water. (He uses A.R. grade chemicals.)

Slowly, and with constant stirring, the sodium carbonate solution is added to the thorium nitrate solution. Using a mechanical stirrer to stir the thorium nitrate solution is best, but a glass rod also works.

Thorium nitrate reacts with sodium carbonate to make thorium carbonate and sodium nitrate. Thorium carbonate does not dissolve in water, so it forms a white precipitate. Sodium nitrate stays dissolved in water. The stirring is continued for a couple of minutes after all the sodium carbonate has been added, then it is allowed to settle. The thorium carbonate settles into a gooey gunk at the bottom of the beaker. As much of the water as possible is poured off. Then 600 ml of distilled water is added to the thorium carbonate and stirred around with a clean glass rod. The chemist makes sure that all the thorium carbonate gets into contact with the clean water, and that any lumps are broken up. This dissolves any remaining sodium nitrate.

The thorium carbonate is allowed to settle again, then as much of the water as possible is poured off. Small amounts of distilled water are added and stirred in until a fairly thick paste is formed. Now the purified pumice pebbles are added and stirred around until they are all evenly coated with thorium carbonate.

A Pyrex glass cake pan is placed on the electric hot plate. The heat is turned on to ¼ maximum and about $^1/_8$ of the coated pumice pebbles are added to the glass pan. They are heated there with constant stirring with a thick glass rod, so that the pieces dry out evenly. When the coated pumice pebbles no longer stick together, they are dry enough. They are transferred to a clean sheet and an equal amount of wet pumice pebbles is put in the cake pan. They are dried out like the first group of pebbles. This process is repeated until all the coated pumice pebbles are dry. Any white powder that failed to stick to the pumice is collected and saved in a glass jar. If it is later necessary to change the catalyst bed, this material is wetted and used to coat new pumice pebbles.

A plug of the purified glass wool about 3 cm long is put into the combustion tube about 15 cm from the male end. This will hold the catalyst bed in place. The tube is filled up with the coated pumice pebbles for a length of 70 cm or so. A small plug of purified glass wool about 1 cm in length is put every 15 cm. This reduces the danger that tar building up on the pumice pebbles will block the tube.

The tube is put inside the furnace. If two Hoskins tube furnaces are used end-to-end, the tube is insulated in the space between the two furnaces with several layers of asbestos paper or cloth. In this space, the tube is filled with loose glass wool. This space is not counted as part of the necessary 70 cm of catalyst bed.

The apparatus is set up as shown in Figure 23. It is tilted at an angle of about 20 degrees, the end with the sep funnel being higher than the end with the collecting flask. The sep funnel has a one-hole stopper with a piece of glass tubing running through it almost all the way to the valve of the sep funnel. This is a constant pressure device that causes the contents of the sep funnel to drip into the tube at a constant rate, no matter what the level of the acids in the sep funnel are at a particular instant.

The sep funnel is connected to the female end of the vacuum adapter. The male end of the vacuum adapter is inserted into the female end of the combustion tube. The male end of the combustion tube is connected to a condenser. The condenser is connected to a vacuum adapter, and the vacuum adapter leads to a 500 ml round bottom flask. The glass joints are lightly greased and wired together where possible. The furnace must be supported to prevent its weight from bending the soon-to-become-soft hot glass tube. Clamps connected to ringstands are used to hold the other pieces in place.

The vacuum adapter connected to the sep funnel is the nitrogen gas inlet. The underground chemist gets a tank of nitrogen at a welding supply shop. He has to make sure that he knows how to use the regulators. He runs a line of tubing from the tank to the "bubbler." The bubbler is shown in Figure 24. It is a bottle with a 2-hole stopper in the top. One hole has a section of glass tubing reaching nearly to the bottom of the bottle. The bottle has about an inch and a half of concentrated sulfuric acid in it. The purpose of the sulfuric acid is to dry the nitrogen gas and to show how fast it is bubbling into the apparatus. The other hole has a short section of glass tubing. Plastic tubing is attached to this tubing and leads to the vacuum nipple of the vacuum adapter.

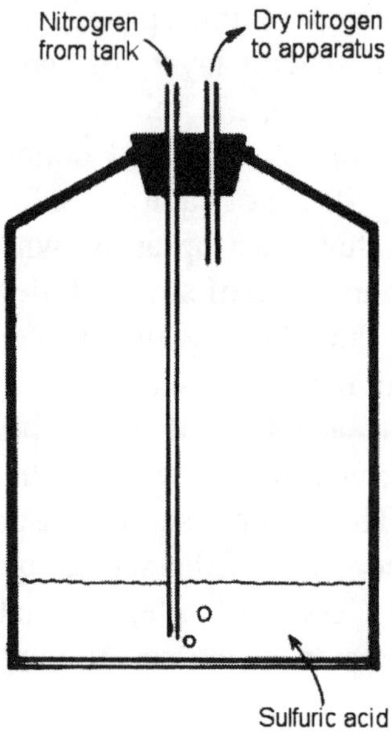

Figure 24

And now the time has come for the underground chemist to fire up the furnace. He places a thermometer capable of reading up to 450° C, or, better yet, a thermocouple in the furnace against the outside of the glass tubing. (If his thermocouple did not come with wiring instructions, he can find the wiring diagram in the *Encyclopedia Britannica* and in many college-level physics textbooks.) The thermometer or thermocouple extends into the central regions of the furnace. The space at the end of the furnace between the outside of the glass tubing and the inside of the furnace's iron tubing is plugged up with pieces of asbestos paper or cloth to hold in the heat.

He turns on the electricity to the furnace, and begins a slow stream of nitrogen (about one bubble per second) through the tube. He keeps a sheet listing the temperatures his furnace gets at various voltage settings on the transformer. Of course, it takes a while for the furnace to heat up to its true temperature at a given setting.

Now the tube furnace is heated to 425-450° C, while the slow stream of nitrogen continues through the tube. The heat turns the thorium carbonate into thorium oxide. The heating continues for 12 hours, after which the catalyst is ready to produce phenylacetone.

The chemist mixes 200 grams of phenylacetic acid with 250 ml of glacial acetic acid. He mixes them thoroughly, the phenylacetic acid dissolving easily in the glacial acetic acid. (Glacial acetic acid is the name for pure acetic acid; it is so called because it freezes at a little below room temperature.) Next add about 10 ml of water to this acid mixture and shake or stir it to get the water evenly dissolved throughout the solution. This small amount of water helps to prevent the formation of tar on the hot catalyst bed while it is working. Tar represents lost product, and a tar covered catalyst can't participate in the reaction we want to perform

This acid mixture is poured into the sep funnel and the funnel is stoppered with the one-hole stopper with the glass tubing constant pressure device. The temperature of the furnace is 425-450° C, and a one-bubble-per-second stream of nitrogen has been flowing through the tube for at least 12 hours. The valve on the sep funnel is opened so that about 20 drops of the acid mixture drip into the tube from the sep funnel every 30 seconds.

A slow flow of water is put through the condenser to condense the ketones as they leave the furnace. The product collects in the 500 ml flask and the nitrogen gas exits through the vacuum nipple of the vacuum adapter connected to the condenser. If there is trouble condensing all the acetone, the 500 ml flask is packed in ice.

It takes about 5 hours for all the acid to drip into the tube. When all the acid mixture has dripped in, 25 ml of acetic acid is added to the sep funnel and dripped in. This flushes the last of the product out of the catalyst bed.

The product in the 500 ml flask consists of a lower water layer and a brown-colored organic layer on top. The latter is poured into a 1000 ml sep funnel; the water layer is then drained off into a clean beaker, and the organic layer is poured into another beaker. The water layer is put back into the sep funnel along with 50 ml of toluene, and the funnel is shaken. It is allowed to sit for a few minutes, then the lower water layer is drained off and thrown away. The toluene layer is poured in with the organic layer in the other beaker.

The chemist is now ready to clean up the phenylacetone so that it can be distilled. He mixes up a supply of 10% sodium hydroxide solution by adding 10 ounces of lye to ¾ of a gallon of water in a glass jug. He pours the organic layer into the sep funnel, adds 400 ml of the sodium hydroxide solution and shakes. The water layer is drained off into a clean beaker and the organic layer is poured into another beaker. The water layer is returned to the sep funnel and 75 ml of toluene added. The funnel is shaken, then the water layer is drained off and thrown away. The toluene layer is poured in with the organic layer. This is repeated three more times, then the phenylacetone is distilled as described in Chapter Three. The yield of phenylacetone is about 100 ml.

The temperature of the furnace is raised to about 525° C, and a slow stream of air is drawn through the tube for two hours. The air is drawn through by turning off the nitrogen flow, opening up the valve of the sep funnel and attaching a vacuum hose to the vacuum nipple of the vacuum adapter on the 500 ml flask side of the apparatus. This air flow burns off built up crud and tar on the catalyst and charges it up for another run. It is done after the first run, and then after every few batches.

The furnace temperature is set at 425-450° C again and the flow of nitrogen through the tube is resumed. It is flushed out for a couple of hours, then the sep funnel is filled with acid mix for another run. It is dripped in as before to get another batch of phenylacetone. In this way, phenylacetone can be produced on a continuous basis.

If the homemade furnace has trouble reaching the necessary temperature, the chemist wraps it with more insulation. If that does not do enough, a lower temperature process can be used by replacing the thorium-oxide-coated pumice pebbles with manganous-oxide-coated pumice pebbles. The process goes as follows:

The pumice pebbles are made and purified with nitric acid as described earlier. In a 1000 ml beaker, 70 grams of manganous chloride ($MnCl_2$) is dissolved in 300 ml of distilled water. In another beaker, 38 grams of anhydrous sodium carbonate is dissolved in 500 ml of distilled water. The sodium carbonate solution is slowly added to the manganous chloride solution with constant stirring. Manganous chloride reacts to form manganous carbonate, which does not dissolve in water and precipitates out. The manganous carbonate is filtered out in a Buchner funnel as described in Chapter Five. The crystals are rinsed with distilled water.

The manganous carbonate is returned to a clean beaker and enough distilled water is added to make it into a fairly thick paste. If too much water is added, it does not stick well to the pumice. The pumice pebbles are stirred in until they are evenly coated. The beaker is heated on a hot plate while the pumice stones are vigorously stirred. Local overheating must be avoided or the catalyst will be ruined. When most of the water is evaporated, the catalyst is transferred to a Pyrex cake pan and gently heated on a hot plate. The pumice chips are stirred constantly to get even drying. When they no longer stick together, they are transferred to a clean sheet of paper.

The chemist fills the combustion tube with the catalyst as before and sets up the apparatus. He heats the furnace to 360-400° C while passing a stream of nitrogen through the tube. This converts the manganous carbonate to manganous oxide (MnO). This heating is continued for 8 hours. Then the heat is reduced to 350° C, while the stream of nitrogen is continued at a rate of one bubble per second. When 350° C is reached, he drips in the same phenylacetic acid-acetic acid mixture used earlier in this chapter. The correct rate is 20 drops every 30 seconds. When it has all dripped in, he adds 25 ml of acetic acid to the sep funnel and drips it in. He then either adds more acid mix to the sep funnel for another run, or shuts down the furnace. If he shuts down the furnace, he must continue the flow of nitrogen through the tube until it has cooled off. This prevents the MnO catalyst from being oxidized to MnO_2, etc. When he turns it back on, he must immediately start the nitrogen flow for the same reason. The product is purified in the same way as described earlier in this chapter.

Since no air is sucked through the tube at high temperature, gunk builds up on the catalyst and eventually puts it out of commission. When this happens, the catalyst bed is changed. The yield using the manganous oxide catalyst bed is not as good as that using the thorium oxide catalyst bed. Thorium oxide is used, unless the chemist has no choice.

A somewhat more complicated way to do this reaction is to use what is called a thorium oxide "aerogel" catalyst. A lower temperature and a higher rate of production are possible. For more information about it, see *Industrial and Engineering Chemistry*, published in 1934, Volume 20, pages 388 and 1014.

Before leaving this topic, we should address the subject of smells, in particular the smells which can accompany this process. The tube furnace is not something which should be set up in an apartment building or in a rental house that the landlord stops by to visit on more than rare occasions.

Phenylacetic acid smells just like cat piss, and the smell gets on surfaces and just lingers and lingers. The heat knows what this smell is associated with, and neighbors smelling it all the time will just get mad. They will think you have a herd of screaming tom cats pissing on the walls or a bunch of ferrets gone bonkers! There are certain sections of cities where using the living room as a bathroom is nothing unusual, but we'll assume that you don't want to set up shop there.

One might say, "Hey, I won't spill any, so I won't stink up the place." Let me assure you, stuff happens, especially with a furnace. All it takes is a loose joint on the glassware to drip phenylacetic acid solution all over the place. Pouring spills happen, and there is the glassware to clean. The stuff gets around.

The best way to clean up phenylacetic acid is to mix some lye in with rubbing alcohol. While wearing gloves, you can sponge this solution into the spilled phenylacetic acid. It will be converted to the sodium salt of phenylacetic acid, which doesn't stink. The sodium salt is also soluble in water, so further cleanup with soap and water will be successful.

Another source of smelly emanations from the tube furnace is due to the form in which the product tends to exit from the furnace. There is a strong tendency for the product mixture to form a fog as it exits the furnace and gets cooled down in the condenser. This fog is pretty hard to get condensed completely down to liquid with just a trip through a condenser. *Organic Syntheses* goes so far as to suggest that the vapors be passed through a bed of marbles to break the fog. Your Uncle suggests that the upper part of the vacuum adapter be packed with glass wool to give a lot of surface for the fog to condense upon before reaching the exit of the apparatus. Fog escaping the apparatus results in a lower yield, and a lot of stink. Both are undesirable.

References

Journal of the Chemistry Society, page 612 (1948); page 171 (1940).

Organic Syntheses, Collective Volume Two.

Chapter Nine
Other Methods of Making Phenylacetone

There are many other methods of making phenylacetone described in the scientific literature. Most of them are dogs, not worth the time and effort. But there are some good methods of making phenylacetone that I have not yet described.

An acceptable method is to oxidize methyl benzyl carbinol (1-phenyl-2-propanol) to phenylacetone (methyl benzyl ketone) with chrome trioxide (CrO_3) in pyridine solvent. The problem with this is that methyl benzyl carbinol is not commercially available, and so must be made from benzyl chloride grignard reagent and acetaldehyde. This grignard works well, although there can be a problem getting unreacted benzyl chloride out of the product. Their boiling points are very close, so distillation does not separate them completely.

For a good typical example recipe using chrome trioxide in pyridine to oxidize an alcohol to a ketone, see *Journal of Organic Chemistry,* Vol. 47, page 2643 (1982). This is the best reagent for this purpose, considerably better than Cr^{+6} and sulfuric acid combinations in either water or acetone solvent. It is, however, stinky and expensive to use pyridine. You won't find that material down at the hardware store either.

The real obstacle to using this possible route to phenylacetone isn't getting the pyridine. It is getting the materials for and doing the grignard reaction that yields the methyl benzyl carbinol. Benzyl chloride is on the less restrictive List II of "hot chemicals." It's also nasty to work with as it's a reasonably potent tear gas. The anhydrous ether needed to do the grignard reaction has long been associated with drug manufacture, smells a lot, and is very prone to ignition of the vapors by sparks and such. Magnesium turnings are also on the California list of controlled chemicals, and may well be on other state reporting lists. If you are in an industrial situation where methyl benzyl carbinol is produced as an intermediate and a few gallons can be carried home without anyone noticing, then this is a fine route to phenylacetone. In other situations, it's not practical.

Another two-step method of making phenylacetone is to make benzyl cyanide from benzyl chloride and sodium cyanide, and then make the benzyl cyanide into phenylacetone by the method described in Chapter Seven. The way to make benzyl cyanide can be found in *Organic Syntheses,* Collective Volume 1. Benzyl cyanide is listed in the table of contents.

This route to phenylacetone has its possibilities. The List II chemical benzyl chloride can be obtained through regular commercial sources, or driven across the Mexican border in amounts under a few kilos. See the latest printing of *21 CFR* parts 1310 onward, for just how much benzyl chloride can be imported at one time without being subject to felony charges.

The cyanide needed for this reaction can be pretty easily obtained in large amounts from chemical outlets that serve the industry. A typical container will hold 100 to 200 pounds of the cyanide in briquette form at a price of a dollar or two per pound. The most common industrial usage of this material is electroplating. Zinc, copper, brass, cadmium and silver are all electroplated from solutions that contain roughly a pound per gallon of cyanide. Crafting a company and cover story consistent with these facts would be advisable.

If one should choose to scale up this process beyond using a few pounds of cyanide, the subject of waste disposal should be addressed. Pretty much any city served by a sewer system monitors the water coming into the sewage treatment plant for the presence of heavy metals and cyanide. If they are finding hundreds of pounds of cyanide coming at them, they will look for the source! For this reason, large amounts of cyanide shouldn't just be poured down the sewer. It shouldn't be dumped into the ground either! Simply pour the waste cyanide into a drum, and when cooking is done, pay the couple of thousand dollars to have it carted away by a hazardous waste disposal company. They won't be looking for incriminating evidence in the waste, and, for a large operation like the one using hundreds of pounds of cyanide, the couple thousand dollars is chicken feed.

An alternative method for disposal of cyanide waste is to destroy it with bleach. Then the water solution which is left can be safely poured down the sewer. To do this, add an equal weight of sodium hydroxide to the cyanide waste and add water to bring the two ingredients into solution. Then slowly and with constant stirring add about 4 volumes of bleach to the cyanide solution. The best bleach is the highly concentrated 15% hypochlorite solution available from pool suppliers, although 5% household bleach can be used by tripling the amount added. The reaction creates a lot of heat, so when the solution starts steaming just take a break and let it cool down.

Another way to come at the production of phenylacetone is to start with common materials that will react to produce phenylacetic acid easily and in high yield. There are more than a couple of such materials, and none of them are subject to reporting requirements. I am ashamed that my government could be so sloppy in compiling their List I chemicals! This situation must be corrected, because it is my lifelong goal to get every chemical under the sun on some type of reporting list.

Probably the simplest and easiest precursor to phenylacetic acid is its ethyl ester, ethyl phenylacetate. This substance is used in fairly large amounts in perfumes, so by disguising yourself as a perfumer, pails or drums of this very useful chemical could be had by tapping into the perfumer's supply lines. Ethyl phenylacetate is also used in artificial fruit flavorings and to produce an aroma of honey. One would first be well advised to become at least passably informed about perfume formulations. Books such as *Common Fragrance and Flavor Materials* by Bauer, *Perfume and Flavor Chemicals* by Arctander and *The Perfume Handbook* by Groom are a great place to start one's education on the topic. Then one goes to trade journals such as *Soap, Perfumes and Cosmetics* to look for advertisements from chemical suppliers to the perfume trade. These suppliers are greatly preferable to doing business with scientific supply houses, who will also stock ethyl phenylacetate. The rule of thumb is that one should try to avoid any supplier who also sells List I chemicals. To sell List I chemicals, one must be a trusted snitch and lackey. I also must regretfully inform you that ethyl phenylacetate has recently been placed on the List I. Perfumers are still using it though, so one may be able to hide among the trees in the forest. If one orders the chemical in solution with other ingredients, then it isn't subject to reporting. The simplest solution of all would be ethyl phenylacetate in alcohol solution. Then this purchase would certainly appear to be from a perfumer seeking to impart its fragrance to his product.

The reaction to convert ethyl phenylacetate to phenylacetic acid is just a simple ester hydrolysis. Alkaline hydrolysis, also called saponification, using NaOH or KOH, is superior to acid catalyzed hydrolysis in this case. The following reactions bring about the transformation:

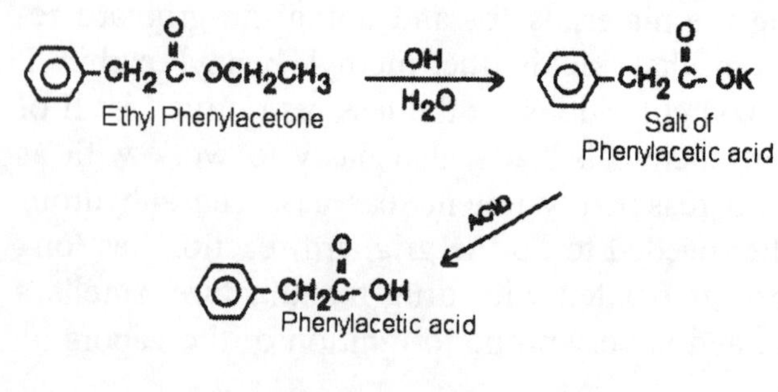

For example, take a 2000 ml round bottom flask, or a copper or stainless steel replica. Into the flask put 200 ml of ethyl phenylacetate, 1000 ml of 190 proof vodka, and 140 grams of KOH (potassium hydroxide). Attach a reflux condenser, and heat the pot to boiling using a hot plate and a pan of cooking oil as a heating bath. Reflux at a moderate rate for 2 hours.

Now distill off most of the alcohol solvent. A vacuum assist can be used to speed up the distillation of the vodka if desired. The volume should be reduced to around 500 ml.

Next pour the residue into a 1500 ml beaker. Rinse the flask with some water to get all the reaction mixture out and into the beaker. Then mix around 200 ml of sulfuric acid into around 400 ml of water. Once the acid solution has cooled down, slowly add it to the reaction mixture in the beaker with strong stirring. Add it at the rate that the heat created by the neutralization of the KOH and the potassium salt of phenylacetic acid will allow. Putting the beaker in some cool water will help. Add acid until a pH of around 1 is reached. Then cool down the mixture to about 5-10° C.

A large amount of phenylacetic acid crystals will form. Filter them out. They should next be recrystallized by putting them into a beaker, and adding about 700 ml of hot water. Heat the mixture with stirring until all the phenylacetic acid dissolves, then cool down the beaker as before. Filter out these recrystallized phenylacetic acid crystals, and preferably dry them off by applying a vacuum to them either in a dessicator or large filtering flask. The yield should be around 140 grams.

The horrors go on and on! Your Uncle thinks that saving the world from the evils of drug abuse is becoming just a hopeless task! See *Journal of Organic Chemistry*, Volume 62, pages 234-35 (1997).

In this recipe, they mix 10 millimoles of phenethyl bromide with 30 millimoles of sodium nitrate and 100 millimoles of glacial acetic acid in 20 ml of DMSO.

They stirred this mixture at 35° C for 6 hours, then they diluted the mixture with several volumes of 10% HCl. They next extracted out the product — phenylacetic acid — with toluene, washed the extract with some water, and distilled it to get 78% yield of phenylacetic acid.

That's a pretty simple recipe. I'll bet it scales up easily. It is high time that phenethyl bromide joins the List I chemicals, or at least, List II. I'll sleep much better at night when something is finally done about this!

Another convenient precursor to phenylacetic acid is mandelic acid. It, too, has been neglected as to placement as a listed chemical under the Chemical Diversion Act. This situation has to be changed! How else can the world be made safe?

Mandelic acid is presently available at a price ranging from $25 to $45 per pound, with discounts for larger quantities. Mandelonitrile is more expensive. A mainstream legitimate use of mandelic acid is as an acidifier for the urinary tract, while mandelonitrile is associated with laetrile through its source from nature, amygdalin.

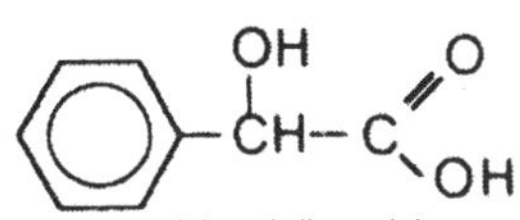

Mandelic acid

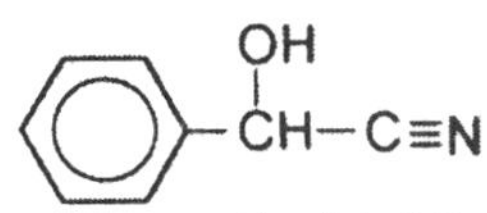

Mandelonitrile

Mandelic acid is easily reduced to phenylacetic acid by the procedure found in *Helvetica Chimica Acta,* Volume 22, page 601 (1939). Mandelonitrile is very simply converted to mandelic acid by the procedure found in *Organic Syntheses,* Collective Volume 1. Look in the table of contents for mandelic acid.

Note here that mandelic acid is a benzyl alcohol, just as ephedrine, pseudoephedrine, and phenylpropanolamine are benzyl alcohols. As a consequence, the direct reduction methods given in Chapter Fifteen can also be used in this case to reduce mandelic acid to phenylacetic acid. The method most suited to reducing large quantities would be to adapt the so-called "Nazi Meth" recipe (Method Three in the section on "Direct Reductions"). Some mandelic acid dissolved in glacial acetic acid with some 70% perchloric acid and palladium catalyst will hydrogenate quickly and efficiently at 80° C and a hydrogen pressure of about 30 pounds.

One could also use The Advanced Fester Formula (Method Five in the section on "Direct Reductions") to do the same hydrogenation by first making the acetic acid ester of the mandelic acid, and then pouring the ester reaction mixture into the electrochemical cell. See Volume 7 of *Organic Reactions* for a good review of this class of reactions, termed hydrogenolysis of benzyl esters.

Another method to convert mandelic acid to phenylacetic acid is to use HI and red phosphorus, just as in Method Four in the section on "Direct Reductions." One simply uses mandelic acid in the reaction instead of ephedrine, pseudoephedrine or phenylpropanolamine. Iodine and red phosphorus are difficult and dangerous materials to buy from scientific supply houses who also sell List I chemicals, even though they have not been formally placed on the reporting list. As a consequence, the HI and red phosphorus option is probably best left alone.

Still another good precursor to phenylacetic acid is out there for the taking. Oh, the sloppiness of those people who compiled the List I chemicals! Something must be done about this, and done soon! Your Uncle can't believe the holes that have been left in federal reporting requirements. This precursor is ethylbenzene.

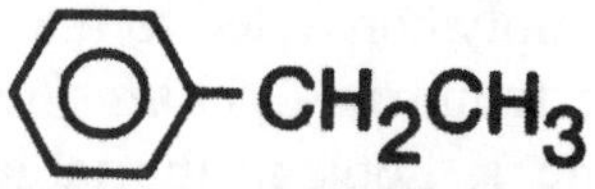

Ethylbenzene

Ethylbenzene is made by the thousands of tons as a feed material for the production of styrene, and as a solvent for plastic resins. By disguising yourself as someone doing work in the latter area, pails and drums of ethylbenzene will become easily available from chemical outlets serving the industry. These are much better people to deal with than scientific supply houses.

Ethylbenzene is converted to phenylacetic acid in 90% yield by a fairly simple chromate oxidation in water solution. The reaction must be done inside a sealed steel pipe, as a temperature of 275° C is required. See *Advanced Techniques of Clandestine Psychedelic & Amphetamine Manufacture* for a complete discussion of this reaction.

Finally, still another good precursor to phenylacetic acid is benzyl cyanide, discussed earlier in this chapter. By heating benzyl cyanide with sulfuric acid solution, phenylacetic acid is produced. See *Organic Syntheses*, Collective Volume 1, for details. Look in the table of contents for phenylacetic acid.

I should note here that in all of these recipes, the phenylacetic acid is recovered from solution in the opposite manner that meth is. Meth is a base, and so dissolves in acid water solutions. To get it out of water solution, the water must be made strongly (pH 13+) alkaline. With phenylacetic acid, it works just the opposite, because it is an acid. Phenylacetic acid dissolves in sodium hydroxide solutions, forming the sodium salt. To get it out of water solution, the water is made acid, generally with dilute sulfuric acid. The crystals of phenylacetic acid are then filtered out or extracted with solvent. The details are given in the recipes cited. It is a good idea to recrystallize the phenylacetic acid made by these methods from water. This will remove traces of sulfuric acid from the crystals. Phenylacetic acid will dissolve in hot water, but hardly at all in cold water. This is how the recrystallization is done. See *Organic Syntheses*, Collective Volume I, under phenylacetic acid, for the details. See also the discussion in the tube furnace chapter for some hints in cleaning up spills and traces of phenylacetic acid, so that lingering smell doesn't leave the finger of blame pointing your way.

Once one has phenylacetic acid, making phenylacetone is pretty easy. See Chapter Three and the tube furnace chapter for the most convenient and practical methods, but there are plenty of others.

In what is actually the best method of making phenylacetone, two molecules of methyllithium react with phenylacetic acid to produce phenylacetone, or one molecule of methyllithium reacts with one molecule of the lithium salt of phenylacetic acid to produce phenylacetone. This reaction is done in anhydrous ethyl ether under an atmosphere of nitrogen. Organolithium reagents burst into flame upon contact with air. Although methyllithium is not so bad in this respect as t-butyllithium, organolithium reagents are dangerous to handle. But, apart from the element of danger, this is the best way to make phenylacetone. The high cost of lithium is offset by the high yields of product. Go to *Organic Syntheses*, Collective Volume 5, page 777, for a sample recipe, and some commentary.

Lithium metal obtained from batteries is probably of high enough quality to give good results in this reaction. See *Advanced Techniques of Clandestine Psychedelic & Amphetamine Manufacture* for a discussion of the method for obtaining lithium metal from batteries. This still leaves the problems associated with obtaining and using the anhydrous ether or THF needed as solvent for the reaction. These problems aside, this reaction produces high yields of a clean product.

Other methods of making phenylacetone call for the use of phenylacetyl chloride, the acid chloride of phenylacetic acid.

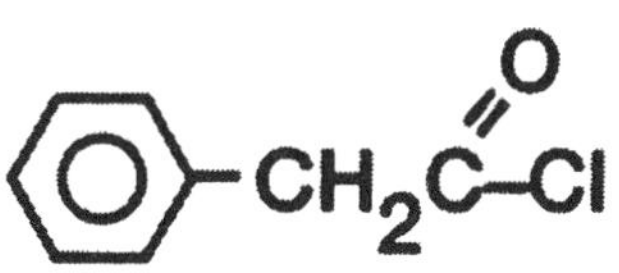

Phenylacetyl Chloride

Can you believe that this chemical isn't on the federal reporting lists? Saints preserve us! Don't get all excited though, as this chemical is only available from those scientific supply houses. They probably have some idea as to what it's good for. I don't know of any common industrial uses for this chemical.

With phenylacetyl chloride, a good way to make phenylacetone is to react methyl zinc reagent with phenylacetyl chloride. The claimed yield is around 70%. Methyl zinc reagent is made by adding zinc chloride to methyl grignard reagent. It is not an especially difficult reaction to do, and the yields are very good. The problem is that phenylacetyl chloride is expensive and hard to find, although it can be made from phenylacetic acid and thionyl chloride ($SOCl_2$). This reaction is described in the *Journal of the American Chemical Society,* Volume 70, page 4214, (1948). This can be found in any good college library.

The production reaction using the phenylacetyl chloride to make phenylacetone was first reported in *Berichte der Deutschen Chemischen Gesellschaft* Volume 5, page 500 (1872). Other references can be found in the *Journal of the American Chemical Society,* Volume 41, page 411 (1919) and Volume 69, pages 2350-54 (1947). Also see the discussion in *Grignard Reactions of Nonmetallic Substances* starting on page 713. This method has its possibilities, but the need for anhydrous ether to do the grignard reaction is a problem, as is getting the phenylacetyl chloride. See *Chemical Abstracts,* Volume 48, column 8261 for a good sample recipe.

Another good way to make phenylacetone is to react phenylacetyl chloride with the ethoxymagnesium derivative of dimethyl malonate. Hydrolysis with acid then produces phenylacetone. This reaction is covered in detail in *Advanced Techniques of Clandestine Psychedelic & Amphetamine Manufacture.*

Another material which can be converted into phenylacetone is cinnamaldehyde:

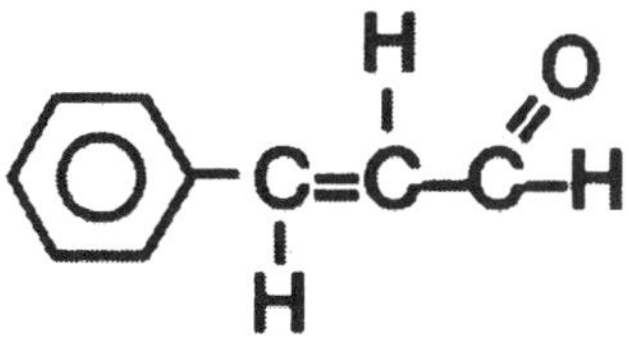

Cinnamaldehyde

Cinnamaldehyde is covered by no type of sales reporting control, and is used very heavily in flavorings and perfumes to give either the taste or smell of cinnamon. It also is present to the extent of 80-90% in oil of cinnamon, which is sold by the thousands of tons to flavor everything from toothpaste to candy to Coca-Cola. This cheap material is out there for the asking. Oh, the humanity! This intolerable situation can't be allowed to continue for a single minute longer!

How does one make phenylacetone from cinnamaldehyde? Looking at the structure of this molecule, what one wants to do is reduce the aldehyde grouping at the end of the propyl side chain all the way down to a methyl group (-CH_3), while still preserving the double bond on that side chain. With that double bond there, conversion to phenylacetone is easy by the methods given in Chapter Ten.

This selective reduction is a tall order! See *Reductions in Organic Chemistry* by Milos Hudlicky, page 102, for a table listing the products obtained by reducing cinnamaldehyde with common reducers. Reducers strong enough to reduce the aldehyde to a methyl group will also reduce the double bond or cause a cyclopropane to be produced. A derivative of cinnamaldehyde must be used, one that allows the aldehyde group to be reduced without losing the double bond on the side chain. This derivative is the tosylhydrazone.

A number of papers have been published on this reaction, and they all get similar and good results, so we can be assured that this method is valid and quite useful. Cinnamaldehyde or cinnamon oil reacts with p-toluenesulfonhydrazide in alcohol solvent to yield the tosylhydrazone derivative, which is filtered out. This derivative is then reduced with sodium borohydride or sodium cyanoborohydride to give allylbenzene in good to excellent yields.

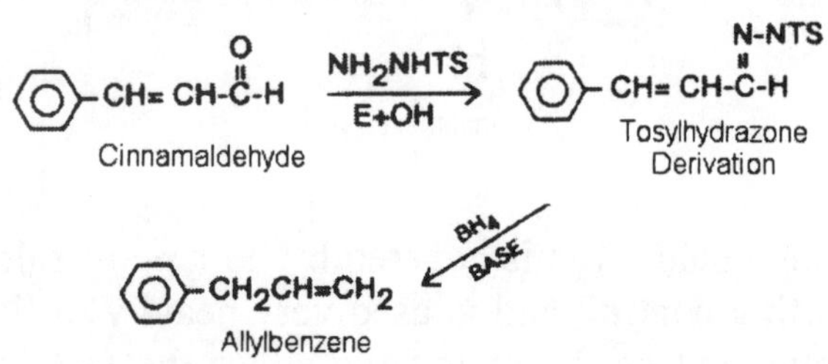

An interesting feature of this reduction is the movement of the double bond, giving allylbenzene instead of the propenylbenzene one would expect. Allylbenzene is some useful and versatile stuff! Either of two pathways to meth can be used with allylbenzene:

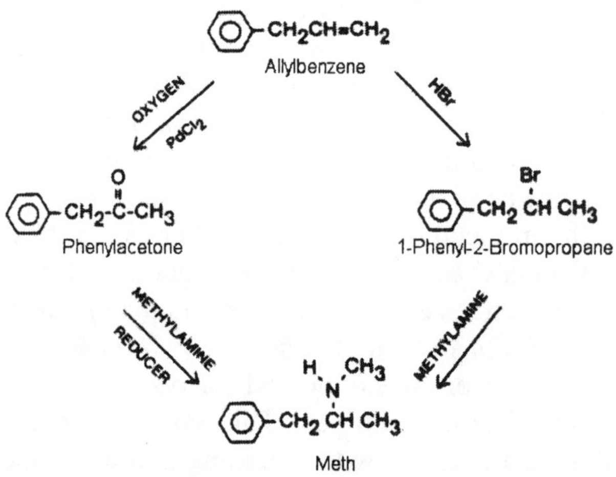

These two paths to amphetamines starting with allylbenzene become of extreme importance in making psychedelic amphetamines such as MDA or MDMA from commonly available essential oils which often contain large amounts of appropriately substituted allylbenzenes. For example, sassafras oil contains 80-90% safrole, which is 3,4-methylenedioxyallylbenzene. One has the choice then of converting the allylbenzene to either the phenylacetone or bromosafrole. Reaction then with methylamine gives MDMA. Substituting ammonia for the methylamine gives instead MDA. The so-called Wacker oxidations of allylbenzenes to phenylacetones are covered in the next chapter. The reductive amination of the phenylacetone to the amphetamine is covered in Chapters Five, Eleven, and Twelve. The other pathway using HBr is discussed in Chapters Seventeen and Twenty. Both are quite workable, and which one to choose is largely a matter of taste or availability of chemicals. Distilling the products is quite important with the second pathway, to get a pure product.

Now to get allylbenzene from cinnamaldehyde or cinnamon oil (aka oil of cassia, Chinese cinnamon oil, or cinnamon bark oil; avoid cinnamon leaf oil), put 130 ml of the cinnamaldehyde or cinnamon oil into a 1000 ml flask. Heating will be required for a fairly short period, so a round bottom flask equipped with reflux condenser is best. A volumetric flask could also be used by wrapping a cold rag around the long neck of the flask to act as a reflux condenser. Next add about 260 ml of 95% ethanol (190 proof vodka). The recipe given in *Organic Syntheses*, Collective Volume 6, page 293, uses 95% ethanol and gets a 90% yield of the tosylhydrazone derivative. Other recipes given in *Journal of Organic Chemistry*, Volume 40, page 923 (1975) and Volume 43, page 2299 (1978) and Volume 46, page 1217 (1981) use absolute alcohol. One-hundred-ninety-proof vodka is available at the liquor store, so this is the way to go. Finally, add about 220 grams of p-toluenesulfonylhydrazide. This material is listed in my Eastman catalog, selling for 30¢ a gram, and is available in bulk at a discount. It can also be made in high yield from p-sulfonyl chloride and hydrazine by the procedure given in *Organic Syntheses*, Collective Volume 5, page 1055. P-toluenesulfonylhydrazide is covered by no reporting requirements, and hasn't been the focus of suspicion. The list of potentially incriminating chemicals is now getting so long as to be almost useless anyway. I would suggest getting some in bulk, and doing no further business with whomever the chemical is obtained from.

Swirl around the contents of the flask to mix them together, then heat the flask on a steam bath or in some boiling hot water with either magnetic stirring or swirling. Once the mixture starts to get hot, the p-toluenesulfonylhydrazide will react with the cinnamaldehyde, and the mixture will become clear, or at least transparent, if oil is being used. A reaction time of about 15 minutes is needed, once the mixture becomes hot. If a makeshift reaction apparatus like a volumetric flask is being used, take care that vapors of alcohol don't come out of the vessel and into contact with sources of ignition, such as a hardware store hotplate. Keep the neck of the flask cool, and attach some tubing to the top to lead escaping vapors away.

Once a clear or transparent reaction mixture has been achieved, stop the heating, and allow the mixture to begin to cool. As it cools, crystals of the tosylhydrazone derivative will begin coming out of solution. This is a good point to pour the mixture into a 600 ml beaker or glass measuring cup. The mass of crystals which will form are much more easily poured and scraped into a filter from this container than from a round bottom flask or volumetric flask.

When the mixture has cooled down to room temperature, the mass of crystals should be filtered out using a Buchner funnel and vacuum flask, just as described for filtering out meth crystals in Chapter Five. The filtered out crystals of the tosylhydrazone derivative should next be recrystallized from 190 proof vodka. This is done by putting the crystals back into a clean round bottom flask or volumetric flask, adding about 200 to 250 ml of 190 proof vodka, and heating the mixture with steam or hot water to redissolve the crystals. When they have redissolved, the heating is stopped, and the liquid poured once again into a beaker or measuring cup once crystals begin to reappear. When the mixture gets cooled to room temperature or below, the mass of crystals is filtered out as before, then spread out on a plate to air dry.

The yield of dried crystals from pure cinnamaldehyde will be about 270 grams. The yield from oil will be correspondingly lower, depending upon what percentage of cinnamaldehyde it contains. Once that number is known, the amount of oil used can be increased proportionally to get maximum use out of the p-toluenesulfonylhydrazide.

With the tosylhydrazone obtained, reduction of it to allylbenzene can be done. There are three good methods, with yields of allylbenzene obtained ranging from 42 to 98%, depending upon the reaction conditions used. The first two methods use sodium borohydride as the reducer. This is a fairly common industrial chemical, and can be obtained in bulk from industrial chemical outlets. A common industrial use of sodium borohydride is to remove tightly complexed heavy metals from water prior to sewering the water. For instance, let's say that one plates copper or nickel, and the rinse water gets contaminated with ammonia or some other complexor for these metals. Before pouring this water down the sewer, the metal must be removed. Borohydride will reduce the complexed metal ion to metal, which will fall out of solution. Similarly, metal stripping baths can be treated in the same way to remove the metals built up in them. Also, electroless nickel-plating baths can be waste treated in the same manner to allow their disposal by pouring the bath down the sewer once the metal is removed from solution. Environmental protection is a good and believable story to use at the industrial chemical outlet.

The first method of reduction of tosylhydrazone to allylbenzene is taken from *Journal of Organic Chemistry*, Volume 43, page 2301, and is labeled Method A in that article. A glass or Teflon-coated container of approximately one gallon capacity is filled first with 2600 ml of glacial acetic acid. This very common industrial chemical is easily available in 55 gallon drums at a price of a dollar or so per gallon. The flask and its contents are cooled down in ice, then 75 grams of sodium borohydride is slowly added with stirring at such a rate that the temperature of the solution is kept in the 15-20° C range.

Once the borohydride has been added to the solution, then add 226 grams of the tosylhydrazone derivative of cinnamaldehyde, slowly with stirring. Continue the stirring for one hour at room temperature, then with continued stirring, raise the temperature of the solution to 70° C for 2½ hours. The solution will gas off nitrogen as the tosylhydrazone gets reduced to allylbenzene.

Next allow the solution to cool, and pour it onto about 10 pounds of crushed ice. When the ice has melted, add a 10% solution of lye in water, slowly and with stirring, until the water solution is basic to litmus or pH paper.

The allylbenzene is then extracted out of the water with two or three 200 ml portions of toluene, hexane, or even Coleman camper fuel. The combined extracts are then distilled. The solvent first distills off as the milky-looking water azeotrope, then as the pure solvent. When the solvent is gone, pure allylbenzene will distill at about 156° C. A reasonably efficient fractionating column will separate these materials with no problem. Vacuum distillation can also be used to collect the allylbenzene once the solvent has been distilled off. A good aspirator pulling about 20 torr with cold water will distill allylbenzene at about 50° C or so. Very cold water flow through the condenser will be required to catch all the allylbenzene at the low temperature that a good aspirator will cause it to boil at. The yield by this method will be about 40 ml or so of allylbenzene.

By increasing the amount of sodium borohydride used, and changing the order of addition, the yield of allylbenzene can be increased to over 50%. This is Method B in the same cited *Journal of Organic Chemistry* article.

Once again we start with a glass or Teflon-lined container of about one gallon capacity. Into it we place about 2500 ml of glacial acetic acid and 225 grams of the tosylhydrazone derivative of cinnamic acid. These ingredients are stirred to make a slurry, then at room temperature, slowly add about 280 grams of sodium borohydride at such a rate that the foaming of the reaction mixture doesn't cause the mixture to froth out of the container. This will take about an hour or so. In this variation, a lot more hydrogen gas gets produced than in the first method, so some ventilation should be provided to avoid explosion hazards.

After the borohydride is added, continue stirring for about an hour at room temperature, then heat the mixture to about 70° C for about an hour and a half. Then allow the mixture to cool down, and pour it onto crushed ice as in the first example, then add sodium hydroxide solution as before to make the solution alkaline, extract the mixture with toluene, hexane or Coleman camper fuel, and distill to get in this case around 60 ml of allylbenzene.

By using sodium cyanoborohydride as the reducer, the yield of allylbenzene can be increased to nearly 100%. Your Uncle doesn't recommend this variation because sodium cyanoborohydride isn't the common industrial chemical that sodium borohydride is. Sodium cyanoborohydride is available from those snitch-filled scientific supply houses, and the notation by the chemical generally says something along the lines of "selective reducer for Schiff's bases." Yes, it's a really convenient chemical to reduce phenylacetone-methylamine mixtures to meth in high yield. This chemical would be too scary for me to order from those people if I wanted to cook some crank. My advice is to stay away from buying this chemical.

This recipe also uses some fairly uncommon solvents, as opposed to the glacial acetic acid used in the previous two methods. This method is provided for educational purposes.

Into a container of around 10,000 ml capacity put 2500 ml of dimethylformamide (DMF) and 2500 ml of sulfolane. Mix them together, then add 300 grams of the tosylhydrazone derivative of cinnamaldehyde and 250 grams of sodium cyanoborohydride. Stir them into solution, then add a few tenths of a gram of the acid-base indicator bromocresol green. When the indicator dissolves, it will color the solution blue.

Next heat the solution to 105° C, then dropwise add concentrated hydrochloric acid until the pH drops below 3.8 as indicated by the color change of the solution from blue to tan. Then add 2000 ml of cyclohexane to the solution, and continue the heating with stirring for about an hour. I would think that toluene could be substituted for the cyclohexane. A little more bromocresol green is added during the heating period, and after one hour of heating, a little more hydrochloric acid is added dropwise to keep the pH below 3.8 as indicated by the tan color. Continue heating for an additional 1½ hours, for a total of around 2½ hours of heating.

After the solution has cooled, pour it into about 7000 ml of water, and shake the mixture around for a bit. Then separate off the organic layer floating on top of the water layer. Take the water layer, and extract it a couple of times with 200 ml portions of cyclohexane or toluene. Add these extracts to the organic layer which had been separated off. Your product is in there, heavily diluted with solvent.

Take the combined organic layer and solvent extracts, and wash this with about 1000 ml of water. Drain off the water layer, and repeat the water wash a couple more times.

The solvent-allylbenzene solution is next put into a large distilling flask, and the solvent is distilled off. Then the allylbenzene can be distilled as in the previous examples. The yield will be over 100 ml of allylbenzene.

Another good method of making phenylacetone is to use a method called the Knoevenagel reaction. In this method, the starting material is benzaldehyde. The advantages to being able to use a wide variety of starting materials to produce phenylacetone are obvious.

There's just so much stuff to "watch" that the "watchers" can't do a proper job of policing transactions. In this case, we can all sleep better at night knowing that benzaldehyde, and the other chemical used in this synthetic route, nitroethane, are both on the List I of reportable chemicals. However, don't let that warm fuzzy feeling of comfort and safety grip you too deeply. Technical grades of benzaldehyde used for flavorings in food are exempt from reporting. Oil of bitter almonds contains 95% benzaldehyde, and it too is exempt from control. The outrage against civilized values continues with the simple electric oxidation of toluene to benzaldehyde given later in this chapter.

Nitroethane is a more difficult matter. It finds use in the varnish industry as a solvent for nitrocellulose, so finding it at a waste exchange is possible. Sending letters to varnish makers similar to the one given in *Advanced Techniques of Clandestine Psychedelic & Amphetamine Manufacture* may get you a castoff drum or pail of it. A fairly easy recipe for nitroethane is also given later in this chapter.

This reaction is fairly easy to do, and is pretty hard to mess up, so long as some basic precautions are taken. The underground chemist does his best to make sure that his glassware is dry, and the alcohol used is absolute (100% with no water). He must also do the processing of this material quickly, because the nitroalkene which is formed in the first phase of this reaction will not keep.

The reaction goes like this:

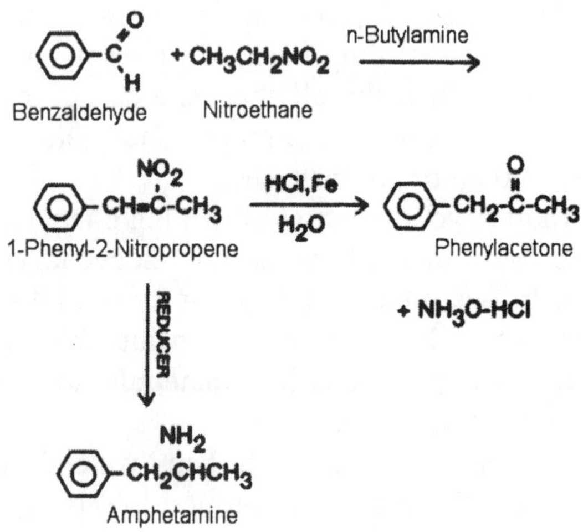

Benzaldehyde reacts with nitroethane in an alcohol solution with n-butylamine catalyst to produce a crystalline substance called a nitroalkene. This nitroalkene can then be reduced by means of iron and HCl to produce phenylacetone. The reduction is similar to the use of activated aluminum in the reaction to produce methamphetamine without the bomb, in that the metal, in this case iron, dissolves and produces hydrogen which reduces the nitroalkene. It is not as complicated as it sounds, and is pretty easy to do. The nitroalkene is first reduced to phenylacetone oxime, which is then hydrolyzed to phenylacetone.

You may wonder, looking at the structure of the nitroalkene molecule, if it is not possible to reduce it directly to the prototype amphetamine, benzedrine. The answer is yes. In fact, one method of making the psychedelic amphetamines such as MDA is to get the properly substituted benzaldehyde (in the case of MDA the proper benzaldehyde is called piperonal) and reduce it using a hydrogenation bomb and Raney nickel, or by use of lithium aluminum hydride. Another good method for reducing the nitroalkene directly to amphetamine is to use zinc amalgam and hydrochloric acid in alcohol solvent. A still better method for direct reduction of the nitroalkene to amphetamine is to use palladium black on charcoal in the champagne bottle hydrogenation bomb seen in Figure ***** in Chapter Eleven. Directions for making palladium black on charcoal are found in Chapter Fifteen. A few grams of catalyst per hundred grams of nitroalkene works nicely. Reaction conditions are room temp at a hydrogen pressure of 30 pounds. Hydrogenation is complete in 5 to 10 hours, and the solvent is 190 proof vodka. Best results are obtained if the nitroalkene is purified by recrystallizing the crude product from alcohol prior to reduction.

We'll go into some of these direct reductions to amphetamine in some more detail at the end of this upcoming recipe. The main point to get now is that racemic amphetamine, aka benzedrine, isn't a bad buzz. Direct reduction to amphetamine rather than proceeding through phenylacetone to get meth eliminates the need to get or make methylamine because the nitrogen atom is already there in place. Amphetamine also carries a great deal lesser federal penalty if one should happen to get busted. The direct reduction techniques are just as applicable to the substituted nitroalkenes one so often comes across in making psychedelic amphetamines as they are to nitroalkene proper.

This reaction is done as follows: Into a clean, dry 3000 ml round bottom flask is placed 400 ml of absolute alcohol, 20 ml of n-butylamine, 428 grams (407 ml) of benzaldehyde, and 300 grams (286 ml) of nitroethene. The underground chemist sets up the glassware for refluxing as shown in Figure 10 in Chapter Three. He includes the drying tube with Drierite as shown in Figure 9. He swirls around the flask to mix the contents, then sets the flask on a hot plate and begins heating it. The water flowing through the condenser should be fairly cool, to be sure of condensing the alcohol vapors. A good, gentle rate of boiling is what he aims for. He continues the boiling for 8 hours. The solution will turn yellow.

He makes sure that his chemicals, especially the nitroethane, are of a good grade. Nitroethane is widely used in the paint and varnish industry as a solvent for cellulose acetate lacquers, vinyl resins, nitrocellulose, waxes and dyes. If he has the industrial grade, he first distills it before use. Benzaldehyde smells like bitter oil of almonds and should be clear. Benzaldehyde is used in flavorings and perfumes.

When the eight hours of boiling is done, he turns off the heat and lets the flask cool down. Once crystals begin to appear, he takes off the condenser and begins stirring the solution with a glass rod. He continues the stirring, and transfers the flask to a sink of cool water to help speed the cooling. He continues the stirring until the mass of crystals becomes too thick to stir, or the flask is cooled off. The idea of the stirring is to prevent the batch from setting into one solid mass of crystals. The crystals should be yellow in color. Be aware that these crystals may form slowly or with great difficulty. Just because the whole mass doesn't immediately turn to slush doesn't mean the batch is a bust. Addition of a seed crystal greatly speeds crystallization. Alternatively, cooling the mixture and allowing it to stand for a couple of days in the cold will generally do the trick. For stubborn cases, adding some water to the reaction mixture will force crystallization.

He now proceeds to purify this 1-phenyl-2-nitropropene. The simplest way to do this is to add ice cold alcohol to the crystals until a slurry is formed (about 200 ml) and then break up any lumps of crystals with a glass rod. He then filters the slurry through a large coffee filter and squeezes the mass to force out as much of the alcohol as possible. Along with the alcohol, he will be removing most of the unreacted benzaldehyde and nitroethene. The crystals will still be yellow, but they will no longer be sticky and gooey. If he still smells n-butylamine on them, he may rinse them with alcohol again.

A better way to clean up these crystals is to recrystallize them. In large batches like this one, it is a lot of work and he must make provisions for exhausting the fumes to the outside to prevent the danger of explosion, but he will get a cleaner product.

It is done as follows: To the crystals which have been rinsed off with alcohol and returned to a cleaned, dry 2000 ml round bottom flask, he adds just enough alcohol to dissolve the crystals. This takes in the neighborhood of 400 ml of alcohol. Any type of alcohol will do. If he has access to methanol or isopropyl alcohol from some industrial source, that will do fine. Alcohol is flammable, so the way he makes the alcohol hot is to place the flask with the crystals into a pan of hot water, and to begin adding the alcohol to it. He swirls it around while adding the alcohol and keeps adding alcohol until the crystals are dissolved. The result will be a clear yellow solution. Now he records how much alcohol he added and places the flask on the hot plate and sets up the glassware for simple distillation as shown in Figure 11 in Chapter Three. A 500 ml flask is fine for the receiving flask. He turns on the heat to the solution, begins water flow through the condenser and distills off about $^1/_3$ of the alcohol he added to the crystals to dissolve them. When $^1/_3$ of the alcohol is distilled off, he removes the flask from the heat, and cools it off in cool water, followed by ice water. He doesn't want to place the flask immediately into ice water, because it might crack.

Now, as the alcohol cools off, it will no longer be able to dissolve the crystals, and they will reform in much cleaner shape because the garbage which is polluting them will stay dissolved in the alcohol. Once the alcohol is cold, he filters the crystals through a filtering funnel the same way it was described in Chapter Five. He places the crystals out to dry on a glass or china plate, and returns the yellow alcohol solution which filtered through to the distilling flask. This solution still contains a good deal of crystals dissolved in it.

He sets up the glassware as before and distills off another $^1/_3$ of the alcohol, then cools off the flask as before. Once again, crystals will form, although they will not be of as high quality as the first crop. He filters them as before, and returns the alcohol to the distilling flask. Now he distills off about ½ as much alcohol as before, then cools off the flask and waits for the crystals to form. This will be his last crop of crystals. He filters them and sets them out to dry. The total amount of crystals he will get will be about 420 grams.

The underground chemist must now proceed to reduce these crystals of 1-phenyl-2-nitropropene to phenylacetone. If he lets them sit around, they will begin to polymerize into a black, gooey mess (though he can delay them going bad by putting them in the freezer).

Into a clean 3000 ml flask, he places 164 grams of the nitroalkene crystals he just made. To that he adds 750 ml of distilled water, 400 grams of cast iron turnings about 1/40 inch in size, and four grams of iron chloride ($FeCl_3$). The flask is placed in a glass dish or metal pan large enough to hold it, and cooking oil is added to the dish so that it reaches about half way up the sides of the flask. He places the flask with the dish of oil onto a hot plate, and heats the oil to about 90° C. He puts a mechanical stirrer into the flask with a glass rod and Teflon-stirring paddle, and begins stirring the mixture in the flask. Once the temperature of the contents of the flask nears 80° C, he measures out 150 ml concentrated hydrochloric acid. He adds it slowly to the flask over a period of 5 hours. The iron will slowly react with the acid and dissolve, producing hydrogen which will reduce the nitroalkene to phenylacetone oxime. The oxime then reacts with more water and HCl to give phenylacetone.

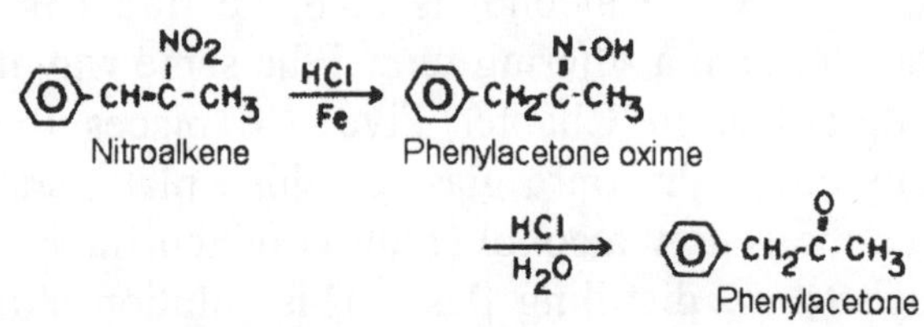

When the acid has all been added, he removes the flask from the heat and lets it cool down. Then he mixes up 80 grams of sodium hydroxide or lye in 300 ml of water. Once they have both cooled down, he adds the sodium hydroxide solution to the 3000 ml flask and swirls it around.

This makes a whale of a mess, as the sodium hydroxide will react with the dissolved iron salts to make iron hydroxide sludge. This sludge will be blue-black, but can turn reddish-orange if the iron salts oxidize to the ferric state. This sludge is a real bitch to try to filter, but it doesn't cause any problem with steam distillation. Steam distillation is the way to go to get very pure phenylacetone out of this reaction.

To steam distill out the phenylacetone, rig up the glassware for simple distillation, bore a hole through a cork and stick some glass tubing through it. Place the cork in the top of the still-head. Some glass tubing should protrude out the top of the cork, and the other end of the glass tubing should reach almost to the bottom of the flask.

Heat the flask to boiling using a pan of hot oil on a hotplate as the heating source. Then attach a line of plastic tubing to the top of the glass tubing, and run the other end of the tubing to the steam escape vent of a pressure cooker. Fill the pressure cooker half full of water, and heat it also to boiling, and let the steam it produces blow through the tubing into the flask. Run the boiling pots at about the maximum rate that your condenser can convert the steam back to water. Collect the steam distillate in glass gallon jugs. It will take at least a couple of gallons of water put through the system as steam to steam distill all the phenylacetone out. Continue steam distilling until no more oily droplets of phenylacetone are noticed coming over with the steam.

Then take the steam distillate in the glass jugs, add a couple hundred ml of toluene to each, and shake to extract out all the phenylacetone into the toluene. Using a sep funnel, separate the top layer of toluene and phenylacetone from the water underneath.

Put this toluene extract containing the phenylacetone into a distilling flask, rig for fractional distillation, and as in the examples given in Chapter Three, distill first the toluene-water azeotrope, then pure toluene, and when that is about gone, apply a vacuum and vacuum distill the phenylacetone. The yield from this size batch will be around 100 ml. The overall yield on the process is about 50%, with around 65% yield in the Knoevenagel reaction to make the nitroalkene, and about 75% yield in the reduction and hydrolysis step using the iron and hydrochloric acid.

The steam distillation can be omitted if a lower grade of phenylacetone is acceptable. In this variation, filter the reaction mixture through a plug of angel's hair to remove the iron metal. Then slowly with strong stirring add the sodium hydroxide solution to the filtered reaction mixture until that nasty iron sludge just begins to form and stay around rather than quickly redissolving in the acid. That is neutralized enough. We don't want a mess. Then extract the reaction mixture with a few hundred mls of toluene. Separate off the toluene-phenylacetone layer (the top layer), and wash it with an equal volume of water, and then with some 5% NaOH solution.

The toluene-phenylacetone solution should then be put in a distilling flask as in the previous example, and fractionally distilled to get about 100 ml of phenylacetone just as in the previous examples. This recipe was taken from *Journal of Organic Chemistry*, Volume 15, pages 8-15 (1950). A good review of this reaction can be found in *Organic Reactions*, Volume 15.

The question which should naturally come up at this point is "Can I use something other than n-butylamine as the catalyst for this reaction?" The answer to that is "yes." See *Journal of the American Chemical Society,* Volume 56, pages 1556-8 (1934) for a nice article on the effect of other catalysts on this reaction. The catalyst has to be a primary amine to get the nitroalkene as the product. Secondary amines give nitro-alcohols if my memory serves me correctly. So other primary amines can be used in place of n-butylamine. Methylamine boils at too low a temperature, and so would be boiled out of the reaction mixture using the procedure given above. It does work, however, and the yields are quite good. To use methylamine as catalyst, the same proportion of ingredients are used as in the previous recipe, except that roughly the same volume of methylamine free base (25 to 40%) dissolved in alcohol or water is used instead of butylamine. Then the mixture is heated on a hot water bath to about 50° C with magnetic stirring. After about an hour or two of reaction, the mixture is cooled. Then mix in about 5 or 6 ml of glacial acetic acid, and pour the product into a glass baking dish and set it in the freezer to crystallize. From there, proceed just as in the above example. Then yield should go up to about 70% for the total process as phenylacetone. Dr. Shulgin in his book *PIHKAL* uses ammonium acetate with good results on most substituted benzaldehydes, but not so good on others. I don't know how well it would work on benzaldehyde. The catalyst has to be soluble in the solution, so trying ammonium acetate in the example just given would be OK, but in the next recipe for the Knoevenagel reaction, which I will give later in this chapter, the solvent is toluene. Ammonium acetate doesn't dissolve in toluene, so it's not a possibility there.

A Russian article dating to the 40s claims that a few drops of ethylenediamine will catalyze this reaction at room temperature and in the dark. Are they being straight, or are they lying through their rotten, stinking teeth? Your Uncle doesn't know. Ethylenediamine is a primary amine, so it should work. The *Journal of the American Chemical Society* article tells of the bad effects of using too much catalyst, so how much ethylenediamine to use? In the above example, 20 ml of n-butylamine is about .2 mole. Since ethylenediamine has two amino groups per molecule, one would guess that it is twice as active as n-butylamine. Then .1 mole of ethylenediamine is about 5½ ml.

Being able to substitute ethylenediamine into the recipe would be nice, as it is commonly used industrially. For example, ethylenediamine is a major component of certain types of nickel stripper. This type of nickel stripper will come as a two-component package. Component number one is a yellow powder, m-nitrobenzenesulfonic acid. This is the oxidizer for the nickel metal, which dissolves the nickel and puts it into solution. Component number two is ethylenediamine, which acts as complexor for the nickel ions, keeping them in solution rather than sludging out. This type of nickel stripper is most favored for stripping nickel from a copper or brass substrate. A good cover then for getting ethylenediamine from industrial outlets is to say one is in the business of refinishing jewelry. The gold plate has mostly worn off this old jewelry, so you strip off the underlying nickel plate to get down to the copper underneath, then you replate this jewelry with a nice shiny new plate of nickel and gold. People are well pleased with these refurbished heirlooms.

To get the ethylenediamine, go to the *Metal Finishing Guidebook and Directory*. Look in the back of the book and look under suppliers of metal stripping solutions. Give them a call and see if they will send you a sample. If not, then buy a pail of the ethylenediamine component. There may be some water in the industrial grade ethylenediamine. That's not a problem in the variation of the Knoevenagel reaction which I'll be describing later, as water in the solution is removed as the azeotrope in that example. If using the first method given above, then it would be wise to check for water content by drying the industrial ethylenediamine over NaOH or KOH pellets and fractionally distilling the material.

The recipe given above allows for the option of direct reduction of the nitroalkene to the amphetamine, because the nitroalkene is isolated and purified as part of the reaction procedure. There are a variety of methods which can be used to reduce the nitroalkene to racemic amphetamine, also called benzedrine. The best and highest yielding procedure uses lithium aluminum hydride in dry ether or THF solvent. If these materials are available to you, quite a number of examples of this process can be found in *PIHKAL*. It was Dr. Shulgin's favorite method for making substituted amphetamines from substituted nitroalkenes, and ultimately from substituted benzaldehydes. Most of us don't have access to these materials, so let's just leave that subject right there.

Hydrogenation methods can also be used to do this reduction. For the use of Raney nickel, see *Chemical Abstracts,* Volume 48, column 2771, and US Patent 2,636,901. For the use of palladium or platinum, see *Chemical Abstracts*, Volume 71, entry 91049 and US Patent 3,458,576. All of these methods get around 50% yield of amphetamine from nitroalkene. An 80% yield is claimed in reducing a substituted nitroalkene to the substituted amphetamine in *Helvetica Chimica Acta,* Volume 51, page 1965 (1968), using palladium catalyst with hydrochloric acid as solvent. My German's not too good, so I can't give you any details other than that they heated the mixture to 85° C and used 10% Pd on C at a pressure of 35 atmospheres. Would the same procedure work on other nitroalkenes? You got me; it's worth a try.

Since all the hydrogenation methods give around 50% yield, one might as well forget about the hassle of hydrogenation, and go with the clandestine choice for this reduction. That choice is electric reduction at a copper or lead cathode. The yield in this procedure is about 50% as well, and is much simpler than the hydrogenation.

The procedure is to be found in US Patent 1,879,003. In an appropriate size beaker, cup, plastic pail or plastic drum, depending upon the batch size, the electric cell shown in Figure 25 should be put together:

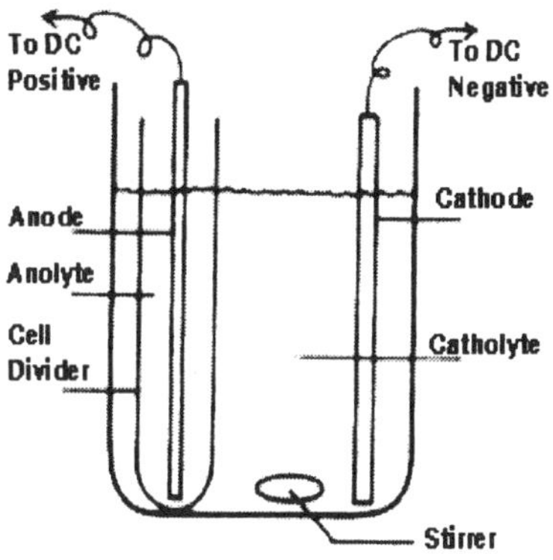

Figure 25

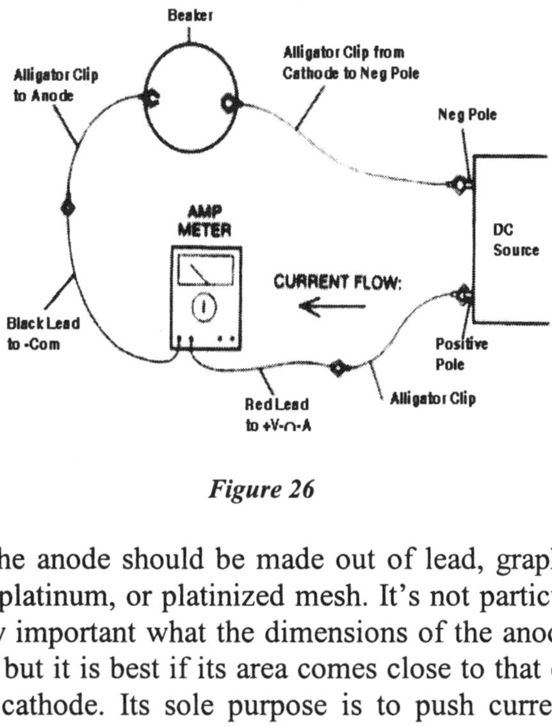

Figure 26

The anode should be made out of lead, graphite, platinum, or platinized mesh. It's not particularly important what the dimensions of the anode are, but it is best if its area comes close to that of the cathode. Its sole purpose is to push current into the solution. The anolyte serves to carry current from the anode to the cell divider. The cell divider in this solution should be made of ceramic such as porcelain. Any glazing on the surface of the ceramic must be removed by sanding or sandblasting, as glaze will prevent the passage of current. What to use as the cell divider will depend upon batch size. For beaker-size batches, a porcelain figurine with any glaze and paint removed will work well. I have a little Frosty the Snowman with porcelain walls about 1 mm thick that works just fine in small cells. The paint that decorated this figure was easily removed to leave what is basically a white hollow cylinder. For larger batches, a porcelain cup with the glaze removed would work. For still larger batches, a flower pot with the holes on the bottom plugged and the glaze removed would work. For drum batches, a toilet reservoir water tank would function, once the glaze is removed. The thicker the walls of the ceramic, the more voltage that will be required to make the cell divider pass current from the anolyte to the catholyte and ultimately the cathode where the reduction takes place. The cell divider is required because contact of the product amphetamine with the anode would result in its destruction. The reduction solution, called the catholyte must not contact the anode.

The cathode should be as large as will conveniently fit in the reaction container. The cathode should be made of lead or copper. It must be clean, and free of dirt. The metal should shine! A "bright dip" is a good way to clean these metals. It is dilute nitric acid. Brief immersion of the cathode in dilute nitric followed by a good water rinse will assure that the surface is clean and active. A solution is made up by mixing two parts ethanol with one part acetic acid and one part sulfuric acid. This is your catholyte. Enough should be made to make the catholyte reach nearly to the top of the cathode. The nitropropene (1-phenyl-2-nitropropene) should then be added to this catholyte. It is added such that one mole (164 grams of the nitropropene) is dissolved in two liters of the catholyte. Do not premix the nitropropene into the catholyte until the reduction is ready to be done.

Anolyte is made by diluting concentrated sulfuric acid to 5% by volume with water. For example, 5 ml of concentrated sulfuric is diluted to 100 ml volume with water. The anolyte is poured into the cell divider until its liquid level roughly matches the catholyte.

Then stirring is begun and current (DC current, of high quality) pushed through the solution. How much current? The surface area of the cathode is measured. Only the area on the side facing the anode is counted. For each square centimeter of surface, .2 amp (200 milliamps) is caused to flow through the solution. If one has a DC rectifier, the voltage is simply increased until the desired amount of amps flows through the cell. If using a battery, choose a battery or batteries hooked in a series to give a voltage sufficient to move the current needed. Increasing voltage increases current flow through the cell. E=IR.

Heat will be produced by pushing current through the cell divider and by reaction at the cathode. The temperature of the solution should be watched, and not allowed to go over 40° C. External cooling is good enough for beakers and pails. For a plastic drum, copper pipe run through the solution as a heat exchanger might be needed as well.

Let's take the example of a one mole batch done in a 2500-3000 ml beaker. A reasonable size cathode in this case would measure at least 12 cm by 6 cm, for a face area of 72 square centimeters. 200 milliamps (.2 amp) per square centimeter of face gives a required current flow of 14.4 amps, shown on the current meter. The current meter is wired in on the line serving the anode. Fairly heavy wiring, well beyond that in a set of alligator clips, would be needed in this case to carry that much current without overheating the wiring. For still larger batches, jumper cables would be needed to carry the current.

The reduction of 1-phenyl-2-nitropropene to amphetamine is an 8 electron reduction. So the theoretical amount of current needed to do the reduction is 8 faradays per mole of the nitroalkene in solution in the catholyte. One faraday is 96,500 coulombs, and a coulomb equals an amp-second. At 14.4 amps, one faraday passes into solution in 6700 seconds. This is 1.86 hours. Eight faradays flow in just under 15 hours. The process isn't 100% efficient, however, so at least 12, and up to 16 faradays will be needed to do the reduction to completion. This will take 24-30 hours. Increasing the electrode area will allow for greater currents to be passed, with a proportional decrease in the amount of time needed to do the reduction. The anolyte may decrease in volume during the course of the reduction, as the water in the anolyte gets converted to oxygen electrically. Water should be added to that compartment of the cell occasionally to make up for such losses.

When the required amount of current has flowed through the solution, the electrodes can be removed from the beaker, pail or whatever, and the catholyte, which now contains amphetamine, can be poured into a distilling flask. The ethanol is then distilled out of the solution. Ethyl acetate is also formed during the distillation by reaction between the acetic acid and ethanol. This distills off easily as well. A vacuum assist during the distillation makes the process go faster, and at a lower temperature.

The residue left in the flask after the alcohol and ethylacetate have distilled off consists of a solution of amphetamine in sulfuric acid and acetic acid. This should be allowed to cool down, then chilled in some ice. Add to this acid solution sodium hydroxide solution to neutralize the acid. 10-20% NaOH or lye in water is about the proper strength for this sodium hydroxide solution. Adding this to the acid produces a violent reaction, generating a lot of heat. It should be added slowly with stirring or swirling, taking breaks to allow the acid solution to cool down during the course of this neutralization. It's going to take in the neighborhood of a couple of pounds of NaOH or lye added to neutralize all the sulfuric and acetic acid in the flask, and make the solution strongly alkaline (pH 13+ to pH papers). As the neutralization nears completion, a layer of amphetamine will start to float on top of the water solution in the flask. When this starts to happen, shaking between adds of sodium hydroxide solution should be done instead of just stirring or swirling. Then check the pH of the water layer. When it stays pH 13+ to pH papers, enough sodium hydroxide has been added. Add water to dissolve any salts which may have precipitated out of the solution.

Now to the cooled down solution in the flask add around 200 ml of toluene, and shake some more to extract the amphetamine into the toluene. Allow the mixture to sit, then separate off the toluene layer floating on top of the water with a sep funnel. Pour it into a distilling flask, and first distill off the toluene-water azeotrope, then pure toluene, and finally vacuum distill the amphetamine as done in the previous examples. Amphetamine will distill under a vacuum a few degrees cooler than methamphetamine. The yield should be in the neighborhood of 70 ml free base, which is made into the hydrochloride salt by the usual procedure of bubbling dry HCl through a toluene solution of the free base. This will yield maybe 70-80 grams of amphetamine hydrochloride.

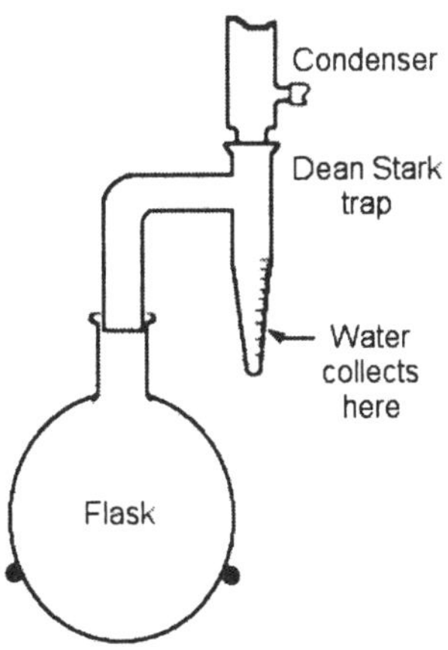

Figure 27

The following variation of the Knoevenagel reaction will give higher yields of product than the first method. The reason for the higher yield is the use in this method of toluene as solvent, and the placement of a Dean Stark trap above the flask to remove water from the mixture as it is formed. Removal of water favors the formation of greater quantities of nitroalkene.

To do the reaction, a 1000 ml round bottom flask is filled, in this order, with 200 ml of toluene, 100 ml of benzaldehyde, 90 grams (86 ml) of nitroethane, and 20 ml of butylamine. It is a good idea to swirl the flask after adding each ingredient to prevent layers from forming. Next the flask is placed on a one burner electric buffet range with infinite control, and the glassware is set up as shown in Figure 27.

The Dean Stark trap is attached to the flask, and a condenser is attached to the Dean Stark trap. Then the buffet range is turned on at a heat setting high enough to produce a rapid boiling of the toluene, and cold water is flowed through the condenser. As the reaction is progressing, the vapors of toluene carry water along with them, and when they turn back to liquids in the condenser, the water will settle in the trap portion of the Dean Stark trap because water is heavier than toluene. You will also note a milky appearance to the toluene due to suspended water in it. The trap portion of the Dean Stark trap is graduated in milliliters. This allows you to keep track of how much water has been collected. Half of the water is collected in the first hour, and the full amount (18 ml) is collected after five hours of boiling. When this is done, the heat is removed, and the flask allowed to cool. This phase of the reaction has just made the nitroalkene. If one should wish to collect the nitroalkene for direct reduction to amphetamine, one just needs to remove the Dean Stark trap, rig the flask for simple distillation as shown in Figure 11, and remove the toluene under a vacuum from an aspirator, using gentle heating from a hot water bath. It should be noted that the nitroalkene has a slight tear gassing effect upon the eyes, and also irritates the skin. Do not use the stuff as a body balm.

If phenylacetone is desired from the nitroalkene, the toluene solution produced in the reaction is used directly in the next step. Once it has cooled down, it is poured into a 2000 ml 3-necked flask. Then into the 3-necked flask is added 500 ml of water, 200 grams of iron powder (40 to 100 mesh), and 4 grams of ferric chloride ($FeCl_3$). Into the center neck of the flask is put a mechanical stirrer reaching almost to the bottom of the flask. There should be a tight seal so that the ensuing vapors of toluene when the flask is heated do not escape by this route. A good condenser is attached to one of the other necks, and a sep funnel, or dropping funnel with matching ground glass joint is put into the remaining neck. With vigorous stirring, the contents of the flask are heated to about 75° C, and 360 ml of concentrated hydrochloric acid is added to the flask by means of dripping it into the mix through the sep funnel over a two hour period. The reaction mixture will boil vigorously. The heating and stirring are continued for an additional half hour after the last of the hydrochloric acid has been added.

Next it is time to get the phenylacetone out of the reaction mixture. Once the flask has cooled down, the iron is filtered out by pouring it through the plug of angel hair described earlier in this chapter. It is a good idea to rinse down the trapped iron powder with a dash of toluene to get any clinging phenylacetone off of it. Then the toluene layer is separated using a sep funnel. It is poured into a round bottom flask. The water layer has about 100 ml of toluene added to it, and this is shaken to draw suspended phenylacetone into the toluene. The toluene layer is then separated and added to the aforementioned round bottom flask. It is then rigged for fractional distillation as shown in Figure 13. The toluene distills off first as the toluene-water azeotrope at 85° C, and then as pure toluene at 110° C. Once the toluene is mostly gone, vacuum is applied, and phenylacetone is collected at the usual temperature range. The yield is about 120 ml of phenylacetone.

This recipe was taken from *Organic Syntheses*, Collective Volume 4. Look in the table of contents for o-methoxyphenylacetone.

Making Shitloads of Your Own Benzaldehyde

As was mentioned earlier in this chapter, small to moderate amounts of that List I controlled chemical benzaldehyde are most conveniently obtained by vacuum distilling oil of bitter almonds, which contains about 95% benzaldehyde, or by similarly distilling technical grade mixtures containing benzaldehyde used for flavorings and fragrances. When larger amounts are wanted, such as a gallon or more, then the recipe which I will give here is the way to go. It is an old industrial process for making benzaldehyde that gives high yields of product, and produces next to no waste to be disposed of. This process uses the electrode-generated reagent, Mn^{+3}, to oxidize toluene to benzaldehyde in very high yield with very few byproducts. The Mn^{+3} is electrically made from the easily available Mn^{+2} salts. During the reaction, the Mn^{+3} is reduced to Mn^{+2}, which can be returned to the process and used over and over again just by electrically oxidizing it again to Mn^{+3}. I suspect from my correspondence with various sources that this method is being used with great success by clandestine cookers out there to make loads of benzaldehyde easily and quickly.

This method is taken from US Patent 808,095, which dates to 1905. This process is no longer used industrially to make benzaldehyde because it has been replaced by the air oxidation of toluene using various catalysts. That's cheaper for those big manufacturers, but for our purposes here, this method is better.

The first thing which must be done is to make mangano-ammonium sulfate. This is the substance that the patent prefers to use as the Mn^{+2} compound. Other, later work published simply used $MnSO_4$ as the Mn^{+2} compound, but they talk about using divided cells whereas this procedure can be done in a simple, undivided cell.

Take two 55-gallon plastic drums, well cleaned with the tops cut off them. Fill each $^{2}/_{3}$ full with hot water. That's roughly 30 gallons of hot water in each. Next add around 1½ gallons of concentrated sulfuric acid to each drum slowly with stirring. The addition of the concentrated sulfuric will make even more heat, so beware of splattering as it goes into the water. That will produce a roughly 1 molar solution of the acid in water.

Next, while the water is still hot, take 85 pounds of manganese sulfate, split it into two equal portions, and add them to the two drums of hot acid solution. Stir until dissolved. Then take about 40 pounds of ammonium sulfate, divide it into 20 pound portions, and slowly with stirring add the ammonium sulfate to each drum. Roughly 50 kilos of the mangano-ammonium sulfate will precipitate out of the hot acid solution as the ammonium sulfate is added and goes into solution. Continue stirring.

Both manganese sulfate and ammonium sulfate are easily available at industrial chemical outlets, and the technical grade of each chemical is good enough for this purpose. Manganese sulfate generally comes as the tetra and pentahydrate mixture, and the amount used in this example assumes that this is the material used. It will come in 50-pound bags for use as a fertilizer ingredient. Ammonium sulfate also comes in 50-pound bags, and it too is used as a fertilizer ingredient.

Next filter out the crystals of mangano-ammonium sulfate by pouring the acid solution through cotton pillow cases. As each pillow case gets full of crystals, hang them up to drip dry.

Now an electric cell should be constructed. The most suitable container would once again be a plastic 55-gallon drum, thoroughly cleaned, with the top cut off. The inner vertical wall of the drum should be lined with a sheet of lead. It's pretty easy to get lead sheets about ¼ inch thick. It is sold by suppliers to electroplaters of chrome. See the *Metal Finishing Guidebook and Directory* for a large selection of such suppliers. Look under "Anodes — Lead" in the back of the book. Your library should be able to get this guidebook for you through inter-library loan. One could also page through the trade magazines *Plating and Surface Finishing* or *Metal Finishing* for ads featuring lead anodes. Typical sheets will measure 4 feet by 8 feet. One sheet will be plenty, and weigh plenty as well.

The standard 55-gallon drum stands 3 feet tall, and has a diameter of about 22 inches. The inner circumference is then about 69 inches. Cut a section from the lead sheet measuring a little under 3 feet by 69 inches, put it in the drum, and make this sheet of soft metal line the inside surface of the drum. This will be your anode, the place at which the Mn^{+2} gets oxidized to Mn^{+3}.

To get current to flow through this cell, we need a cathode. Start with a copper ring about one foot in diameter. It should be heavily constructed enough to support the cathode, which in this case is also made of lead. Cut strips of lead roughly an inch wide and long enough to almost make it down to the bottom of the drum, or a little under three feet long. You will need ten or twelve of them. These are your cathodes. Clamp the top of each of them to the copper ring with even spacing. Next take two sections of steel rod. Each one is two feet long. Wrap them in some heavy plastic, using rubber bands to keep the plastic tightly wrapped around the rods. These are the supports for your cathode ring. Lay them on top of the drum, then rest the rings of cathodes on top of these rods. Finally, stick a mechanical stirrer down the center of the drum, almost reaching the bottom. When done it should look like Figure 28.

Then into the barrel put 45 liters of water, followed by 79 kilos (43 liters) of concentrated sulfuric acid. The industrial grade of sulfuric acid so easily available in 55-gallon drums is good enough for use in this process. If the drum says that the concentrated sulfuric acid is 66 degrees balme, it is roughly 98% sulfuric acid. The acid, of course, should be added slowly with stirring, and good ventilation should be provided as the dilution of the acid will make a lot of heat and steam that carries some acid up with it. Once the acid solution has cooled down to a little above room temperature, then 47.5 kilos of the mangano-ammonium sulfate made earlier should be added to the drum.

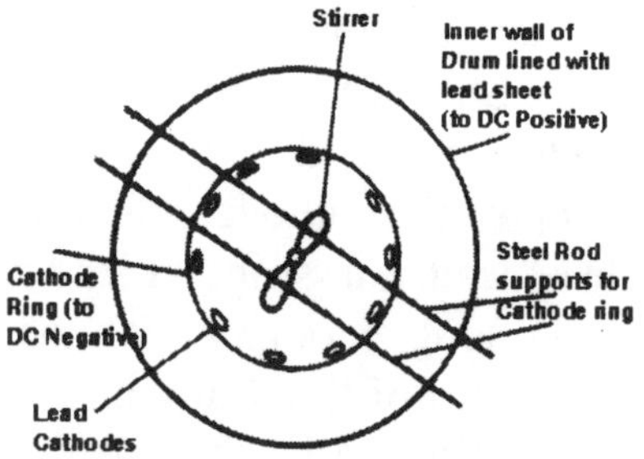

Figure 28

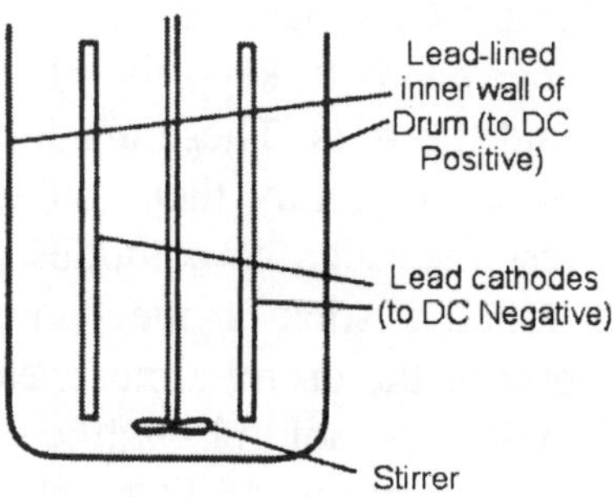

Figure 29

Now begin moderate stirring of the drum contents, and do a calculation. It is time to calculate how much of the lead sheet anode is submerged by the liquid in the drum. In a typical 55-gallon drum with a diameter of 22 inches, the inner wall of the drum has a circumference of about 69 inches. The amount of ingredients added will fill the drum about $^2/_3$ full, or to a depth of about 2 feet. This gives roughly 1600 square inches of lead sheet facing the cathodes. The backside of the lead sheet, of course, shouldn't be counted. This process works well when around 3.5 amps per square decimeter (or about .23 amps per square inch) of lead anode face in contact with the solution is applied. In this example, around 375 amps DC should be used. To provide this much DC current, a rectifier must be purchased. Go once again to the *Metal Finishing Guidebook and Directory* or to the trade journals *Plating and Surface Finishing* or *Metal Finishing* for ads galore by dealers of rectifiers to electroplaters. The general rule of thumb is that a rectifier will cost about $1 for each amp of capacity. For this application, a rectifier able to supply at least 500 amps should be purchased. One doesn't want to continually max out the rectifier. A rectifier this size will run on 240 volt 3 phase AC. Fairly heavy wiring will be needed to serve it with power.

The cables going from the positive pole of the rectifier to the anode, and from the cathode ring to the negative pole of the rectifier, will have to be heavy as well to carry the nearly 400 amps of DC current. The points at which the cables make electrical contact with the anode and cathode should be clean, and clamps should be used to tightly make contact at these points. This is important to prevent sparking at these points, which could ignite the hydrogen gas made at the cathode and the oxygen gas made at the anode.

With stirring, turn on the rectifier and increase the voltage until the required amount of DC current is flowing through the cell. Rectifiers used for electroplating have fairly accurate current meters built into them. Along with the oxygen produced at the anode, the other reaction going on there is oxidation of Mn^{+2} to Mn^{+3}. Continue the flow of current until about 6000 amp hours have passed through the cell. With a current flow of 375 amps, it will take 16 hours to pump 6000 amp hours through the electrochemical cell. A dark red salt, ammonium managanese alum, ($Mn_2(SO_4)3(NH_4)_2SO_4$), forms as a result, and falls out of solution. This is the oxidizer for the toluene.

During the course of passing current through the electric cell, a fair amount of the water in the solution gets converted to hydrogen and oxygen. Water should be periodically added to the cell during the passage of electricity to maintain the original water level. Another point that should be made is that the acid mist which rises from the anode and cathodes as a result of their fizzing off of hydrogen and oxygen may coat the sides of the drum and the plastic-coated metal rods supporting the cathode ring. This conductive film could lead to a short, which would blow a fuse in your rectifier. Placing a few rags under the metal rods where they touch the drum will prevent this. It is even better to support the cathode ring from the ceiling. That will eliminate this danger. Contact between the cathodes and the anode must similarly be prevented. Don't stir too rapidly and cause the easily bent lead cathodes to touch the anode.

Once the required amount of current has flowed through the cell, turn off the current, then pull the cathode ring out of the drum. Then add 4 kilos (4.6 liters) of toluene. The toluene available at the hardware store in the paint section is good enough for this reaction. Turn up the speed of agitation to get the toluene, which would naturally like to float on top of the mixture, down into contact with the Mn^{+3} oxidizer.

The preferred temperature for the oxidation is 50° C. The reaction mixture is likely to be nearly that warm at the end of the electrolysis, but it will cool down once the current is turned off. The directions call for the addition of about 8 liters of additional water during the early stages of the oxidation, so adding boiling hot water will help to maintain warm reaction conditions. Additional heating may have to be supplied.

At the end of two to three hours of stirring, the oxidation of the toluene is complete. The red Mn^{+3} salt will have redissolved, and a clear solution will be produced once again. One's eye can be the guide to when the reaction is complete.

The droplets of benzaldehyde churning through the reaction mixture can be extracted by adding about three gallons of toluene, and continuing the strong stirring. Then turn off the stirrer, and allow the toluene-benzaldehyde droplets to rise to the top of the reaction mixture. This top layer can then be separated off, and distilled in the same manner as all those previous examples. Benzaldehyde boils at 178° C at atmospheric pressure, and at roughly 80° C under a typical good aspirator vacuum of 25 torr or so. The yield will be around a gallon and a half. The toluene which distills can be reused, of course, in the next batch.

To do the next batch, one simply reinserts the cathode ring, checks to make sure that the liquid level in the drum is about what it was originally, and repeats the electrolysis of the mixture to reoxidize the Mn^{+2} back up to Mn^{+3}. In this way, batch after batch can be run with the only feed material needed being toluene. This is a good way to make lots of benzaldehyde.

To make lots of nitroethane, see the simple directions given in the *Journal of the Chemistry Society of London*, 1944, pages 24-25. An alternative procedure is the reaction between nitric acid and ethane found in *Oil Gas Journal* Volume 54 No. 36 pages 110-114 (1956). These processes are clandestine suitable, but getting a pail or drum ready-made would be quite a bit easier.

Phenylacetone from Hydratropic Aldehyde

There is another phenylacetone recipe that looks pretty good. I haven't heard of a lot of use of this method in clandestine chemistry, but it is simple and creates phenylacetone in one step from commercially available materials. The biggest problem with this procedure is that the starting material from which the phenylacetone is made, hydratropic aldehyde, isn't very widely used commercially. It finds some use in perfumery, but massive amounts of it aren't used. A five-gallon pail might be the largest quantity that one could reasonably obtain through normal commercial channels. To obtain this chemical, one would probably be best advised to go to one of the perfumery trade publications mentioned throughout this book and in *Advanced Techniques of Clandestine Psychedelic & Amphetamine Manufacture* and look through the trade journals for ads from chemical suppliers. Then disguise yourself as a perfumer, and order some.

To do this reaction, one only needs a beaker, or some other glass container. Into the beaker put 400 ml of concentrated sulfuric acid. Then measure out 90 grams of hydratropic aldehyde, also known as 2-phenylpropanol, or as methyl-phenylacetaldehyde. Cool the sulfuric acid down to -16° C using an ice bath to which either salt or alcohol has been added to further chill down the ice. When the sulfuric acid has cooled down to -16° C, slowly with stirring add the hydratropic aldehyde over a period of about one hour. Maintain the cold temperature of the mixture throughout the addition of the aldehyde, then let the mixture react for an additional 15 minutes after all the aldehyde has been added. Don't let the temperature fall much below -16° C, or the hydratropaldehyde will freeze in the reaction mixture and will fail to convert to phenylacetone. The following reaction, which is just a rearrangement of the aldehyde molecule, takes place:

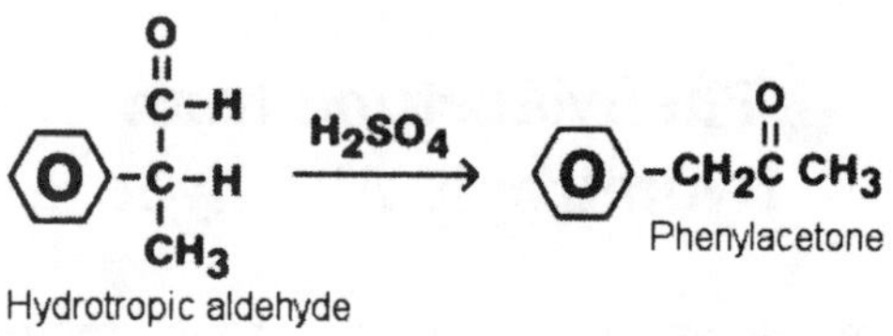

After the additional 15 minutes of reaction, the sulfuric acid solution is then poured onto about 5 pounds of cracked ice. The whole mixture is then allowed to come to room temperature. As the ice melts, an oily liquid and some gummy goo are formed. Extract the water with about 300 ml of toluene. This will pull the phenylacetone into the toluene. Separate the toluene layer from the diluted sulfuric acid. A large sep funnel would be the best tool for this job. The toluene will be floating on top of the water. Wash the toluene/phenylacetone solution with some water, and then with some bicarbonate dissolved in water. This will remove any sulfuric acid which has been carried over. Finally distill off the toluene, and vacuum distill the phenylacetone. The yield will be around 60 ml of phenylacetone. Not a bad procedure if you can get the hydratropic aldehyde.

Should hydratropic aldehyde become a "listed chemical," it would still be pretty easy to set up large scale production of this substance using commonly available chemicals. Hydratropaldehyde can be made in fairly good yield from cumene, which is a cheap and readily available commercial solvent and chromyl chloride. The procedure is called the Etard Reaction, and can be done in a variety of solvents such as carbon tetrachloride or chloroform. For a review of this reaction, see *Chemistry Reviews,* Volume 58, pages 1-60 (1958). Chromyl chloride is pretty easily made from chromium trioxide and hydrochloric acid. For details on this method, see *Inorganic Syntheses*, Volume 2 page 205 (1946). I must forewarn you that chromyl chloride is a nasty and fuming chemical, which must be handled with respect. In spite of that, putting together a large-scale phenylacetone factory by this method wouldn't be very hard to do.

Chapter Ten Psychedelic Phenylacetones From Essential Oils

Question: "Hey Fester, are they putting some kind of inhibitor in the sassafras oil? I've tried the nitrite oxidation of safrole you give in *Secrets of Methamphetamine Manufacture* and *Practical LSD Manufacture*. I've used both palladium chloride and the mixed catalyst of palladium chloride and copper chloride, and I've tried both ethyl and methyl nitrite as oxidant. My yield of methylenedioxyphenyl-acetone is miniscule. This is frustrating and costly. I hate sticking my neck out risking a bust and not getting the phenylacetone I need to make that X. I can read the patent you cite as well as you, and I see that you have faithfully presented the patent contents. What gives?" — The Iceworm

For years in my books, a recipe taken from US Patent 4,638,094 issued to Ube Laboratories in Japan has been presented. In the patent, the inventors claim a 90% yield of phenylacetones from allylbenzenes using Pd chloride catalyst, and methyl or ethyl nitrite in alcohol solution.

Since allylbenzene isn't a very common chemical, this procedure wasn't of much significance to meth cookers. However, substituted allylbenzenes are very common in essential oils, and these substituted allylbenzenes can be converted to the phenylacetones to be used to make psychedelic amphetamines like MDA or MDMA. Boy, was that a good patent!

One problem, however. When people tried putting the patent into practice, the 90% yields claimed turned into yields of about 5% in reality. "Oh, what an abortion!" to quote from the old guy who taught me the art of electroplating. We used to say that when a part failed to plate properly. Now, like that great old guy, it's time for that recipe to be laid to rest.

So what's up with the people at Ube Laboratories and their bogus patent claims? Your guess is as good as mine. It cost them a boatload of cash to come into a foreign country and file a patent roughly 40 pages long with fifteen claims. I see that they are still publishing papers in the scientific literature dealing with related reactions. Was this part of some nefarious disinformation scheme? Your Uncle hopes that researchers are not so easily bent to such purposes.

Luckily, there are at least a couple of related procedures that do give the desired phenylacetone in yields that make the process useful. My hat goes off to the folks out there who took the general methods published, and demonstrated their effectiveness. Method number one is to be found in *Organic Syntheses*, Volume 62, in a recipe titled "A General Synthetic Method for the Preparation of Methyl Ketones From Terminal Olefins: 2 Decanone."

Here we just take the general method published, and substitute in our allylbenzene for the terminal olefin used in the recipe. Other than that, the procedure is followed word for word. This process can be scaled up or down as desired to make batches of whatever size. This procedure requires chemical glassware, and getting that equipment is getting to be more and more of a challenge. In particular, a dropping or addition funnel with pressure-equalizing side arm is needed. This bit of equipment looks like your typical sep funnel, but some glass tubing runs from below the valve of the sep funnel up into the reservoir of the sep funnel. This bit of equipment is needed because the reaction is done under pressure, and to get the allylbenzene such as safrole obtained from sassafras oil to drip down into the reaction mixture, its flow can't be fighting against the pressure inside the reaction vessel. Such a dropping funnel will cost about $150 for a 500 ml capacity, and it is prone to breakage if handled roughly. Replacing it could be a hassle, so take care.

Dimethylformamide is used as the solvent for this variation of the reaction. This is not at present a particularly risky chemical to obtain. It also has a variety of industrial uses, such as a solvent for Orlon, and it's also added as solvent in paints and similar coatings that require a bake cure cycle. In particular, it's a major component of the thinner for Xylan, the Teflon-based paint whose application and use is detailed in my book *Vestbusters*. Listen up clandestine cookers! Having trouble getting glassware? Coating your own homemade metal reaction vessels with Teflon or Teflon-based paint will give chemical resistance similar to glass, while avoiding all the heat the enemy has associated with buying glassware.

One also needs palladium chloride as catalyst for the reaction. This is an expensive chemical to buy, and is no doubt the subject of ever increasing scrutiny by the heat. This is especially true since $PdCl_2$ is used to make palladium catalyst for the hydrogenation of ephedrine, pseudoephedrine or phenylpropanolamine to amphetamine or methamphetamine. Hydrogenation is looking like the way clandestine cookers should be moving now that iodine and red phosphorus have gotten so "hot," so I'd be very wary of buying palladium chloride from lab chemical suppliers. See *Advanced Techniques of Clandestine Psychedelic & Amphetamine Manufacture* for a very simple and good procedure for making your own palladium chloride from an ingot of palladium. It's much cheaper, and safer. The local photo shop may also have palladium chloride, as it is used to produce specialized photographic effects.

The other chemicals used in this process are no problem to obtain. Cuprous chloride (CuCl) or cupric chloride (CCl_2) is used as a co-catalyst in the reaction. Getting either one of them will present no problem. Cuprous chloride is better in the reaction than cupric chloride because it has less tendency to chlorinate the product. It costs a lot more than cupric chloride, though, and is less commonly used than is cupric chloride, so cupric chloride may be the best choice. Oxygen gas used to keep the catalyst active is easily obtained in cylinders at the welding supply shop, or even at the pharmacy. Say you are getting it for your Grandfather, who was too winded to make the trip down to the store himself. To do a convenient and productive one mole batch, one starts with a three-necked flask of about 2000 ml volume. Yes, I know how difficult they are to obtain these days. A copper fabricated replica coated on the inside with Teflon or Teflon-based paint like Xylan 1006 will give good service. See *Vestbusters* to learn how to order and apply these materials. After application and bake cure, check the inside with a dentist's mirror to make sure that the coating is in good shape. Copper metal has been chosen here to allow the use of magnetic stirring. There are color changes in the solution to be watched and followed, so glassware is very certainly preferred.

This batch size is about the upper limit of what can be done with the usual magnetic stirrer and commercially available stir bars. For larger batches, one of those super strong Edmund Scientific magnets coated with some Teflon or Teflon-based paint will allow good stirring of the contents. For still larger batches, replacing the magnet in the stirrer with one of those super strong Edmund Scientific magnets will push the limit up some more.

Now into the three-necked flask, put 700 ml of dimethylformamide and 100 ml of water. Follow that with 17.5 grams of $PdCl_2$ and either 98 grams of cuprous chloride or 170 grams of cupric chloride dihydrate. Put the addition funnel with pressure-equalizing side arm in the center neck of the flask. Attach a gas line from the oxygen cylinder to one of the outer necks, and to the other outer neck, attach a large toy balloon. Into the addition funnel, put one mole of allylbenzene. For example, 150 ml of safrole is one mole. Safrole is 80-

90% of sassafras oil, and if one really wanted to be lazy, one could just use the sassafras oil as is. It's better to fractionally distill the oil to get a nice clear safrole feed for the reaction. The impurities in the oil interfere with the reaction and lower the yield. Wire a stopper in place on the addition funnel so that it doesn't blow its lid when pressurized. Similarly, wire the connections on the two outer necks of the flask in place, and wire the addition funnel down so that it doesn't blow off.

Now begin stirring, and feed enough oxygen pressure into the vessel so that the balloon inflates. Continue stirring for one hour to allow the solution to absorb oxygen. If the balloon deflates, just pipe more oxygen into the flask.

After one hour of stirring under oxygen, begin slowly dripping the allylbenzene such as safrole into the solution. Stirring must be efficient during the addition. I'm not sure if safrole dissolves well in dimethylformamide. If it does, then cutting the safrole in half with DMF would help to make sure the addition is done slowly enough. For this size batch, the addition of the safrole or other allylbenzene should take 30 minutes to an hour.

As the allylbenzene is added to the solution, it turns black, then after a few hours, it works its way back to green. The balloon should be kept inflated with oxygen during the course of the reaction. Stir the mixture for 24 hours.

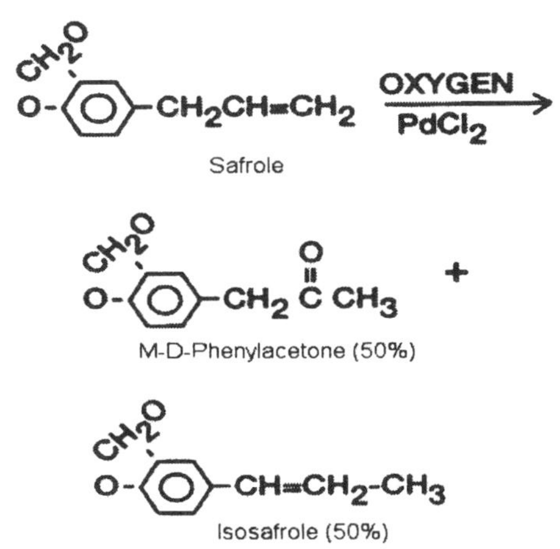

The reaction (at the bottom of the preceding column) takes place, illustrated for the reaction with safrole, but all allylbenzenes behave similarly.

After the 24-hour reaction period, the oxygen pressure can be released, the stirring stopped, everything disassembled, and the reaction mixture is poured into 3000 ml of 3N hydrochloric acid solution (one part concentrated hydrochloric acid, 3 parts water).

The acid solution should then be extracted a few times with 300 ml portions of toluene. The products will extract into the toluene, and the extracts should be pooled together for further purification. The catalyst stays in the dilute hydrochloric and it can be recovered by boiling away the acid. The last part of this boil down should be done under a vacuum to protect the catalyst from burning.

We return to our products. In the toluene extracts, one can expect to be contained about a 40-50% yield of the phenylacetone. If safrole was used in the reaction, that would be methylenedioxyphenylacetone. The other product obtained would be a 40-50% yield of the propenylbenzene. In the same case using safrole, that propenylbenzene would be isosafrole. This propenylbenzene is not a waste. It can be used to make the phenylacetone as well. For instance, one could look in *PIHKAL* on page 734 for the procedure using hydrogen peroxide and formic acid in acetone solution to convert the propenylbenzene to the phenylacetone. I did this reaction once and really hated it. The directions call for removing all the volatiles from the reaction mixture under vacuum when the batch is done. This destroyed my aspirator. Peroxyformic acid is a bitch on metal!

Alternatively, one could go to the second edition of my book *Practical LSD Manufacture*, and choose from the methods given there for converting propenylbenzenes to the phenylacetone. I am partial to the electric cell method, first converting the propenylbenzene to the epoxide, and then rearranging to the phenylacetone.

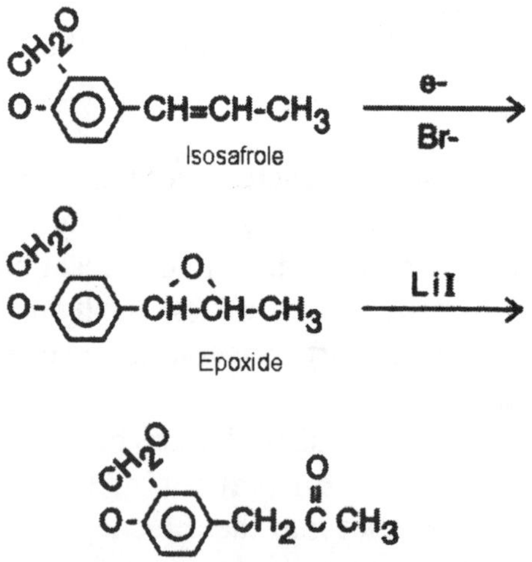

To read more about this reaction method, see *Journal of Organic Chemistry*, Volume 46, pages 3312-15 (1981) and Volume 49, pages 1830-32 (1984). Note that in these papers, they used a platinum cathode instead of stainless steel, and so got partial reduction of the epoxide. Lithium iodide isn't the only substance that will bring about the rearrangement of the epoxide to the phenylacetone.

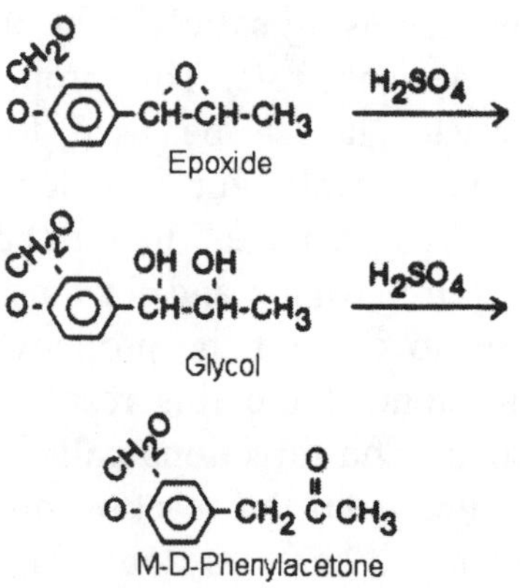

Another way of doing the electric cell method of turning the propenylbenzene into phenylacetone is given in the *Journal of Organic Chemistry* article, Volume 49. If, at the conclusion of passing current through the reaction mixture, a little 1% solution of sulfuric acid is added and stirred for an hour, the product of the cell is 98% yield of the same glycol by the formic acid and peroxide method. Much higher yield, and a quicker and easier reaction. The glycol is then made into phenylacetone by the usual method given by Shulgin in *PIHKAL*, page 734. Heating with a sulfuric acid solution turns the glycol into the phenylacetone.

We digress. We still have the toluene extract of the reaction mixture. It contains the phenylacetone and the propenylbenzene. Both are valuable and must be recovered. How to do it depends upon batch sizes, and how much equipment the clandestine cooker has available.

For reasonable size batches, such as the one given in this example, the simplest way to separate the two substances is by distillation under a vacuum. A claisen adapter packed with broken pieces of glass, and insulated on the outside by wrapping with aluminum foil provides good separation. The toluene-water azeotrope will distill first, thereby drying the solution. Then pure toluene will distill. The toluene can be recycled to extract the next batch. When the toluene is almost all distilled off (there will be a little over 150 ml of high boiling liquid left in the distilling flask, and the rate of toluene distillation will slow and the temperature shown on the thermometer will rise above the 110° C boiling point of toluene), then the heat should be turned off the mixture, and once cooled down a bit, vacuum can be applied. The vacuum should be applied cautiously at first, in spurts, to allow the vacuum induced boiling to cool the contents of the flask further. Then full force vacuum can be applied, and a vacuum distillation commenced. Boiling chips made from a pumice footstone will work on sassafras family compounds, but an oil bath must be used to heat the flask. Direct heating on a hot plate surface will cause bumping during distillation. With meth and regular phenylacetone, even direct heating like that doesn't cause bumping during distillation.

Once the toluene has been removed under a vacuum, the receiving flask should be changed to catch the propenylbenzene, such as isosafrole in this example. A good aspirator using cold water will distill it at about 120° C. The yield of it will be somewhere around 75 ml. After changing receiving flasks, the phenylacetone will then distill around 155-160° C under that same good vacuum. If the water where you live has a temperature akin to urine, then these boiling points will be correspondingly higher. The yield of phenylacetone will also be around 75 ml.

The phenylacetone is then of course used to make the amphetamine. To make meth from phenylacetone, just pick out any of the methods given in this book. Ditto for MDMA. To make MDA from the phenylacetone, the best method is hydrogenation using Raney nickel catalyst as described in US Patent 3,187,047.

If you would like to check out this whole topic being discussed at incredible length, go to www.lycaeum.org/~strike. I hosted the web site there for a month or two in the most foolhardy act anyone has talked me into for ages. While you're there, you might as well pick up Strike's book on this same topic. It's well worth the $25. Just promise your Uncle that you won't even consider using the methods in his book using mercury.

Another way of doing this same reaction dispenses with the need for the special dropping funnel, the chemical glassware, and all the dancing around required to use oxygen under pressure. One can do this reaction in a beaker, or any other glass or Teflon-lined container, if the oxygen for the reaction is supplied by hydrogen peroxide. Hydrogen peroxide is a fairly common industrial chemical. It comes in 55-gallon drums at a price ranging from 40 to 80 cents a pound for 30-35% hydrogen peroxide. A guy named Peter Card used to sell quarts and gallons mail order at a jacked up price for alternative health uses. Check alternative health publications to see if he or somebody else is still in business. In any case, 30-35% hydrogen peroxide is a commonly used substance not requiring one to do business with chemical supply houses such as Aldrich.

To do this variation, let's consider the same one mole size batch. 700 ml of dimethylformamide is placed in the beaker, followed by 100 ml of 30-35% hydrogen peroxide. The catalyst is next added with strong stirring, and a drip of the allylbenzene such as safrole is commenced into the solution. 150 ml of safrole is added over one half to one hour as in the previous example, and the stirring continued for 24 hours as in the previous example. Then, just as before, the reaction mixture is poured into 3N hydrochloric acid solution, and the product extracted out using toluene. I like to use toluene as extractant whenever possible because it is available at the hardware store in the paint thinner section, and it doesn't smell much.

So here once again we have the toluene extract containing the phenylacetone and propenylbenzene. Let's use this opportunity to check out another way to separate the propenylbenzene from the phenylacetone. This way is preferable if really big batches are being done, or if access to chemical glassware for vacuum distilling is denied you. The first thing which should be done is to distill off the toluene as in the previous example. A homemade still could do this job quite well. Be careful of fire hazards when constructing it. Use electric heating to warm the pot. Make sure the toluene vapors get condensed and cooled well. Supply ventilation to keep flammable fumes from binding up. When the toluene has distilled, the propenylbenzene and phenylacetone mixture can be drained from the still by opening a drain or spigot on the bottom of the distilling pot.

Now this mixture can be separated by forming the bisulfite addition product of the phenylacetone. Sodium bisulfite or sodium metabisulfite is widely used in brewing to disinfect the brew vessels, and in vinting to keep the wine from spoiling. A few pounds on up is easily gotten at brew supply shops. It is cheap. 100 to 150 grams will be plenty for this size batch. Add water to the 150 grams of bisulfite with stirring to make a saturated solution. This requires 500 ml of water.

Add the phenylacetone and propenylbenzene mixture to the saturated bisulfite solution, and shake or strongly stir for a few hours. Heavy masses of the bisulfite addition product with phenylacetone form in solution as crystals. After the reaction time is finished, they can be filtered out and rinsed down with some toluene. The filtrate will consist of a layer of the propenylbenzene in toluene from the rinse floating on top of the bisulfite solution. This top layer should be separated off, washed with some water and the toluene evaporated away to give the propenylbenzene in about 75 ml yield. If safrole was used in the reaction, then the isosafrole thereby obtained will smell like licorice.

The crystals of bisulfite addition product should be placed in a boiling flask rigged for reflux along with about a quart of a saturated solution of sodium carbonate or bicarbonate. Arm and Hammer baking soda or baking powder will work just fine. Reflux for a couple of hours to regenerate the phenylacetone from the bisulfite addition product. Cool, and separate off the phenylacetone. Extracting the bicarb solution with toluene will prevent any loss of product. Mix the toluene extract with the phenylacetone, distill off the toluene, and the residue is pure phenylacetone. If you are familiar with the smell of phenylacetone, you will instantly recognize it as product. Methylenedioxyphenylacetone smells almost exactly like regular phenylacetone in its pure state. Sometimes scams are run, selling people the MD-phenylacetone, telling them it's regular phenylacetone. In this case, the phenylacetone is likely to smell a little like the candy shop stuff it was made from. The yield in this case of phenylacetone will be around 75 ml.

Obtaining dimethylformamide can be an obstacle to people contemplating this route of production. If that is the case, it is easily replaced with alcohol. The preferred alcohols to use in place of the formamide are 190 proof vodka and 91% isopropyl alcohol. The latter material can be picked up off the shelf at the drug store. There needs to be some water in the mixture for the reaction to go, and these two alcohols are the most easily available and non-toxic choices.

To do a run using the alcohol solvent instead of formamide, the same equipment and procedures as used in the last two examples are followed, except that the formamide is replaced with the alcohol solvent, and only cupric chloride can be used as co-catalyst. From there, it works exactly the same, except that the yield of product is closer to 70%. The reaction also seems to go faster in alcohol.

What would seem to be a much better method of doing this conversion of the allylbenzene to the phenylacetone is to be found in *Journal of Organic Chemistry,* Volume 45, pages 5390-92 (1980). In this paper they found that t-butyl alcohol is a much better solvent for doing the palladium catalyzed oxidation of the terminal olefin to methyl ketone. This is good, as t-butyl alcohol is a common and cheap industrial chemical. Fifty-five gallon drums of it are no problem to obtain at all.

In this solvent, palladium chloride can't be used; rather palladium diacetate is the preferred catalyst, and just a tiny amount of it does wonders. Later, I'll tell you how to make your own palladium diacetate from an ingot of palladium metal.

In their procedure, they mix a liter or so of t-butyl alcohol with the terminal alkene, such as, for instance, safrole. About one mole of safrole (150 ml) in the liter of t-butyl alcohol would be a good mix. This mixture is heated to about 80° C in any glass or Teflon-coated reaction vessel. Then a gram and a half of palladium diacetate is added to the reaction mixture. After adding that, about 500 ml of 30-35% hydrogen peroxide is added dropwise over a 30-minute period with good stirring. The temperature of the reaction mixture is maintained at around 80° C for about six hours, and an orange-colored solution should result. If during the addition of the peroxide, it looks like the safrole or product phenylacetone is coming out of solution, then just add some more t-butyl alcohol solvent.

After the six hour reaction time, cool the mixture and pour it into 5000 or so ml of water. The product phenylacetone should float to the top as a yellow layer. Separate it off, and extract the t-butyl alcohol and water solution with a few portions of toluene. Add the toluene extracts to the main body of the phenylacetone, and wash a few times with 5% NaOH solution to remove t-butyl alcohol. Distilling should give over 80% to near 90% yield of phenylacetone. I know the first three methods given here in this chapter work. This one should too.

Making Palladium Diacetate

The best ways to make palladium diacetate involve dissolving some of a palladium ingot in nitric acid to make a solution of palladium nitrate in nitric acid or anodically dissolving some of a palladium ingot in hydrochloric acid, as described in *Advanced Techniques of Clandestine Psychedelic & Amphetamine Manufacture,* to make palladium chloride in hydrochloric acid. The acid is neutralized by slowly adding sodium hydroxide solution with good stirring, then the addition of some more sodium hydroxide solution precipitates the palladium as hydroxide. The hydroxide sludge is separated from the liquid and then reacted with acetic acid to make palladium diacetate. See *Chemistry and Industry* (London) 1964, page 544 and US Patent 3,318,891.

For example, let's take that convenient and productive one mole size batch which uses about 1½ grams of palladium acetate as catalyst. To prepare the catalyst for this batch, one would take an ingot of palladium (obtained from the neighborhood coin and precious metal dealer), and dissolve away about .75 grams of the ingot either by immersing the ingot in nitric acid, or anodically dissolving it in hydrochloric acid.

One then takes the acid solution containing the dissolved palladium, and slowly with strong stirring adds a 5-10% solution of sodium hydroxide to the acid. One must keep the temperature of the solution below 60° C during the addition of the hydroxide, as higher temperatures will produce a polymeric metal-hydroxide sludge which is useless for producing catalyst. Nestle the acid beaker in ice or cold water to keep the temperature of the solution down while the acid is being neutralized.

Using a pH meter, keep track of the pH of the acid solution during the addition of sodium hydroxide solution. Once the pH of the solution nears 7, a sludge of palladium hydroxide will start to form. Continue adding sodium hydroxide solution with stirring until a pH of 10 is reached. It does no harm to overshoot this pH, within reason. For example if a pH of 11 is produced, no harm will be done. No more than 15 minutes should be taken to move the pH of the solution from about 3 or 4 up to 10 or 11. Most of the sodium hydroxide solution will be used in neutralizing the initially strong acid with a starting pH below 1, to get the pH up around 3 or 4. Comparatively little sodium hydroxide will be needed to move the pH from 3 or 4 up to 10. The movement of the pH reading from 3 or 4 up to about 10 must be done quickly to prevent the formation of a polymeric hydroxide sludge.

When a pH of 10 or so is reached, a precipitate of palladium hydroxide will form in the solution. These tiny particles will settle to the bottom of the beaker with time. Decant off the clear water solution, and wash the sludge with four 25 ml portions of distilled water. The sludge is suspended in the clean water, then allowed to settle out, and the water rinse is decanted off. Finally, using a Buchner funnel and source of vacuum, the palladium hydroxide sludge is filtered out of the last water rinse, and sucked to a semi-dry filter cake. A reasonably fine grade of laboratory filter paper will catch all the palladium hydroxide. The water filtrate should be clear.

Next, add the semi-dry palladium hydroxide filter cake to about 20 ml of glacial acetic acid, and stir for about 2 hours at 80° C. This will form a solution of palladium acetate in acetic acid. This solution can be added to the batch as catalyst. If the solution stands around for a while in the cold, yellow needle-like crystals of palladium acetate will form. This too can be used as catalyst, so long as all the crystals get rinsed into the reaction solution.

Electric Oxidation of Propenylbenzenes to Phenylacetones

Question: "I've tried the electric cell procedure you gave in *Practical LSD Manufacture* a number of times. My yield of phenylacetone is OK, but no better than what I can get by the old formic acid and peroxide in acetone method. I took a look at the European patent you cited in the book, and it seems I'm following the directions. Why am I not getting the near 100% yields claimed in the patent?" — Puzzled in Holland

Answer: This isn't a case of a bogus patent. This procedure is the real McCoy. See also *Journal of Organic Chemistry*, Volume 46, pages 3312-3315 (1981) and Volume 49, pages 1830-32 (1984) for confirmation of the excellent results. Sigeru Torii, the author of the *Journal of Organic Chemistry* articles is no lightweight either. He is highly respected in the field of electrochemistry.

Let's start by getting you some good quality graphite rods or bar stock to use as anodes for this reaction. Those welding supply shop carbon rods aren't the best for doing electrochemistry with. The following list of suppliers taken from the *Metal Finishing Guidebook* is only a partial listing of all the suppliers of these good quality graphite anodes. If they should happen to ask, say you use these graphite anodes in a Wood's nickel strike, which you use to electroplate on top of stainless steel. Plating suppliers aren't a suspicious lot, so no problems should result from buying graphite anodes.

Atotech 1-800-752-8464
MGP 216-459-0817
Plating Supplies 413-786-2020
Sifco 216-524-0099
Technic 401-728-7081
Unique Industries 203-735-8751
Univertical 313-491-3000

The next thing to line you up with is a good source of DC current. A toy train transformer or other cheap transformer just doesn't turn out a "clean" enough DC output. These cheapie DC sources have what is called "AC ripple." This is AC current superimposed upon the DC output. If you hook such a cheap DC source to an oscilloscope, you will see on the screen the so-called "picket fence" output. This won't do.

A good source of DC current for electric reactions is a Hull Cell rectifier. These are very commonly used by electroplaters to do test plating runs in the lab, and to troubleshoot their plating baths. The output is easily varied from zero to over 20 volts DC, and will commonly put out up to around 15 amps. They generally have pretty accurate gauges on them which measure the voltage and amperage flowing at any given time. New, they go for around $500 to $600. Some major suppliers of these rectifiers are:

Larry King Corp. 718-481-8741
Kocour 773-847-1111
Plating Test Cell Supply 216-486-8400
Technic 401-728-7081

A cheaper and quite good source of high quality DC current comes from batteries. To use batteries as a power source, one chooses a battery or batteries hooked in series such that the voltage delivered is sufficient to move the needed amount of current through the solution. An amp meter, such as the $50 Radio Shack multi-tester, must also be wired into the line leading to the anode. See Figure 29:

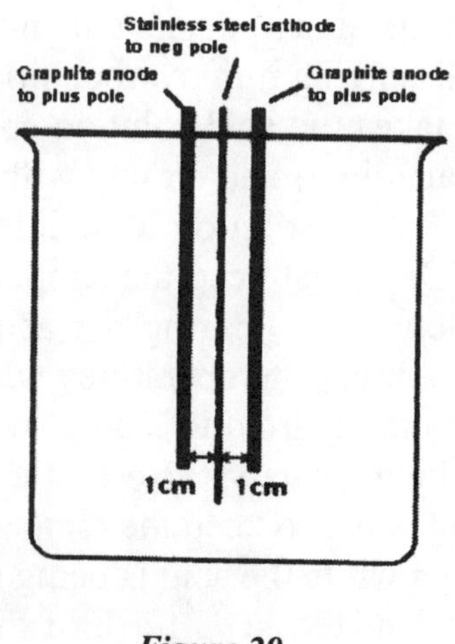

Figure 29

Now let's look more closely at the procedures given in both the patent and the *Journal of Organic Chemistry* articles. The first step is to dissolve about 25 grams of sodium bromide in 100 ml of water. It must be completely dissolved, no crystals of solid sodium bromide left floating around. This is added to the beaker first. This provides about 2.5 moles of NaBr for each mole of the propenylbenzene, such as isosafrole or asarone, used in the reaction. In the *Journal of Organic Chemistry* procedure, they found that anything over 1.5 to 2 moles of NaBr for each mole of isosafrole gave good results. The 2.5 moles used here allows higher current densities to be used, and results in more efficient production of the epoxide from the propenylbenzene.

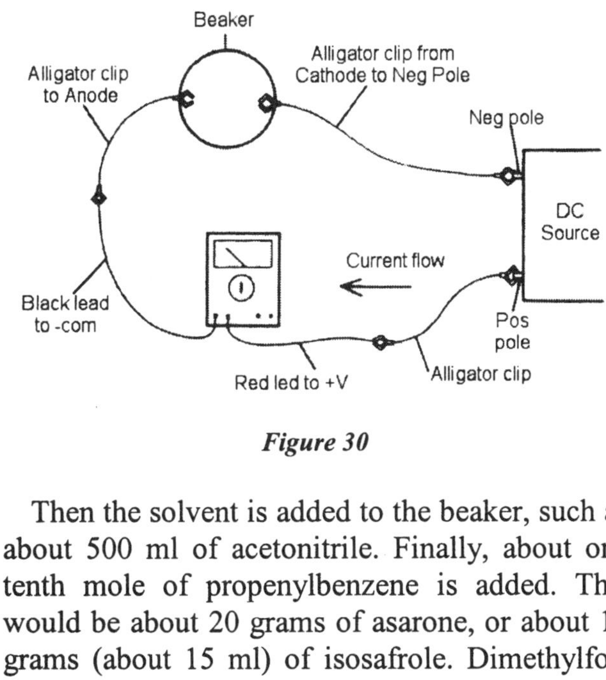

Figure 30

Then the solvent is added to the beaker, such as about 500 ml of acetonitrile. Finally, about one tenth mole of propenylbenzene is added. This would be about 20 grams of asarone, or about 16 grams (about 15 ml) of isosafrole. Dimethylformamide can also be used as solvent by varying somewhat the amount of solvent and water used. This procedure can be scaled up as desired. It was patented as an industrial process.

In the *Journal of Organic Chemistry* article, a platinum anode and platinum cathode of approximately equal size were used. Using platinum as an anode allowed a current density of only about 7 milliamps per square centimeter of the anode surface facing the cathode. The back area can't be counted because the current can't reach there. When graphite is used as an anode, a current density of around 100 milliamps (one tenth amp) per square centimeter of the anode surface facing the cathode can be used. The electric oxidation is also much more efficient using graphite. Only about 2 to 2.5 faradays per mole of propenylbenzene needs be passed through the solution using a graphite anode to get a near quantitative yield of the epoxide. This compares to 4 to 4.5 faradays per mole of propenylbenzene using a platinum anode to get similar results. Using graphite anodes allows therefore a much faster reaction. Graphite is a hell of a lot cheaper than platinum to boot.

The use of a platinum cathode in the *Journal of Organic Chemistry* procedure also seems to be greatly inferior to the stainless steel cathode used in the patent procedure. Besides being much more expensive, when platinum cathodes were used the product was a mixture of mostly epoxide, along with some glycol. It would seem that the epoxide undergoes partial hydrolysis at the platinum cathode to a glycol.

So let's follow the patent procedure. For a one tenth mole size batch, a central stainless steel sheet having an area of 100 square centimeters immersed into the solution is connected to the negative pole of the DC power source. On both sides of this central stainless steel sheet, graphite anodes with a total surface area of about 70 square centimeters in contact with the solution are placed at a distance of one centimeter from the cathode sheet. Since only half of the anode will be facing the stainless steel cathode, this gives about 35 square centimeters of anode surface which can pump current into the solution. It is this face area that we base the current density calculations upon. One can choose to use a larger anode and cathode array, so long as the proportions between anode and cathode area are maintained, and the current applied increased to maintain about 100 milliamps per square centimeter of anode facing the cathode.

So then strong stirring is begun, and the DC power source turned on. The applied voltage is adjusted until the amp meter shows a current flow of 3.5 amps. This is 100 milliamps per square centimeter of the graphite anode face. Continue passing current at this rate until about 24,000 coulombs have passed into the solution. At 3.5 amps this takes about 2 hours.

What to do next to this solution? The simplest and most direct procedure is the one given in the patent, and in *Practical LSD Manufacture*. The acetonitrile is evaporated away under a vacuum, and caught in a cold trap for reuse. The residue left in the evaporating flask, which is the epoxide, is dissolved in some ethyl acetate, and then refluxed with a little lithium iodide to isomerize the epoxide to the desired phenylacetone.

An alternative is to use the procedure that Torii used when he got the mixture of epoxide and glycol in his reaction product. He converted the whole of the mixture to the glycol. To do this, the electrodes were removed from the reaction beaker at the end of the electrolysis, and about 80 ml of a 1% aqueous solution of sulfuric acid was added to the reaction mixture. This was stirred for about an hour, then the sulfuric acid was neutralized by adding sodium bicarbonate (Arm & Hammer). After that the reaction mixture was transferred to an evaporating flask, and the solvent removed under a vacuum. You can pretty much forget about reusing the solvent in this variation, as water from the sulfuric acid solution will form an azeotrope with the acetonitrile. It will be tough to get out.

The sulfuric acid solution causes the following reaction to take place:

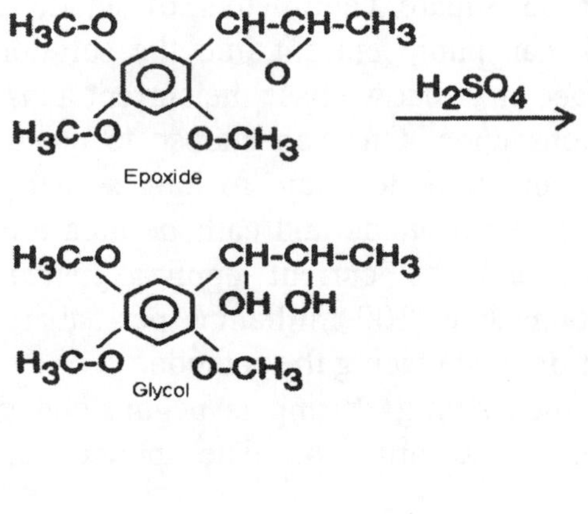

After the solvent is evaporated under a vacuum, the residue in the flask, which consists of close to a 100% yield of the glycol mixed in with crystals of sodium carbonate, is extracted with some ethyl acetate. The ethyl acetate solution of the glycol is next washed with some saturated salt solution. Finally, the ethyl acetate solution is dried with some anhydrous sodium sulfate, then the ethyl acetate is evaporated away under a vacuum to leave a residue of nearly pure glycol. It will be an oil that may crystallize over time. One should not wait around for that to happen. Rather, moving right on to making the phenylacetone should be done.

The procedure for doing this is found on page 734 of *PIHKAL*. For a batch of the size given in this example, the residue of nearly pure glycol is mixed with about 40 ml of methanol (Heet gas line de-icer). Then about 250 ml of a 15% solution of sulfuric acid in water is added, and after thorough mixing, the solution is heated for about 3 hours on a steam bath or in a pan of boiling hot water. After cooling this reaction mixture down, extract three times with 50 ml portions of toluene. The combined extracts are washed first with water, then with a 5% solution of sodium hydroxide or lye in water.

The toluene solution containing the phenylacetone should then be fractionally distilled. The toluene-water azeotrope will distill first as a milky-looking mixture, then pure toluene will distill, looking clear. When the toluene is mostly all gone, the heat should be removed from the distilling flask, and vacuum applied cautiously at first, as the hot mixture will boil furiously at first when vacuum is applied. Then full strength vacuum can be used, and the pure substituted phenylacetone will distill as a clear liquid. If cold water is used with a good aspirator, methylenedioxyphenylacetone will distill around 150° C to 160° C. This is the product from isosafrole, and is used to make MDA or MDMA (X or Ecstasy). If asarone (or B-asarone) was used as the starting material, the boiling point is considerably higher. It's probably better in this case just to remove the toluene under a vacuum rather than try to distill the 2,4,5-trimethoxyphenylacetone. Further purification can be done by the bisulfite addition method given in *Practical LSD Manufacture*. This phenylacetone is used to make TMA-2.

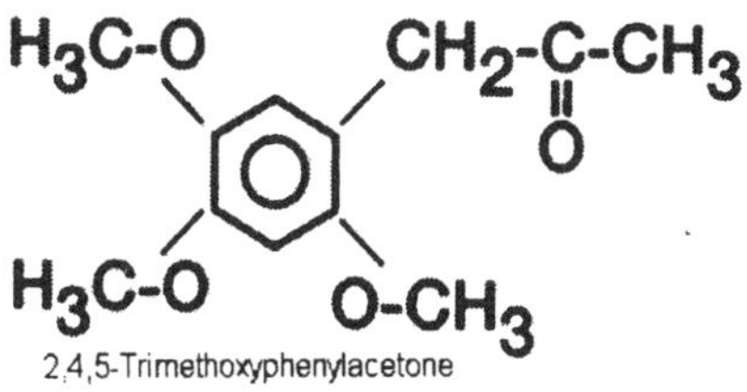

2,4,5-Trimethoxyphenylacetone

That should take care of your problems, Puzzled. The most important points are to use good grade graphite anode material, and use an AC-ripple-free DC power source. The electrodes, of course, can be used over and over. A good cleaning after each batch run should be all they need to keep them in working order.

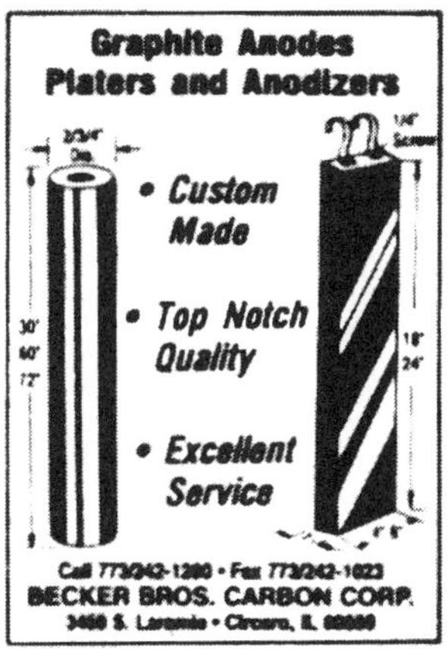

Figure 31

Chapter Eleven
The Way of the Bomb

"Blessed be the bomb... and all its work." —
the mutants of *Beneath the Planet of the Apes*

When underground chemists move up to industrial-scale manufacture of methamphetamine, it soon becomes obvious that the Leuckardt-Wallach reaction is not suitable for making large amounts. There are two reasons for this. N-methylformamide distills slowly, because of its high latent heat of vaporization. This makes the production of large amounts of N-methylformamide a very time-consuming process. Secondly, the Leuckardt-Wallach reaction can take up to 48 hours to complete.

It's also a rather finicky and kind of unpredictable reaction, as was noted earlier in Chapter Five. I still prefer it, however, because of the warm memories it brings back to me.

To increase production, a faster method of turning phenylacetone into methamphetamine is necessary. Reacting phenylacetone with methylamine and hydrogen in an apparatus called a "bomb" is such a method. A bomb is a chemical pressure cooker where hydrogen gas is piped under pressure to react with the phenylacetone and methylamine. It is called a bomb because sometimes reactions like this are done under thousands of pounds of pressure, and occasionally the bomb will blow up. This reaction is done under a pressure of only 3 atmospheres, 30 pounds per square inch greater than normal air pressure, so there's no danger of the hydrogenation bomb going off. This reaction is called reductive amination. It is not especially difficult to do, but it is necessary to have the hardware in proper working condition and to keep out materials that would poison the catalyst. Reductive amination is a quick, very clean and high-yield process.

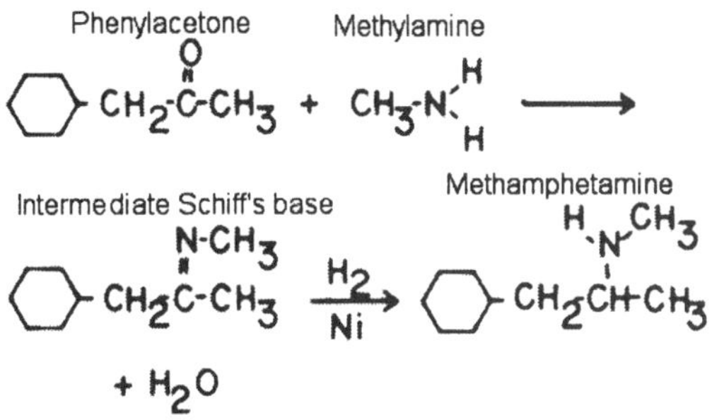

Phenylacetone reacts with methylamine to produce a Schiff's base and a molecule of water. This Schiff's base then reacts with hydrogen and platinum catalyst and gets reduced to methamphetamine. To encourage the formation of this Schiff's base, the amount of water in the reaction mixture is held to less than 10%; 5% is even better. If the underground chemist is able to get methylamine gas in a cylinder, it is easy to control the amount of water in the reaction mixture, but 40% methylamine in water can be made to work with little effort.

Two main side reactions interfere with the production of methamphetamine in the hydrogenation bomb. They are both controlled by properly adjusting the conditions inside the bomb. The first side reaction is the reduction of the phenylacetone.

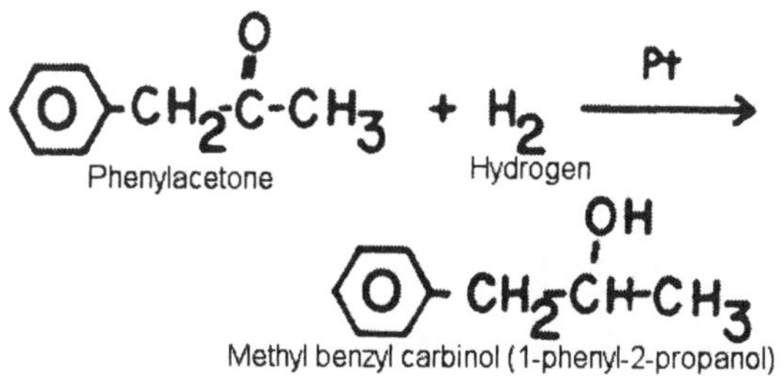

The phenylacetone can react with hydrogen and platinum instead of with methylamine. This side reaction is held to a minimum by not letting the hydrogen gas pressure get much above 30 psi. It is also controlled by encouraging the phenylacetone to react with methylamine instead. This is done by keeping the amount of water in the reaction mixture small, having enough methylamine around for it to react with, and running the reaction at the right temperature.

The other side reaction that can be a problem is phenylacetone reacting with methamphetamine to produce a tertiary amine.

This reaction is held to a minimum by having enough methylamine in the reaction mixture to tie up the phenylacetone, and by keeping the solution fairly diluted, so that they are less likely to bump into one another.

The psychedelic amphetamine MDMA can similarly be made cleanly and on a large scale by the same method. One simply uses methylenedioxyphenylacetone, made according to the directions in the previous chapter, in place of phenylacetone. It reacts in exactly the same way to give around 90% yields of MDMA from the m-d-phenylacetone.

When it is desired to make benzedrine from the phenylacetone, or a psychedelic amphetamine such as MDA or TMA-2 from the appropriately substituted phenylacetone, then the preferred catalyst for the reduction is Raney nickel rather than platinum. In this case, ammonia is used to form the Schiff's base with the phenylacetone, which then reduces to the desired amphetamine through the absorption of hydrogen from the catalyst. Even better results are obtained when the ammonia is used in the form of ammonium acetate, because ammonia in this form is much more efficient at forming the Schiff's base with the phenylacetone than straight ammonia.

Raney nickel is a much less convenient catalyst to use than platinum. It is attracted towards magnets, just like iron filings, so magnetic stirring inside a hydrogenation bomb isn't possible with Raney nickel. Hydrogenations done with this catalyst also generally require much higher hydrogen pressures be applied inside the bomb, and that heating of the mixture be done. Platinum, on the other hand, works just fine at relatively low hydrogen pressures and at room temperature. Magnetic fields don't affect it, so magnetic stirring can be used with this catalyst. It is much more convenient to use.

Reductive alkylation using platinum catalyst can be done inside a very easily constructed apparatus. For example, a champagne bottle of about 1.5 liter capacity can be used. Champagne bottles are built to withstand pressure, and have no problem holding up to the approximately 30 pounds of pressure used in the reduction. The glass is quite inert to chemicals, so it is desirable from that point of view as well. On the downside for this reaction vessel, there is no obviously easy way to attach pressurized lines to the bottle other than using hose clamps, or inserting a stopper into the neck and wiring it into place. Glass is brittle, and may break if the clamps are tightened down too much. Heating isn't possible using this glass container, as this will likely cause it to crack, and then burst under the 30 pounds of pressure. Coating the bottle with fiberglass resin helps avert this danger.

An alternative is "The Poor Man's Hydrogenation Device." It is an aluminum fire extinguisher which has been emptied of contents, washed, and then coated on the inside with either Teflon or a Teflon paint such as Xylan 1006. See *Advanced Techniques of Clandestine Psychedelic & Amphetamine Manufacture* for more on this topic. The fire extinguisher has the advantages of greater strength, easy attachment of pressurized hydrogen hose lines, and its own built-in pressure gauge.

To do a ½ pound batch in a hydrogenation bomb, start with a 1000 or 1500 ml beaker. Put 500 ml of 190 proof vodka in the beaker, followed by 200 ml of 40% methylamine solution in water. While stirring slowly add 200 ml of phenylacetone. The Schiff's base will form right there, and generate a fair amount of heat. Pour this mixture into the hydrogenation bottle, then add 5 grams of platinum oxide catalyst (Adam's Catalyst). Using anhydrous methylamine from a cylinder would be even better. By tipping the cylinder upside down, liquid methylamine can be added to the alcohol. In this case around 100 ml of methylamine would be enough.

For most of us, both 40% methylamine in water or the anhydrous methylamine are unavailable, as they are List I chemicals. Home brew methylamine is generally made in the form of methylamine hydrochloride. See the methylamine recipes in this book, and a couple more good ones in *Advanced Techniques of Clandestine Psychedelic & Amphetamine Manufacture.* To use methylamine hydrochloride, I would think that it would have to be free based, because methylamine hydrochloride isn't very soluble in alcohol at room temperature.

To do this variation using methylamine hydrochloride, one again starts with 500 ml of 190 proof vodka, and places it in a 1000 or 1500 ml beaker. To this add 180 grams (4.5 moles) of sodium hydroxide (NaOH). Stir to dissolve it to the extent possible. Then with cooling and stirring add 300 grams (4.5 moles) of methylamine hydrochloride. Add the methylamine slowly, so that the heat made by neutralizing the hydrochloric acid doesn't make the methylamine free base fume off. Stir until the NaOH pellets have disappeared, and have been replaced by a heavy precipitate of salt.

I don't know if free NaOH would be harmful to the efficiency of the catalytic hydrogenation, but I think it would be better to play it on the safe side. It is then better to be a little bit light on the usage of NaOH rather than adding an excess of it.

When the NaOH pellets have completely dissolved, then the salt formed by the neutralization should be filtered out. The clear, filtered methylamine solution in alcohol can then be used just as in the previous example. Two hundred ml of phenylacetone or methylenedioxyphenylacetone if MDMA is being made, is then added to the methylamine solution. After stirring the mixture and allowing it to cool, it can be poured into the hydrogenation apparatus shown in Figure 32.

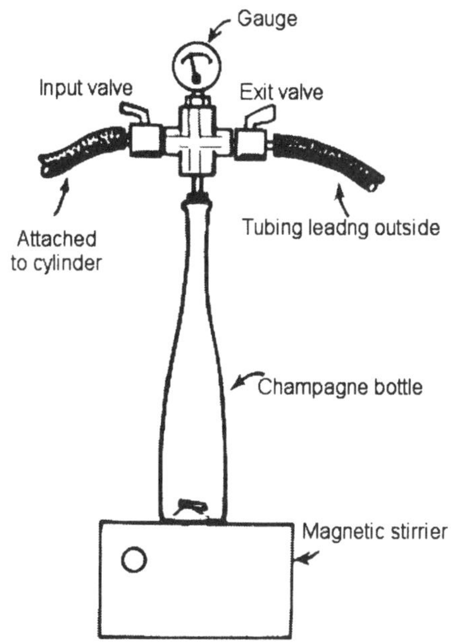

Figure 32

A magnetic stir bar is slid down into the bottle, and regardless of the variation used, about 5 grams of platinum black or platinum oxide (Adam's Catalyst) will be enough for the purpose.

The apparatus shown in Figure 32 can be constructed by anyone with access to machinist's tools. Alternatively, the clandestine operator can have it made for him with little or no chance of anyone suspecting its real purpose. The threads are fine, and coated with Form A Gasket immediately before assembly. The valves are of the swage-lock type.

Before beginning production using this device, the joints are checked for leakage by brushing soapy water on them and looking for the tell-tale bubbles.

The chief danger in using the hydrogenation apparatus is from fire due to leaking hydrogen coming into contact with a spark or flame. The magnetic stirrer is a possible source of static-induced sparks. To eliminate this danger, it is wrapped in a sturdy bread or garbage bag. This prevents hydrogen from coming into contact with it. Good ventilation in the production area likewise prevents hydrogen from building up in the room.

To begin production using this device, the champagne bottle is attached to the rig immediately after filling with the reactants. The air is sucked out of the bottle by attaching the exit valve, a vacuum line leading to an aspirator. After sucking out the air for 30 seconds, this valve is closed, and hydrogen is fed into the bottle from the cylinder until it has pressurized to a few pounds above normal air pressure (i.e., a few pounds show on the gauge). Then the input valve is closed, and the bottle is vacuumed out once more. Now the bottle is practically free of air. The exit valve is closed once again, and hydrogen is let into the bottle until the gauge shows 30 pounds of pressure. This is 3 atmospheres of pressure, counting the 15 pounds needed to equal air pressure. Magnetic stirring is now started, and set at such a rate that a nice whirlpool forms in the liquid inside the bottle.

The hydrogen used in this reaction is of the purest grade available. Cylinders of hydrogen are obtained at welding supply shops, which generally have or can easily get electrolytically produced hydrogen. This is the purest grade. The cylinder must have a regulator on it to control the pressure of hydrogen being delivered to the bomb. The regulator must have two gauges on it, one showing the pressure in the cylinder, the other showing the pressure being fed into the line to the bomb.

With "The Poor Man's Hydrogenation Device," all the construction is omitted. After filling the extinguisher about half full of reaction mixture, the top is screwed back on. A vacuum line is attached to the spray nozzle of the fire extinguisher, the valve is opened by squeezing the spray lever, and the air sucked out. Then the hydrogen line is attached, and about 30 pounds of hydrogen gas pressure put into the extinguisher tank.

After beginning stirring the contents of the bomb, an induction period of about an hour or so usually follows during which nothing happens. No hydrogen is absorbed by the solution during this period. It is not known just why this is the case, but nothing can be done about it. Use of pre-reduced platinum catalyst does not eliminate this delay. (Pre-reducing is a procedure whereby the platinum catalyst is added first, and then contacted with hydrogen to convert the oxide of platinum to the active metal.)

In an hour or so, hydrogen begins to be absorbed by the solution, indicating production of methamphetamine. The pressure goes down on the gauge. More hydrogen is let in to maintain the pressure in the 30-pound range. Within 2 to 4 hours after uptake of hydrogen begins, the absorption stops. This indicates the end of the reaction.

The valve on the cylinder is now closed, and the exit valve slowly opened to vent the hydrogen gas outside. Now the bottle is removed from the apparatus, and the platinum is recovered for reuse by filtering the solution. The platinum is stored in absolute alcohol until the next batch. Many batches can be run on the same load of platinum catalyst, but it eventually loses its punch. It is then reworked in the manner described later.

The filtered reaction mixture is then poured into a 2000 ml round bottom flask, along with 3 or 4 boiling chips. The glassware is set up as shown in Figure 11 in Chapter Three. The chemist heats the oil no hotter than 110° C, and distills off the alcohol and water. When the volume of the mixture gets down to near 500 ml, he turns off the heat and transfers the reaction mixture to a 1000 ml round bottom flask with 4 boiling chips. He sets up the glassware for fractional distillation as shown in Figure 13 in Chapter Three, and continues distilling off the alcohol. The temperature shown on the thermometer should be about 80° C. When the volume of the reaction mixture gets down to about 300 ml, he turns off the heat and lets it cool off. He attaches a 250 ml round bottom flask as the collecting flask and begins a vacuum distillation. The last remnants of alcohol are soon gone, and the temperature shown on the thermometer climbs. If he is using an aspirator, when the temperature reaches 80° C, he changes the collecting flask to a 500 ml round bottom flask and distills the methamphetamine under a vacuum. If he is using a vacuum pump, he begins collecting methamphetamine at 70° C. He does not turn the heat setting on the buffet range above $^1/_3$ of the maximum. Virtually all of the material distilled is methamphetamine. He will get between 200-250 ml of clear to pale yellow methamphetamine, leaving about 20 ml of residue in the flask. A milky color to the distillate is caused by water being mixed with it. This is ignored, or removed by gentle heating under a vacuum.

The distilled methamphetamine is made into crystals of methamphetamine hydrochloride in the same way as described in Chapter Five. He puts about 75 ml of methamphetamine in each Erlenmeyer flask and adds ether or toluene until its volume reaches 300 ml. Then he bubbles dry hydrogen chloride gas through it and filters out the crystals formed. The yield will be close to ½ pound of pure methamphetamine.

Making and Reworking Worn Out Platinum Catalyst

Ready-made platinum oxide (Adam's Catalyst) is available from scientific supply houses at the steep price one would expect for platinum compounds. It's not on any reporting lists, and I don't think it's a particularly "hot" item at the time I am writing this. However, being able to bypass these snitch dens is always of value. Platinum catalyst can be made either from platinum metal, or from the chloride salts of platinum. The metal is easily picked up at dealers of precious metals and coins in the form of ingots and coins. The chloride salts of platinum can be easily obtained from suppliers of precious metal salts to electroplaters. Go to the *Metal Finishing Handbook and Directory* and look in the index under platinum plating baths for a list of suppliers. For a really thorough discussion on making platinum catalyst, see *Organic Syntheses*, Collective Volume I. Look in the table of contents under Platinum Catalyst for Hydrogenation.

The process used to turn platinum metal into active catalyst is identical to the method used to recycle worn out platinum catalyst into reborn material. The first step is to dissolve the metal in aqua regia. Aqua regia is a mixture of three parts hydrochloric acid, and one part nitric acid. Only laboratory grade acids in their concentrated forms are used for this process. Lower grades may well introduce catalytic poisons into the precious metal. The nitric acid is the 70% material. The hydrochloric acid is the 37% laboratory material. About a pint of mixed acid serves well to dissolve the few grams of platinum needed to run man-sized batches of methamphetamine. The acids are simply mixed, and then the platinum metal is added. A few fumes of NO_2 are given off in the dissolution process. Occasional swirling and some heating speeds the process of dissolving the platinum. The dissolution converts the platinum to chloroplatinic acid H_2PtCl_6. This substance is the starting point for both of the alternative pathways to active platinum catalyst.

When all of the platinum metal has disappeared into solution, heat is applied to boil away the acid mixture. Then some concentrated hydrochloric acid is added, and this too is evaporated away to dryness. The addition and evaporation of hydrochloric acid is repeated several times until the residue is free of nitrates.

With chloroplatinic acid thusly obtained, the manufacturing chemist has two alternative methods with which to convert it into active material ready for use. The first method is the classical route involving a fusion of the chloroplatinic acid, or preferably its ammonium salt, with sodium nitrate at a temperature of about 450° C. This method entails the obvious difficulty of accurately measuring and controlling such a high temperature. One can read all about this method in *Organic Syntheses*, Collective Volume I, pages 463 to 470.

The second method uses sodium borohydride to convert the acid directly into platinum black. This method is simpler and produces a much more active catalyst. The procedure is based on the method given by Brown and Brown in the *Journal of the American Chemical Society*, Volume 84, pages 1493 to 1495 (1962). The yield is about 3 grams of the extra high activity catalyst, and does the job of 5 grams of the catalyst prepared by the classical method.

To prepare this catalyst, 8 grams of chloroplatinic acid is dissolved in 100 ml of absolute alcohol. Then, in another beaker, .8 grams of laboratory grade sodium hydroxide is dissolved in 10 ml of distilled water. This is diluted to 200 ml of total volume by adding absolute alcohol, and then 10 grams of sodium borohydride is added. The alcohol-NaOH-water-sodium borohydride solution is stirred until the borohydride is dissolved. The borohydride solution is now added to the chloroplatinic acid solution with vigorous stirring. It is added as quickly as possible without letting the contents foam over. A large amount of hydrogen gas is given off while the borohydride reduces the chloroplatinic acid to platinum black. This process is done in a fume hood or outside to prevent hydrogen explosions.

About one minute after all the borohydride solution has been added, the excess borohydride is destroyed by adding 160 ml of glacial acetic acid or concentrated hydrochloric acid. The solution is then filtered to collect the platinum black. It is rinsed with a little absolute alcohol, with added filter paper and all (to prevent loss of catalyst sticking to the paper), directly into the champagne bottle for immediate use. If it must be stored before use, it is put in a tightly stoppered bottle filled with absolute alcohol.

As an alternative, one could simply add the jet black suspension of the platinum catalyst in the alcohol and borohydride reduction mixture directly to the hydrogenation bomb without bothering to destroy the excess borohydride left in the solution or trying to filter out the very small particles of platinum black catalyst.

Two other methods exist for converting chloroplatinic acid to PtO_2 catalyst. One is slowly adding, with stirring, the chloroplatinic acid to a water solution containing an excess of NaOH, then filtering out the catalyst and rinsing it off. See *Zeitschrift der Anorganische Chemie*, Vol. 40, page 434 (1904), and Vol. 44, page 171 (1905). The other method employs formaldehyde to reduce the chloroplatinic acid. See *Berichte der Deutschen Chemischen Gesellschaft*, Vol. 54, page 113 and page 360 (1921).

Making Benzedrine or Psychedelic Amphetamines From Phenylacetone(s)

In this procedure, methylamine isn't needed. That's one hassle avoided. That hassle is replaced with the need for bulky and unwieldy equipment to be constructed. As was mentioned earlier, Raney nickel is the preferred catalyst for these reductive aminations. It can't be magnetically stirred because it is attracted to magnets. Much higher pressures of hydrogen are required when using this catalyst, and heating of the hydrogenation mixture to around 80° C should be done.

To construct a hydrogenation device suitable for use in this reaction, one must be able to weld stainless-steel. Let's start with the construction of the hydrogenation bottle, and use as the example a bottle of about one gallon capacity.

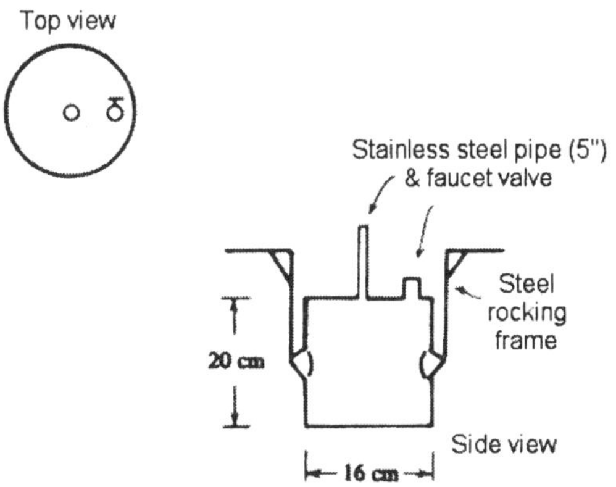

Figure 33

If the underground chemist has to make his own one gallon reaction bottle, he uses stainless steel $^{1}/_{8}$ to $^{3}/_{16}$ inch thick, such as a section of stainless-steel pipe. For a volume of about one gallon, it should be about 16 cm in diameter and 20 cm in height. The bottom is Tig welded on, this process being much easier if it starts out a few millimeters larger in diameter than the pipe section.

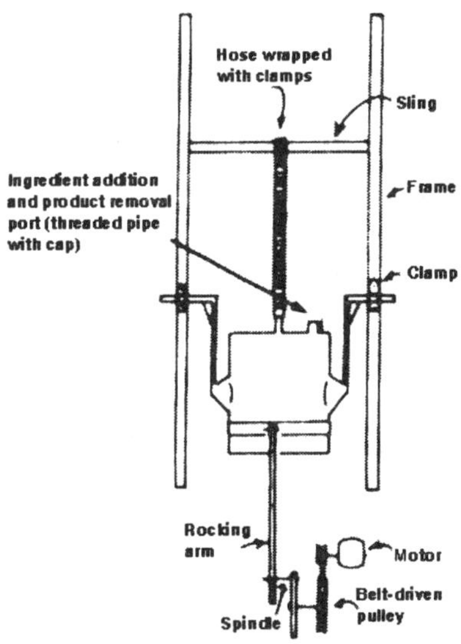

Figure 34

The top of the tank has 2 holes drilled in it. One small one in the center of the tank is an entrance for the hydrogen gas. This has a section of stainless-steel pipe about 5 inches long welded around it. It is usually necessary to melt in some stainless-steel welding rod while making this Tig weld, to get it strong enough. This top section is then welded onto the top to create the reaction vessel shown in Figure 33.

A steel rocking frame is then welded onto the outside of the reaction vessel as shown in Figures 33 and 34. The area where it is welded should be reinforced. All welds are done with a Tig welder.

The chemist can now assemble the bomb. He starts out with heavy wooden planks as the base. This will keep vibration to a minimum. He sets up and bolts down the frame. He attaches some clamps to this frame, then puts sheaths and bearings on the arms of the steel rocking frame, and suspends the reaction vessel about 6 inches off the ground. It should swing back and forth easily.

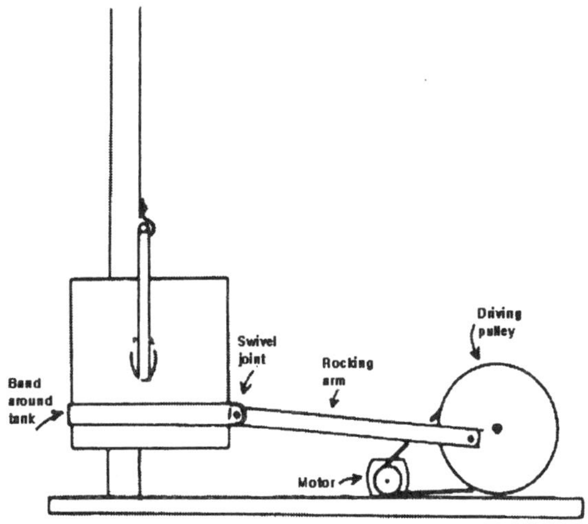

Figure 35

He attaches a band around the reaction vessel, just below where the steel rocking frame is attached to the reaction vessel. The band is attached to the rocking arm, which is attached to a spindle on the driving pulley, as shown in Figures 34 and 35. Both these joints should swivel easily. The driving pulley is about 10 cm in radius. The pulley on the motor has a radius of about 2 cm. The spindle, which extends from the driving pulley to the rocking arm, is about 3 cm from the center of the driving pulley.

The motor is the usual 1760 rpm type of motor, with a power of at least $^{1}/_{30}$ hp. When the motor is turned on, it spins the driving pulley, which moves the rocking arm back and forth, which in turn shakes the reaction vessel.

The chemist is now ready to test the system. He opens up the valve and puts 2000 ml of distilled water in the reaction vessel. He closes the valve and turns on the motor to begin shaking. If any water comes out the top of the stainless-steel pipe, he secures the wooden base to minimize vibration. He shuts it off and opens the valve, then siphons out all the water.

He now runs a line of heavy rubber tubing from the hydrogen cylinder to the stainless-steel pipe. He crimps in the end of the pipe, then pushes the rubber hose down over the pipe, at least halfway to the tank. He superglues it to help hold it in place. Then he covers the entire length of the hose with a series of pipe clamps so that it does not blow out or slip off the pipe. This hose is slung over a sling in the frame so that it leads straight down to the reaction vessel. There must be enough slack to allow for the rocking motion.

If any water came out of the pipe in the test run, the hose must have catalytic poisons removed from it by boiling it in 20% sodium hydroxide solution, then rinsing it off in boiling water.

The chemist closes the valve and begins putting pressure in the tank, starting with a pressure value of 50 psi. He brushes soapy water around the joints to look for any leaks. If there aren't any, he works the pressure up to 300 psi. If leaks are found, he tries brazing over the faulty joint. His welds must be nearly perfect.

First, the chemist must find out how high the heat must be set to get an 80° C temperature in the contents of the bomb. He fills the bomb half-full of isopropyl rubbing alcohol and turns on the heat. He keeps track of the temperature of the alcohol while stirring it with the thermometer. He finds the heat setting needed for an 80° C temperature and how long it takes to reach that temperature. Then he removes the isopropyl alcohol from the bomb and rinses it out with ethyl alcohol.

Heating for the hydrogenation bottle is supplied by wrapping it with heating tape. Either just enough heating tape is used to reach the required 80° C temperature, or preferably a lot of heating tape is used and the wiring for the heating tape is hooked up to an adjustable transformer to control the amount of heat supplied.

That is quite an engineering job, isn't it? In *Advanced Techniques of Clandestine Psychedelic & Amphetamine Manufacture* I propose a convenient tabletop MDA recipe. This uses a Raney nickel cathode to do the hydrogenation by means of electrically generated hydrogen on the nickel cathode surface. I don't know how good the results are using this technique. It certainly is much more convenient than constructing this monstrosity!

A variety of hydrogenation mixtures have been used over the years to convert phenylacetone or substituted phenylacetones to the amphetamine. In the earliest works, strong ammonia solution (28% NH_3, 56% NH_4OH) was simply added to an alcohol solution of the phenylacetone, and then the mixture hydrogenated using Raney nickel catalyst. They claimed pretty good results. Later workers trying to recreate these good results didn't get such satisfying yields. They found that liquid anhydrous ammonia added to the alcohol solution of the phenylacetone in alcohol worked much better. Ammonia is much less willing to form the Schiff's base with phenylacetone than methylamine. Less water in the solution encouraged the Schiff's base to be formed, and yields up to 80% from m-d-phenylacetone were obtained. This method is good enough to be used in production situations. If cylinders of anhydrous ammonia aren't easily available to you, see *Advanced Techniques of Clandestine Psychedelic & Amphetamine Manufacture* for the chapter detailing a convenient method for making alcohol saturated with ammonia. This method produces a fairly anhydrous solution.

The preferred hydrogenation solution for the reductive amination of phenylacetone and substituted phenylacetones is given in US Patent 3,187,047. The solution proportions are:

3 kilos of phenylacetone or substituted phenylacetone

1.2 kilos of ammonium acetate

180 ml of glacial acetic acid

9.5 liters of methanol

300 ml of water

500 grams of Raney nickel catalyst

A one-gallon hydrogenation bottle will hold approximately one-seventh of the above-stated hydrogenation solution. That would fill the bottle approximately half full, which is about the maximum amount one would want to fill the hydrogenation bomb. So one would simply use about one-seventh of the ingredients listed above. This mixture shouldn't be mixed up before use, as the Schiff's base will start to polymerize and undergo other unwanted reactions on standing for prolonged periods.

Using a funnel, the reactants are added to the hydrogenation bottle. Raney nickel generally comes commercially as a suspension in alcohol. It should be thoroughly shaken before measuring out the amount of Raney nickel added, so that a homogenous mixture is being measured out.

When the reactants have been added to the bomb, the cap is screwed onto the addition portal, making sure the threads are clean, and that the cap has been tightened enough to prevent leakage. A pressure of at least a couple hundred pounds per square inch of hydrogen is then let into the bomb, heating is begun along with the shaking of the hydrogenation bottle. Hydrogen absorption by the solution starts almost immediately, and speeds up as the mixture warms. The hydrogen pressure is maintained by periodically opening the valve to the hydrogen cylinder.

A few hours after the mixture gets warmed up, the absorption of hydrogen by the solution stops. It is now time to get the product. Heating is stopped, and once the solution has cooled down close to room temperature, the shaking is stopped. All valves from the hydrogen cylinder are checked to make sure they are tightly closed. Then with good ventilation, the cap is first loosened on the hydrogenation bottle to vent off hydrogen pressure. Then the cap is removed, and the contents of the bomb poured or pumped into a one gallon beaker or jug.

The Raney nickel catalyst must first be removed from the product mixture. This can be done by letting the mixture sit and letting the catalyst settle to the bottom of the container. Then the product mixture can be decanted off of the settled catalyst. Another way to remove the catalyst is to filter it out. This is much faster. One must be careful filtering out Raney nickel catalyst because it is pyrophoric. That means that when it dries out and is exposed to air it first starts to smoke, then bursts into flames. This is quite dangerous because the methanol solvent is flammable as hell. When filtering the catalyst, keep it wet by rinsing it with more alcohol solvent. Whichever method of removing the catalyst is used, it should be saved because it should be reusable several times. The catalyst is kept submerged under alcohol in a bottle or jar until its next use.

The product mixture should next be poured into a distilling flask. The apparatus is set up for distillation, then the methanol is distilled off. The residue left inside the distilling flask is allowed to cool down. When it has cooled, a 10-20% solution of sodium hydroxide should be added slowly with occasional shaking of the contents of the flask. When enough sodium hydroxide solution has been added to leave the water layer inside the flask strongly alkaline (pH 13+) then a product layer of amphetamine will float on top of the water. It should be no more than yellow colored. A strong smell of ammonia should also be noted from the free basing of the ammonium acetate. Add about four or five volumes of toluene for each volume of phenylacetone used in the batch, and shake the mixture in the flask to extract all the amphetamine into the toluene. This layer will float on top of the water when shaking is stopped. This layer should be separated off using a sep funnel. This amphetamine-toluene solution can be allowed to sit for a couple hours to settle out entrained water, then after pouring the toluene-amphetamine solution off the settled water, the solution can be poured into a large filtering flask, and a vacuum applied to it. This will cause the ammonia in the solution to be sucked out. Some shaking while under vacuum will help speed along the removal of ammonia. Your nose can be the guide as to when all the ammonia has been removed from the solution. Then bubbling with dry HCl through the solution can be done just as described in Chapter Five to get crystals of amphetamine hydrochloride. These are filtered out as described in Chapter Five.

Alternatively, one can distill the toluene-amphetamine solution, and then dilute the distilled amphetamine with some fresh toluene, and bubble it with HCl as described in Chapter Five. The yield in either case will be around a pound from the roughly 450 ml of phenylacetone used in this example.

Raney Nickel Catalyst

Ready-made Raney nickel catalyst, sold as a suspension in alcohol or water, is available from scientific supply houses at a fairly steep price. I'm not sure just how "hot" this material is to purchase through regular channels. Buying the material ready-made will assure one of having a good and active catalyst.

Scientific supply houses also sell Raney alloy, or aluminum nickel-alloy — the material from which Raney nickel catalyst is made. This sells at a more reasonable price, but I'm also unsure of just how "hot" it is.

In *Advanced Techniques of Clandestine Psychedelic & Amphetamine Manufacture,* I describe melting your own Raney alloy in a high temperature crucible. This isn't very hard to do, and is very low profile indeed! Let me improve the procedure given in that book. If one takes a clean file and files off the required amount of nickel metal from the nickel anode, these small particles of nickel metal melt into the molten aluminum much easier than larger chunks of nickel metal. The result is a more homogenous and more easily made alloy. Nickel is about as hard as iron, so filing chunks of nickel metal down to grit is not very difficult to do.

Making Raney nickel catalyst from the Raney alloy, either purchased as such or homemade isn't very hard to do either. The alloy is a 50-50 mixture of nickel and aluminum. By reacting the finely ground alloy with sodium hydroxide solution, most of the aluminum dissolves out, leaving the Raney nickel catalyst. The catalyst retains a few percent of aluminum mixed in with the nickel, and this few percent of aluminum is very important in its activity. Otherwise one could just use nickel metal and expect to see catalytic activity. Let's make a batch of Raney nickel catalyst from Raney alloy.

Here's how it's done. In a 2000 ml beaker, the chemist dissolves 190 grams of sodium hydroxide pellets in 750 ml distilled water. The solution is cooled down to 10° C by packing the beaker in ice. He adds 150 grams of the nickel aluminum alloy to the sodium hydroxide solution. It is added slowly and with vigorous stirring. The temperature of the solution must not get above 25° C. The sodium hydroxide reacts with the aluminum in the alloy and dissolves it, producing aluminum hydroxide and hydrogen gas. The nickel is left as tiny black crystals. The hydrogen which bubbles out of the solution causes foaming, so the alloy is added slowly enough that the foaming doesn't get out of control. If that fails, 1 ml of n-octyl alcohol helps to break up the foam. It takes about two hours to add all the alloy to the sodium hydroxide. When all of the alloy has been added, the stirring is stopped and the beaker is removed from the ice bath. The bubbling of hydrogen gas from the solution continues as the beaker warms up to room temperature. Hydrogen gas is not poisonous, but it is very flammable. Smoking around it can cause an explosion.

When the bubbling of hydrogen from the solution slows down, the beaker is set in a large pan of hot water. Then the water in the pan is slowly heated to boiling. This will get the hydrogen bubbling again, so it is heated on an electric heater in a well-ventilated area. This heating is continued for 12 hours. Distilled water is added to the beaker to maintain its original volume.

After the 12 hours are up, the chemist removes the beaker from the boiling water bath and stirs it up. Then he allows the black Raney nickel catalyst to settle to the bottom of the beaker, and pours off as much of the sodium hydroxide solution as possible. The nickel is transferred to a 1000 ml graduated cylinder with the help of a little distilled water. If the nickel catalyst is allowed to dry out, it may burst into flames. It must be kept covered with water. Again the chemist pours off as much of the water as possible. Then he adds a solution of 25 grams of sodium hydroxide in 250 ml of distilled water to the nickel in the graduated cylinder. The cylinder is stoppered with a cork or glass stopper (not rubber) and shaken for 15 seconds. Then it is allowed to settle again and as much of the sodium hydroxide solution as possible is poured off.

The catalyst is now ready to have the sodium hydroxide removed from it. All traces must be removed, or it will not work. The chemist adds as much distilled water to the cylinder as it will hold, then shakes it to get the nickel in contact with the clean water. He lets it settle, then shakes it again. When the nickel has settled, he pours off the water and replaces it with fresh distilled water. This washing process is repeated 25 times. It takes that much to remove all the sodium hydroxide from the catalyst.

After the water has been poured off from the last rinse with distilled water, 100 ml of rectified spirit (95% ethyl alcohol) is added to the nickel and shaken. After the nickel has settled, the alcohol is poured off and the washing is repeated two more times with absolute (100%) alcohol. The result is 75 grams of Raney nickel in alcohol. It is transferred to a bottle that it will completely fill up. If necessary, more alcohol (100%) is added to fill up the bottle. Then the bottle is tightly stoppered. When the chemist is ready to use it, he shakes it to suspend the nickel and measures out the catalyst. One ml contains about .6 grams of Raney nickel catalyst.

It has been claimed that a more active catalyst can be made by adding the sodium hydroxide solution to the nickel-aluminum alloy instead of vice versa. But when this is done, care must be taken that the foam doesn't get out of control. Also, the alloy must be stirred into the solution so it can react. Other than that, the catalyst is prepared in exactly the same way.

Raney nickel is a pain in the ass to use, largely because its magnetic properties prevent the use of magnetic stirring during reduction. There are other nickel catalysts that are said to resemble Raney nickel in activity. Even better, they aren't pyrophoric, so that danger is eliminated. They are also said to work at fairly low hydrogen pressures and without the need for heating to be applied. "The Poor Man's Hydrogenation Device" could be used instead of that shaker bottle monstrosity!

These catalysts are made by the reduction of the common nickel salts such as nickel chloride or sulfate by borohydride or by zinc dust. These common nickel salts are widely used in electroplating. For some directions, see *Journal of Organic Chemistry*, Volume 35, page 1900 (1970) and *Chem. Commun.*, 1973, page 553 and *Bull. Chem. Soc. Japan,* Volume 33, page 232 (1960). It's worth checking out!

One final topic must be addressed while on the subject of hydrogenations. That subject is the use of catalytic converters from cars as hydrogenation catalysts. I've been hearing these stories for about a decade now. I don't know how well it would work. It may well be the equivalent of an urban legend. We dope cookers have our own lore and mythology, you know?

The converter contains a mixture of platinum and palladium thinly spread over the surface of a supporting medium. Palladium behaves quite similarly to platinum as a hydrogenation catalyst, except that it remains active in acid solutions.

To use this material as a hydrogenation catalyst, a new catalytic converter should be obtained. An old used one from the junkyard will not do, because the catalyst may have been poisoned from the use of bad fuel, and it will also likely be covered with a sooty film. Cut open this new catalytic converter to get the converter element inside. The catalyst itself is a mass of BB-sized pellets packed inside the converter. These should be removed and washed clean in soapy hot water, followed by a rinse with rubbing alcohol.

There are problems associated with the use of this makeshift hydrogenation catalyst. The amount of catalyst used must be greatly increased because these pellets don't possess the enormous surface area that a finely divided material has. Catalytic hydrogenation takes place at the surface of the metal, so this is an important point. Further, these pellets aren't easily stirred up in the solution to catch a breath of fresh hydrogen, so they can quickly go dead. They are better used in a bomb which is agitated by shaking. Also, their mass makes them likely to break a hydrogenation bomb made of glass, so a bomb constructed of metal is more compatible with their use.

Your Uncle has an idea, however, which would rise above the level of Urban Legend. That is to use these palladium-plated beads as the catalytic cathode in the Fester Formula given in *Advanced Techniques of Clandestine Psychedelic & Amphetamine Manufacture.* In the past couple of years, the amount of palladium used in catalytic converters has been increased. That's why the price of the metal has gone up so much of late. It also means that the converter is even more suitable for use in the Fester Formula. I must add an update here also on the subject of hydrogenation of ephedrine. Check out US Patent 6,399,828. They report that the sulfate salt of ephedrine hydrogenates poorly, giving low yields of product. To get higher yields, the catholyte then shouldn't be dilute sulfuric acid. Use instead dilute hydrochloric acid for the hydrogenation, but keep the separate dilute sulfuric acid solution for anodizing the palladium surface. One would then simply place some of the beads inside a plastic mesh bag finely woven enough to hold them in place, but yet coarse enough to allow free contact with the stirred catholyte. Putting a lead of insulated wire down into the middle of the mass of beads, with the insulation removed at the end, would make electrical contact with the bead mass. It could then be used just like a palladium ingot or sheet of palladium electroplated metal. First it should be anodized in dilute sulfuric acid to coat the surface with the blackened palladium needed for the electrocatalytic hydrogenation. Then, by hooking the lead inside the bead mass to the negative pole of DC, hydrogen is instead generated on the bead surface, and hydrogenation of the acetic acid ester of ephedrine, pseudoephedrine or phenylpropanolamine to amphetamine can be done. It may be difficult to estimate the active surface area using these beads. That will complicate calculating how much current to pass, and how long to continue passing current. It's better to use too low a current density than too high, and it's better to continue passing current for a longer period than suggested than for too short a period.

References

Organic Reactions, Volume 4, page 174.

Journal of the American Chemical Society, Volume 61, pages 3499 and 3566 (1939); Volume 66, page 1516 (1944); Volume 70, pages 1315 and 2811 (1948).

Reductions in Organic Chemistry, by Milos Hudlicky.

Practical Catalytic Hydrogenation, by Freifelder.

Chapter Twelve
Reductive Alkylation Without the Bomb

The process of reductive alkylation using the hydrogenation bomb, as you saw in the previous chapter, is not without difficulties or dangers. Just for starters, consider the danger of hydrogen gas building up in a poorly ventilated workplace. Add to that the danger of the bomb blowing up if the welding of the seams is not done well. There is also sometimes a problem making catalyst of predictable activity so that consistent results are obtained.

All of these problems, except for the hydrogen gas danger, can be eliminated by instead using activated aluminum to convert mixtures of methylamine and phenylacetone to meth. In this method, the aluminum turnings take the place of hydrogen gas and the catalyst in the reductive alkylation process. The yields are very good, the process is very simple, and no special equipment is required. The reaction is also quick enough that it can be used in large-scale production.

Activated aluminum is next to impossible to purchase, but very easy to make. The raw material is aluminum foil. The foil is amalgamated with mercury by using mercury chloride. The result is aluminum amalgam.

To make activated aluminum, the chemist takes 100 grams of the aluminum foil, and cuts it into strips about 2½ cm wide, and 15 cm long. He folds them loosely, and puts them into a 3000 ml glass beaker or similar container. He does not stuff them down the neck of a flask or similar container from whence they would be hasslesome to retrieve. He packs them down lightly so that they are evenly arranged, then covers them with a .1% solution by weight (1 gram in one liter of water) solution of sodium hydroxide.

He warms the mixture by setting it into a hot water bath until a vigorous bubbling of hydrogen gas has taken place for a few minutes. He is careful here that the mixture does not overflow! Then he pours off all the sodium hydroxide solution as quickly as possible, and rinses the strips with distilled water, and then with 190-proof vodka. This preliminary treatment leaves an exceedingly clean surface on the foil for amalgamating with mercury.

While the surface of the strips is still moist with vodka, he adds enough of a 2% by weight solution of mercury (II) chloride (aka mercuric chloride, $HgCl_2$) in distilled water to completely cover the foil. He allows this to react for about 2 minutes, then pours off the mercury solution. He rinses off the strips with distilled water, then with 190 proof vodka, and finally with moist ether. Moist ether is either purchased as is, or made by adding water to anhydrous ether with stirring until a water layer begins to appear at the bottom of the ether. The chemist uses this material immediately after making it.

Most people have found that this method of amalgam preparation to be more work than it's worth. Better to follow the simpler procedure that Dr. Shulgin uses in his work. He takes 100 grams of clean aluminum foil, cuts it into one-inch squares, then covers these foil cuttings with 3000 ml of water containing 2.5 grams of mercuric chloride. The amalgamation is then allowed to proceed until a light grey precipitate is formed, some fine fizzing begins, and an occasional silver streak can be seen on the aluminum. This takes 15 to 30 minutes depending upon conditions like temperature, etc. Then the water solution is drained off the aluminum, and the aluminum is then rinsed twice with 3000 ml portions of water. After draining off the water, the amalgam is used pronto. Method Five and Six's procedures are the best.

Method One

In this method, the activated aluminum reacts with alcohol and water to produce hydrogen gas. This hydrogen then reduces the Schiff's base formed from methylamine and phenylacetone to give methamphetamine.

The chemist needs a magnetic stirrer-hotplate to do this reaction. On top of the stirrer-hotplate, he places a Pyrex bowl or cake dish large enough to hold a 3000 ml flask. The bowl or dish cannot be made of metal, because the magnetic stirrer will not work through it. He places the 3000 ml flask in the dish and fills it with cooking oil until the oil reaches about halfway up the sides of the flask. He must be sure to leave enough room for the oil to expand as it heats up. He puts the magnetic stirring bar in the flask along with 1600 ml of absolute alcohol or 190-proof grain alcohol. Then he adds 340 ml of phenylacetone and 450 ml of 40% methylamine in water. Now he turns on the magnetic stirrer and begins heating the oil in the dish. He keeps track of the temperature of the oil with a thermometer, and does not allow it to go above 100° C. While the oil is heating up, he adds 180 grams of activated aluminum to the flask. He makes sure that the stirring is fast enough that the turnings do not settle to the bottom of the flask. The reaction mixture will quickly begin to turn grey and foamy. The aluminum is added at such a rate that the bubbling and foaminess it produces does not overflow the flask. When all of it has been added, a condenser is fitted to the flask, and water flow is begun through it.

The chemist now lets them react for 8 hours. He keeps the temperature of the oil bath at 100° C, and the stirring strong. The activated aluminum slowly dissolves and produces hydrogen gas. The explosive danger from this gas is eliminated by running a length of tubing from the top of the condenser out the window. The fumes from the reaction are noxious, so this is far better than just staying upwind.

When the 8 hours are up, he removes the flask from the oil bath and wipes the oil off the outside of the flask. He decants the solution to remove the aluminum sludge, then rinses the sludge with some more alcohol to remove the last traces of product from it. The rinse alcohol is added to the rest of the filtered product.

The underground chemist can now distill the product. He pours it in a 3000 ml round bottom flask that is clean and reasonably dry, and adds a few small pieces of pumice. He places the flask on the electric buffet range, then sets up the glassware for fractional distillation, as shown in Chapter 3. He begins heating it. The first thing that distills is a mixture of alcohol, water, and methylamine. This occurs when the temperature shown on the thermometer is about 78-80° C. He collects about 1600 ml of this mixture, then removes the flask from the heat. He lets it cool down, then pours the contents of the 3000 ml flask into a 1000 ml flask, along with a few fresh boiling chips. He puts about 15 ml of alcohol in the 3000 ml flask, swirls it around to dissolve the product left clinging to the insides, then pours it into the 1000 ml flask.

The chemist again sets up the glassware for fractional distillation, with a 250 ml flask as his receiver. He applies a vacuum, preferably from an aspirator, and begins vacuum distillation. When the boiling gets under control, he begins heating the flask. The last remnants of alcohol and water will soon be gone, and the temperature shown on the thermometer will climb. When it reaches about 80° C with an aspirator, or about 70° C with a vacuum pump, he quickly changes the receiving flask to a clean, dry 500 ml flask, and reapplies the vacuum. He will get about 350 ml of clear to pale yellow methamphetamine free base. A few milliliters of tar will be left in the distilling flask. The liquid free base is converted to crystals by dissolving it in ether or toluene and bubbling dry HCl through it, as described in Chapter Five.

The underground chemist gets an even purer product by varying this procedure slightly. Once the 1600 ml of alcohol, water, and methylamine is distilled off, he pours a mixture of 650 ml of 28% hardware store variety hydrochloric acid and 650 ml of water into what remains in the 3000 ml flask, after it has cooled down. A lot of heat is produced in the mixing because the methamphetamine free base is reacting to make the hydrochloride, so he adds it slowly, then swirls it. When it has cooled down, he stoppers the 3000 ml flask with a cork or glass stopper and shakes it vigorously for 3 to 5 minutes. It should pretty much all dissolve in the hydrochloric acid solution. Now he adds 200 ml of ether or toluene to the flask and shakes it up well. The ether or toluene dissolves any unreacted phenylacetone and tar. He lets it sit for a few minutes. The ether or toluene layer floats to the top. He pours it slowly into a 1000 ml sep funnel, so that the top layer all gets into the sep funnel. Now he lets it set, then drains the lower acid layer back into the 3000 ml flask.

The acid must now be neutralized to give back amphetamine free base, so it can be distilled. The chemist mixes up a solution of 350 grams of lye in 400 ml of water. When it has cooled down, he pours it slowly into the acid solution in the 3000 ml flask. A lot of heat is generated from the reaction. When it has cooled down, he stoppers the flask and shakes it strongly for about 5 minutes. When standing, the amphetamine forms a layer on top. He slowly pours it into a 1000 ml sep funnel. He drains the water layer back into the 3000 ml flask. The methamphetamine layer in the sep funnel may have some salt crystals floating around in it. He adds 100 ml of toluene to it plus a couple hundred ml of a dilute lye solution. He stoppers and shakes the mixture. The salt will now be dissolved in the water. He drains the water layer into the 3000 ml flask and pours the methamphetamine-toluene solution into a clean 1000 ml flask. There is still some methamphetamine left in the 3000 ml flask, so he adds a couple hundred ml of toluene to it. If there is a lot of undissolved salt in the flask, he adds some more water to it. Now he shakes the flask to dissolve the meth in the toluene, then lets it set. The toluene comes up to the top. He pours it off into the sep funnel, and drains off the water layer. He pours the toluene layer into the 1000 ml flask with the rest of the product.

He can now begin distilling it. He adds a few boiling chips, sets up for fractional distillation, and proceeds as described in Chapter Five. The yield once again is about 350 ml of free base, which makes close to 400 grams of pure crystal.

Method Two

This method is not as good as the first one. It takes longer, it uses up more chemicals to make a given amount of product, and less can be produced at a time.

The equipment is set up as in Method One. Into the 3000 ml flask is placed 1575 ml of 190 proof alcohol and 150 ml of distilled water. Then the chemist adds 150 ml of phenylacetone and 220 ml of 40% methylamine in water. He begins magnetic stirring and adds 160 grams of activated aluminum. He heats the oil bath to 100° C or so and attaches a condenser to the 3000 ml flask. He begins water flow through the condenser and gently boils the contents of the flask for 16 hours. At the end of this time, he removes the flask from the heat and lets the aluminum sludge settle. He filters the alcohol solution, rinses the sludge with alcohol and adds the filtered alcohol to the rest of the product. Then he proceeds as described in Method One. The yield is about 150 ml of methamphetamine.

Method Three

This method is not as good as Method One either. Ether is used as the reaction solvent, which adds danger and expense. The ether is a chemical which should be rarely used by clandestine chemists. Another problem with this reaction is that it is done so dilute that large amounts can't be made at one time.

In the same set-up used in Methods One and Two, the underground chemist places 1000 ml of absolute ether in a 3000 ml flask. Then he adds 100 ml of phenylacetone and 160 ml of 40% methylamine. He begins stirring and adds 65 grams of activated aluminum. He attaches an efficient condenser, runs cold water through it, and heats the oil bath to 45-50° C. He gently boils the solution for 6 hours. The activated aluminum reacts with the water in the methylamine to produce hydrogen.

When the six hours have passed, he distills off the ether and treats the residue as described in Method One, (i.e., distills it under a vacuum, etc.). The yield is about 90 ml of meth.

For more information on this method, see U.S. Patent Nos. 2,146,474 and 2,344,356.

Method Four

This variation on the activated aluminum method of reductive alkylation has the advantage of using methylamine hydrochloride directly in the reaction soup. Since methylamine is now very dangerous or impossible to obtain commercially, and also since the best method for making methylamine yields methylamine hydrochloride, the usefulness of this variation is obvious.

This method involves the addition of an alcohol solution containing the Schiff's base formed between methylamine and phenylacetone onto the activated aluminum. In the other methods, the opposite order of addition was employed. To maximize yields of product, the competing side reactions are suppressed. In the case of activated aluminum methamphetamine production, the main side reaction is the reduction of phenylacetone into an interesting, but quite useless pinacol. It has the structure shown at the top of the next column:

Pinacol

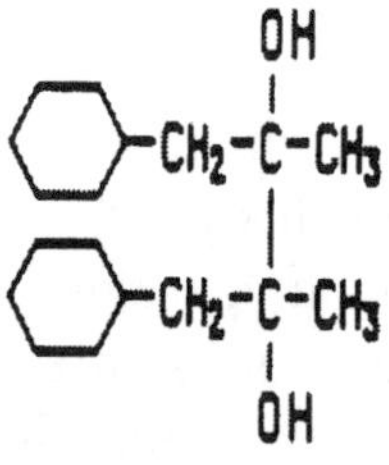

This side reaction is minimized by keeping the amount of water in the reaction mixture to a minimum, and also by using a healthy excess of methylamine. This scheme of things encourages the phenylacetone to tie itself up with methylamine to form the Schiff's base, rather than float around freely in solution where it could be reduced by the aluminum.

To do this reaction, two 2000 ml volumetric flasks are obtained. Volumetric flasks work well for this reaction because the chemist can swirl around their contents quite forcefully without danger of spillage. They also pour pretty well. One volumetric flask is for preparing the activated aluminum, and is also the ultimate reaction vessel. The other volumetric flask is for the preparation of the Schiff's base. The lab work is organized so that both products are ready to react at about the same time.

Into the volumetric flask destined to be the ultimate reaction vessel, the chemist places 108 grams of aluminum foil. It is cut into one inch squares. The best brand of aluminum foil for this purpose is Reynolds Heavy Duty Wrap. It is then treated with sodium hydroxide solution as described in Method 1. After a few good rinses to remove the sodium hydroxide, it is ready to become activated aluminum. To do this, the volumetric flask is filled almost to the neck with distilled water, containing about 2½ grams of $HgCl_2$. The flask is swirled every few minutes for the next 30 minutes. During this time, the water becomes a cloudy grey color, and the aluminum loses its shine. The water is then decanted off the aluminum, and the flask is filled up with fresh distilled water to carry away unreacted mercury. After a period of swirling, the rinse water is poured off, and the rinse repeated with a fresh portion of distilled water. On the last rinse, the chemist makes sure that the water drains off well. This leaves activated aluminum ready to go.

In the second volumetric flask, Schiff's base is made. To do this, 163.5 grams of sodium hydroxide is dissolved in one liter of 190 proof vodka. To this is added 270 grams of methylamine hydrochloride. This methylamine is dry so that the chemist is not weighing water contamination. If this is home brew methylamine hydrochloride, the first crop of crystals is acceptable material, but the second and third batches of crystals are recrystallized as described in *Organic Syntheses*, Collective Volume I. Look in the table of contents for methylamine hydrochloride. The mixture is kept cool during the addition to prevent methylamine gas from escaping. Good stirring is also essential. The result of this operation is an alcohol solution of methylamine. Some salt and water are formed.

To make the Schiff's base, 200 ml of phenylacetone is then added to this solution. The addition produces a fair amount of heat, and some methylamine fumes are driven off as a result. Active swirling of the flask keeps this to a minimum. The chemist also tips the flask during swirling to dissolve any phenylacetone which may be stuck up in the neck of the flask. This is the Schiff's base solution.

To do the reaction, the Schiff's base solution is poured onto the activated aluminum. Once the pouring is complete, they are swirled together energetically for a few seconds, then a thermometer is carefully lowered into the flask. Following this, a section of plastic tubing is stuffed into or over the top of the volumetric flask, and led outside. This is for fume control. The reaction mixture is swirled continuously for the first few minutes. The temperature rises quite rapidly because the reaction is really vigorous. It is necessary to have a bucket of ice water close by to dunk the reaction vessel into to keep it under control. The experimenter strives to keep the reaction mixture in the 50 to 60° C range. After the initial rush, occasional swirling is acceptable, so long as the temperature guidelines are followed. After 90 minutes, the reaction is complete.

At least in the case of MDMA, for meth, one should cook more aggressively. Once the initial rush of the reaction is over, heat should be applied to bring the mixture to reflux as soon as it is safe to do so. Serious experimenters should also note that Dr. Shulgin likes to have some free NaOH floating around the reaction media, along with salt. Even more serious experimenters should note that the original patents for this process stated that best yields were obtained when the hydrogen gas generated was held in by means of a pressure vessel. An emptied fire extinguisher may well be the best reaction vessel.

To process the product, the alcohol solution containing the product is poured off into the distilling flask. The mud-like gunk at the bottom of the flask contains a fair amount of trapped product. This gunk is untreatable as is, but with some lightening up, it can be filtered. A lab product called Celite is added to the gunk until it appears more amenable to filtration. As an alternative, washed white sand, found in the cement section of your friendly neighborhood store, is a good substitute. This is mixed in with the gunk until it lightens up a bit. Then two portions of 200 ml of warm vodka (190 proof) are mixed in and the trapped product is filtered out of the gunk. These gunk filtrates are added to the main product, and the whole mother lode readied for processing.

The first step is to place all the liquid into the distilling flask along with a few boiling chips, and remove the alcohol with a vacuum. A fractional distillation then gives pure methamphetamine free base ready for crystallizing into the hydrochloride.

The same method can be used to give MDMA just by substituting MDA phenylacetone for regular phenylacetone.

Method Five

This is the so called Laboratories Amidos recipe. If you don't feel like reading a French patent, the description of the recipe can be found in *Chemical Abstracts,* Volume 62, column 5228 (1965). This procedure mainly varies from the other processes given in this chapter in that a smaller amount of mercuric chloride is used, but this smaller amount of mercuric chloride isn't rinsed off of the aluminum foil and removed from the reaction mixture. Rather, this smaller amount of mercuric chloride is allowed to remain in the reaction mixture, where presumably it just keeps on working forming freshly amalgamated aluminum during the course of the reductive alkylation. The word out there is that this variation is a good recipe to use for this method of making meth or MDMA from the phenylacetone.

To do the reaction, start with a 1000 ml flask, and place inside that flask 40 ml of phenylacetone or m-d-phenylacetone, followed by 200 ml of 190 proof vodka, and 200 ml of a 25% solution of methylamine in water. This solution can be made from methylamine hydrochloride just by dissolving the required amount of methylamine hydrochloride in water (roughly 100 grams of methylamine hydrochloride diluted to 200 ml with water) and then adding the calculated amount or slightly more NaOH to the water solution of the methylamine hydrochloride with cooling and stirring. In this case that would require about 60 grams of NaOH. The salt formed by this neutralization isn't harmful, and may even be beneficial, so this neutralized methylamine solution can be just poured into the mix as is without any filtering.

Next add 40 grams of aluminum to the flask. The patent specifies aluminum turnings, but I would think that aluminum foil fresh off the roll and cut up into one-inch squares would work just as well. Then finally add .3 gram of mercuric chloride. Swirl or stir this mixture to get the mercuric chloride dissolved, and aluminum amalgam formed on the surface of the aluminum. Mercuric chloride dissolves very slowly even with stirring, so let one's eyes be the guide to when the mercury has formed an amalgam on the surface of the aluminum.

Once the amalgam has formed, rig the flask for reflux, and heat the mixture to boiling for a couple of hours. At the end of the heating period, allow the mixture to cool down, then rig the flask for distillation. Apply a vacuum from an aspirator to boil off the alcohol and the methylamine. Some heat will have to be applied to the flask to make this distillation proceed at a reasonable rate.

When the alcohol and methylamine have distilled off, the residue inside the flask should be poured into about a liter or so of ice water. Rinse the inside of the flask with water and add it also to the ice water.

Then to the ice water mixture, add about 120 grams of potassium hydroxide (KOH) pellets. Stir the mixture until the pellets have dissolved. The KOH will react with the residue of aluminum in the mixture to form aluminum hydroxide. Hydrogen gas will fizz out of the solution during this process and produce some foul mist. One is advised to stay upwind, and add the KOH at such a rate that the mixture doesn't foam over. The KOH will also react with the mercury in solution to form the hydroxide or oxide of mercury. This, too, will form a sludge in the water solution.

When the fizzing of the aluminum has subsided, the mixture should be strongly shaken for a few minutes. Any gas given off should be vented occasionally. Then extract out the meth or MDMA with two 200 ml portions of toluene. After these extracts have been separated off the water layer, they should be filtered if any sludge is seen floating around the extracts. This should completely remove traces of mercury.

Next mix up about a pint of 10% HCl solution. This is made by starting with 30% hardware store hydrochloric (muriatic) acid, and diluting it with two volumes of water. Place the toluene extracts into a large sep funnel, then add the pint of 10% HCl to the sep funnel. Shake strongly for a few minutes. The meth or MDMA will form the hydrochloride salt, and dissolve in the dilute acid. What remains in the toluene solution will be bunk that can be thrown away. One can check the toluene for the presence of phenylacetone, but I doubt that a significant amount will be found.

After the shaking, allow the mixture to sit in the sep funnel for a while. The toluene solution will rise to the top, and the acid solution of the product will be on the bottom. Separate the two, then return the acid solution to the sep funnel. Now to this acid solution, add a 20% solution of lye in water until the mixture is strongly alkaline (pH 13+ to pH papers). This addition should be done slowly with shaking between adds of lye solution. As the mixture gets hot, take time to allow it to cool. Roughly 100 grams of lye should be enough to make this solution strongly alkaline, but let pH papers confirm the situation. As the acid gets neutralized, the free base of meth or MDMA will form and float on top of the water solution.

When the solution has been made strongly alkaline, shake some more, and allow the mixture to cool. When cooled down, extract out the product with two 150 ml portions of toluene. The pooled extracts can be allowed to sit for a while to settle out any entrained water, then poured into a fresh container and bubbled with HCl to get crystals of meth or MDMA. Alternatively, the toluene extracts can be placed into a distilling flask, and fractionally distilled. This is done just as described in Chapter 5. Then the distilled meth or MDMA can be diluted with a couple hundred ml of toluene and bubbled with dry HCl to get crystals of meth or MDMA. In either case, the yield would be around 30 grams of product. Distilling would of course give a purer product.

Method Six: The Racer's Edge

Methylamine is the crucial ingredient required in all methods of producing meth or MDMA from their respective phenylacetones. There are several recipes for cooking one's own methylamine in the next chapter, but if one has chosen the activated aluminum reduction as the meth production route, then it is more than just convenient to also use activated aluminum to make the methylamine. How the gods have smiled upon us to order the laws of chemistry in such a way that activated aluminum is also the most effective way to convert nitromethane to methylamine!

Nitromethane is pretty easily available at hobby shops where it is sold as fuel for model airplanes and race cars. Gallon jugs full of the fuel are likely to be lining the wall in one corner of any store catering to the needs of the model enthusiast at a price of roughly $25 each. This fuel is sold in concentrations of 10% to 50% nitromethane dissolved in methanol solvent. For the purposes of this reaction, any fuel 10% concentration and above will work just fine.

Now to convert the nitromethane model airplane fuel to methylamine, one needs a 2- or 3-necked flask, a good condenser, and a sep funnel. The flask should be at least 2000 ml in capacity. Into the flask put about 150 grams of heavy-duty aluminum foil cut up into one inch squares. Next, measure out enough model fuel so that one has roughly 160 ml of nitromethane. If it is 10% nitromethane, that would mean using roughly 1600 ml of fuel. 20% nitromethane fuel would require the use of half that amount, and stronger fuels would require that even less be used, but they would have to be diluted with methanol to be used. To the fuel mixture, now add 55 ml of water. This is very important in getting the reaction to turn out correctly, as some water is needed to get the activated aluminum to work. Stir or shake the fuel mixture to get the water evenly dissolved in.

Now add just enough of the model fuel mixture to the flask to just barely cover up the aluminum foil. Finally add ½ gram of mercuric chloride to the flask and swirl to get it spread around.

The aluminum foil will begin to amalgamate pretty soon, so quickly attach the condenser to one neck of the flask and the sep funnel or dropping funnel to the other. If there is a third neck on the flask, stopper it. Swirl the flask every couple of minutes to mix the ingredients, and begin a flow of cold water through the condenser. It should be as cold as possible, and siphoned ice water from a pail as described in the N-methyl formamide chapter would be best.

The contents of the flask will begin to get hot as the reaction kicks in. One should be noting significant heating within 15 minutes. It is likely to boil on its own due to the heat of reaction. If a significant heating of the solution isn't noticed within 20 minutes, immerse the flask in boiling hot water to kick start things. It is more likely that the solution will get hot enough that the flask should be wrapped in cold wet towels to calm things down.

Another thing which will be noted as the initial reaction kicks in is that some methylamine gas is escaping out the top of the condenser. You neither want to breath in this stuff, nor lose it to the air. Simply attach some tubing to the top of the condenser and lead it into some dilute hydrochloric acid solution. There the bubbles of methylamine gas will be caught as methylamine hydrochloride. You can later evaporate that solution down and obtain some extra methylamine. It is cleaned up as described in the next chapter.

When the initial rush of the reaction has subsided, take some of the remaining model fuel-water mixture and pour it into the sep funnel attached to the flask. Drip in this solution at such a rate that the boiling of the solution stays manageable. Continue adding this solution until all of the original solution you have measured out has been added. This will take at least half an hour. During the course of the reaction, continue to swirl the flask every couple of minutes to keep the aluminum from getting stuck on the bottom of the flask and covered with goo.

In about an hour, the reaction will begin to slow. When it has stopped boiling, take a pipette and withdraw about 10 ml of the reaction mixture. Squirt it into a beaker half filled with water. Next, take some hardware store 30% hydrochloric acid, and dilute it with two volumes of water. Mix well, then using a pipette, measure out again about 10 ml of the acid solution and with stirring slowly add the acid to the methylamine reaction mixture until a pH of about 7 is reached. A pH meter is very handy for tracking this reaction. Note how much acid was needed to react with the methylamine. If 10% nitromethane fuel was used, the amount of acid required should be 5 to 7 ml. If 20% nitromethane was used, the amount of acid required would be double that. If the amount of acid required is significantly less, then the reaction isn't complete, and the mixture should be boiled for another hour or so. If this still doesn't bring the reaction to completion add about 200 ml of 50% sodium hydroxide to the reaction mixture, and let it boil some more. This last resort shouldn't be done routinely because it makes your product less suitable for direct use in the next stage of the reaction, which is cooking meth or X from the methylamine just made. One might well have to distill a mixture to which sodium hydroxide has been added.

This reaction has just made roughly 3 moles of methylamine dissolved in methanol. That is about the right amount to use to convert one mole of the phenylactone to meth or X. One mole of phenylacetone is roughly 135 ml, and one mole of m-d-phenylacetone is around 160 ml. We now proceed to the next stage, which is using this methylamine to make the product.

If NaOH did not need to be added to complete the reaction, just let the reaction mix cool and settle. Then decant off the product methylamine solution through a filter to remove aluminum debris and keep it in a stoppered bottle while preparing the next stage of the reaction. How to dispose of the mercury amalgamated sludge left over is covered later in this chapter.

Making Product

Into the same three-necked flask used to make the methylamine (once it has been cleaned out and dried) put roughly 40 grams of heavy-duty aluminum foil cut up into one inch squares. Then add around 100 ml of clean methanol to cover up the aluminum foil. Add more methanol if they aren't submerged. Finally, add ¼ gram of mercuric chloride, and swirl it a bit to mix.

As it is dissolving, get the methylamine solution in methanol just made and add it to the flask, and swirl some more. Attach a condenser to the flask. Finally, add one mole of the phenylacetone of your choice and 30 ml of water. Mix some more, then add a magnetic stir bar for stirring action. Allow the reaction to kick in, then supply heat as needed to keep it boiling for a few hours. Your reaction is complete.

Let the debris settle for a bit as it cools, then decant the reaction mixture through a filter. The larger particles can be caught first with a stainless steel screen before going through the finer filter. To the debris left in the flask, add 100 ml of methanol and heat to boiling to get trapped product out. Filter that solution and add it to your main load of product.

Now one has to distill off the methanol from the reaction mixture. Simply pour the filtered reaction mix into the original reaction flask after it has been cleaned out, and rig for simple distillation. Boil off most of the methanol from the product until it seems that there are two layers forming in the distilling flask, or until one is down to roughly 200 ml left in it, whichever comes first. Then let it cool.

It's now time to get the product. When the solution has cooled down, add about 200 ml of 10% sodium hydroxide solution in water to the flask, and shake it for a couple of minutes. Then add about 400 ml of toluene or xylene to the flask and shake some more. When the layers separate upon standing, the product will be in the toluene layer, which will float above the sodium hydroxide and water layer.

Pour the mixture into a separatory funnel, and then drain off the water layer. The toluene layer should then be poured into a distilling flask. One then distills off the toluene or xylene as in the previous examples, and then the product is distilled under a vacuum as in the other examples. One will get roughly an 80% yield of free base meth or MDMA, which is then dissolved in several volumes of toluene and the crystals of product are obtained by bubbling dry HCl gas through this solution of product in toluene, just as in the other examples.

Aluminum Amalgam

Question: "Why won't my aluminum amalgamate?" — someone on the Internet

Answer: The questioner here refers to the methods given earlier in this chapter wherein aluminum metal amalgamated with mercuric chloride solution produces aluminum amalgam, which can be used as a reducer for the Schiff's base produced by mixing methylamine and phenylacetone or methylamine and MD-phenylacetone to yield methamphetamine or "X" respectively.

If we look in *Reagents for Organic Synthesis*, Vol. 1, by Fieser, under aluminum amalgam, we find two procedures for making aluminum amalgam. In the first procedure, the oil and grease free aluminum is first etched with dilute sodium hydroxide to the point of strong hydrogen evolution. Using really dilute sodium hydroxide solution, like the one-half percent solution recommended in the book you are now reading, the solution is going to have to be heated to get a strong hydrogen evolution.

Once the aluminum starts to fizz well, the sodium hydroxide solution is decanted off, and the metal is washed once with water so that the surface still retains some sodium hydroxide solution. Sodium hydroxide clings to surfaces, so this isn't a difficult requirement to meet.

A one-half percent solution of mercuric chloride in water is then poured on top of the aluminum until it is submerged. After about a two-minute reaction time, the mercuric chloride solution is decanted off, and the process repeated with some fresh mercuric chloride solution. The light brown smut which formed on the aluminum metal surface from the sodium hydroxide etch should now all be gone, and replaced with a shiny surface layer of aluminum amalgam. It is now ready to use, once rinsed free of the mercuric chloride solution.

In variation number two given in *Reagents for Organic Synthesis*, the surface of the aluminum foil is first sandpapered, then it is immersed in mercuric chloride solution to form the aluminum amalgam.

Experience has shown that when using clean aluminum foil fresh off the roll, the preliminary sodium hydroxide etch or sanding of the surface can be dispensed with. It just takes considerably longer for the amalgam to form because the surface oxide layer on the aluminum has to be broken before amalgamation can proceed. Let your eyes be your guide to the completion of the process. According to *PIHKAL*, the amalgamation is allowed to proceed until the surface of the aluminum looks grey with occasional silver spots, and fine bubbles of gas form. This can take over 20 minutes using aluminum foil straight off the roll.

Another question which must be asked is, *"Is the mercuric chloride dissolved in the water solution used for amalgamation, or is it still floating around in a crystalline state?"* Mercuric chloride is the most slow-assed dissolving substance I have ever seen! Stirring the solution helps to speed up somewhat the dissolving process, as does grinding the crystals in the water solution with the end of a glass rod, but the dissolving is still a painfully slow operation. Until those crystals of mercuric chloride dissolve, they can't amalgamate the aluminum metal. If you are just tossing the mercuric chloride in water, and then pouring it on top of the aluminum, this is the likely cause of failure to get aluminum amalgam.

Another question is, *"Why are you choosing this method of reduction to get the amphetamine?"* The method is simple, and the yields are respectable, but equally simple and higher yielding alternatives are available. The use of mercury in drug synthesis just gives me the Willies. This is especially true if the cooker is less than highly skilled, and if one doesn't have access to the proper equipment. Getting one's hands on good glassware to use for reactions and for distilling the products is a greater challenge every day. If expedient methods of product purification must be used, the synthetic route should be chosen to produce as clean and desirable crude product as possible.

For example, sodium cyanoborohydride is a very effective reducer for methylamine/-phenylacetone mixtures. It works equally well with methylenedioxyphenylacetone/methylamine mixtures to give "X" or Ecstasy.

To use this method, put 250 grams methylamine hydrochloride into about 1100 ml of methanol or 95% ethanol (190 proof vodka). Then with good stirring, add 66 ml of phenylacetone or methylenedioxyphenylacetone, followed by 30 grams of sodium cyanoborohydride. Adjust the pH of the continuously stirred mixture to pH 6. To measure pH, dampen some pH paper with water, then put a drop of the reaction mixture on the pH paper. To adjust pH, use concentrated HCl or sodium hydroxide. In this case, I believe hydroxide would be called for. Maintain pH at 6 for a day or so until the reaction is complete. This is indicated by the stabilization of pH.

Pour the reaction mixture into about 2½ gallons of water, and make the solution strongly acidic with HCl. Extract the water solution with about 500 ml of toluene. You get this at the paint thinner section of the hardware store. Save this extract, it may contain some phenylacetone.

Then the water solution, which contains the product, is made strongly alkaline (pH 13+ to pH papers) by adding sodium hydroxide or lye slowly with stirring. The free base of the product meth or "X" will come out of solution as small droplets which will over time work their way to the top of the solution.

Once the solution has cooled down, it should be extracted with two 250 ml portions of toluene. This toluene extract will contain the product, and it should be fairly clear, with maybe some yellow coloration. Allow it to sit for several hours, so that any entrained water can settle out of the toluene.

The toluene extract is next poured into a fresh container, taking care to leave any water droplets which settled out behind. Bubbling dry HCl gas through the solution will precipitate out the hydrochloride salt of the product as nice white crystals. They are filtered out, rinsed with some more toluene, sucked dry of solvent and then spread out to dry on a plate. This procedure of crystallization is described in greater detail in Chapter Five.

In Dr. Shulgin's experiments in *PIHKAL*, he got nearly 100% yield of meth and "X" using this procedure. That means 1(one) ml of the phenylacetone will give about one gram of product.

Unused methylamine from the reaction, which is a very valuable commodity, can be recovered from the alkaline water solution by heating it to boiling, passing the vapors up through an ice water cooled condenser, and then passing the methylamine gas into hydrochloric acid to form methylamine hydrochloride again. See Chapter Four for a drawing of the apparatus. After evaporating away the excess hydrochloric acid, crystals of the methylamine hydrochloride are obtained.

The main drawback to this method is the fact that cyanoborohydride is pretty much only available from scientific supply houses such as Aldrich. Those folks were born to be inveterate snitches. Any business that sells List I chemicals has to have an approved snitch plan in place. Of course, mercuric chloride is also mainly available from these same supply houses, but it has many more uses than cyanoborohydride. Cyanoborohydride is almost exclusively used as a reducer for Schiff's bases.

Clandestine chemists have recently found that when making MDMA, the much more common chemical, sodium borohydride, can be used. In *Advanced Techniques* I gave a lot of coverage to this chemical, and how it can be obtained so I won't waste space here to repeat. What they found was that using almost water free reaction conditions, the methylenedioxyphenylactone could be reduced to MDMA in almost 100% yields using methylamine free of water and sodium borohydride as the dry powder. They put their m-d-phenylacetone into a flask, and then added three moles of methylamine in methanol. The most convenient way of doing this is to put three moles of methylamine hydrochloride into 1000 or so ml of methanol, and then adding three moles of KOH to free base it, and after a good mix and reaction of half an hour, filtering out the salt which forms. One could also use the fairly dry methanol solution of methylamine which is made in Method 6 of this chapter. Once the two of them are mixed, slowly add about one-tenth the weight of sodium borohydride as compared to the phenylacetone. The addition should take at least half an hour, and the Schiff's base will be turned into MDMA. Then the MDMA can be recovered just as in the sodium cyanoborohydride method just mentioned. The bad thing about this method is that it doesn't work for making meth from phenylacetone, as even small amounts of water ruin the yield. Check out *Journal of Organic Chemistry* Volume 28, pages 3259-61 (1963) for some background on this reaction route. The free wheeling experimenter looking to break new ground may wish to try dimethylaminoborane instead as the reducing agent. This substance is easily obtained in large amounts from the reducer component of many electroless nickel-plating baths which produce a nickel-boron alloy. Check the MSDS sheet of the electroless nickel product before buying some, and keep in mind that most electroless nickel baths produce a nickel-phosphorus alloy instead by using sodium hypophosphite as the reducing agent.

This leaves catalytic hydrogenation of the methylamine/phenylacetone mixture using platinum catalyst as the non-toxic and non-polluting method of choice. Catalytic hydrogenations using platinum catalyst have been done for over a century. This speaks to the simplicity and effectiveness of the procedure. The yields from catalytic hydrogenation are about the same as using cyanoborohydride.

The best hydrogenation vessel for clandestine cooking is an aluminum fire extinguisher, emptied and cleaned out (see "The Poor Man's Hydrogenation Device" in *Advanced Techniques of Clandestine Psychedelic & Amphetamine Manufacture).* Then a magnetic stir bar is put in the fire extinguisher bottle followed by 150 ml of phenylacetone followed by 300 ml of 190-proof vodka. Then 150 ml of 40% methylamine in water is added. As an alternative, an equivalent amount of methylamine hydrochloride can be put in a beaker with the alcohol, and an equimolar amount of sodium hydroxide added to free base the methylamine. Then this mixture is poured into the extinguisher. Finally, about 3 grams of platinum catalyst is added. The top is screwed on the extinguisher, and most of the air pulled out with an aspirator. Then 30 pounds per square inch pressure of hydrogen is added, and stirring begun. After an induction period of about half an hour or so, uptake of hydrogen begins. The pressure is maintained at 30 psi for the few hours it takes to complete the reaction.

Then the catalyst is filtered out for reuse. The unused methylamine is boiled out and piped into hydrochloric acid for reuse. Then the alcohol is evaporated away, and the residue is then either distilled under a vacuum, or simply dissolved in about 700 ml of toluene. Bubbling dry HCl through this toluene solution will give crystals of the amphetamine hydrochloride as in the last example.

Question: "What do I do with my mercury waste?" — Somebody else on the Internet

Answer: You take the same approach that we industrial chemists have been using for decades now. First of all, you choose to use a process that doesn't make such intractable waste. If you just are cooking a little bit, one can pour the remains of amalgamated aluminum down the toilet and let the metropolitan sewage plant deal with it.

If one is making greater amounts of mercury waste than that, the only responsible thing to do with it is to mix up some concrete, and while it is still wet, stir the mercury waste into it. After it has set up, the rock can be tossed anywhere. The mercury will be trapped for as long as the cement lasts. We should be talking thousands of years there.

The use of mercury in certain reactions needs to be condemned. For example, phenylacetones can be made from allylbenzenes by reaction with mercury acetate followed by oxidation with chromic acid. The waste potential in this route is enormous, without discussing the use of mercury in drug synthesis. Such methods are best left alone.

Chapter Thirteen
Methylamine

Methylamine is very high on the do-not-ever-purchase-through-regular-commercial-channels list. As such, any meth production scheme that uses the phenylacetone route will also have to produce its own methylamine. This is no great challenge. In the days before methylamine became commercially available, researchers and practical cookers in industry always had to make their own. To our benefit, they left good directions for us to follow. See *Organic Syntheses*, Collective Volume 1, pages 347-9.

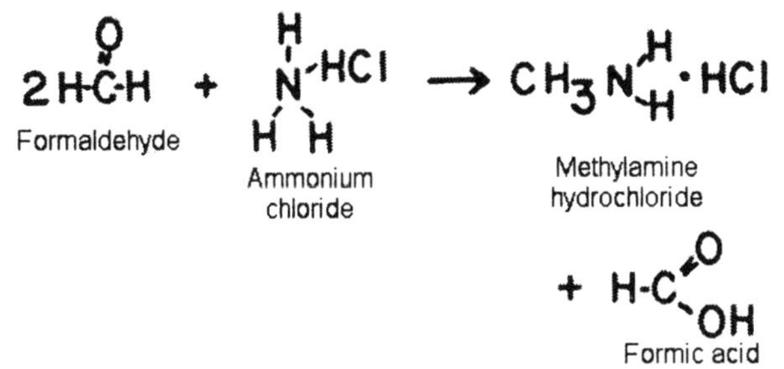

The reaction to produce methylamine is cheap, but requires a lot of labor. Two molecules of formaldehyde react with ammonium chloride to produce a molecule of methylamine hydrochloride and formic acid. Both starting materials are easily obtained in 5-gallon-pail or 50#-bag sizes from commercial chemical outlets serving industry.

The glassware is set up as shown in Figure 11 in Chapter Three. The chemist places 1000 grams of ammonium chloride and 2000 ml of 35-40% formaldehyde in the 5000 ml flask sitting in the pan of oil. (These chemicals need not be a very high grade; technical grade is good enough.) He puts a thermometer in the oil next to the flask and heats the oil to 105° C or so, with the aim of heating the contents of the flask to about 100° C or so. A thermometer inserted into the flask is used to monitor its temperature. A bubbling reaction kicks in, and a condensate made up of formic acid and methyl collects in the receiving flask. When this distillation slows in a couple of hours, raise the temperature inside the flask to 104° C, but no higher. Continue heating at this temp until no more distillate comes over (4 to 6 hours). Periodic applications of aspirator vacuum to the batch will increase yield of methylamine because it pulls the CO_2 out of the reaction mixture.

Then he turns off the heat and removes the flask from the pan of oil. Some liquid will have collected in the 2000 ml flask; he throws it out and rinses the flask with water. The 5000 ml flask is set in a pan of room temperature water to cool it off. A good amount of ammonium chloride crystals precipitate from the solution. He does not want these chemicals, so he filters them out. He returns the filtered reaction mixture to the 5000 ml flask and again sets up the glassware as shown in Figure 11. A 250 ml flask is used as the collecting flask. The reaction mixture should be clear to pale yellow.

He turns on the vacuum source and attaches it to the vacuum nipple of the vacuum adapter. He boils off the water and formic acid in the reaction mixture under a vacuum. Heating the flask in the oil pan speeds up the process, but the oil is not heated above 100° C. When the volume of the contents of the flask is reduced to about 1200-1300 ml, he turns off the vacuum and removes the flask from the oil pan. The flask is put in a pan of room temperature water to cool it off. Some more crystals of ammonium chloride come out of solution. He filters out these crystals and pours the filtered reaction mixture into a 2000 ml flask. He sets up the glassware as before, and again boils off the water and formic acid under a vacuum. He does not heat the oil above 100° C.

When the volume of the reaction mixture has been reduced to about 700 ml, crystals of methylamine hydrochloride begin to form on the surface of the liquid. It looks a lot like a scummy film. When this happens, the vacuum is disconnected and the flask is removed from the oil bath. The flask is placed in a pan of room temperature water to cool it off. As the flask cools down, a lot of methylamine hydrochloride crystals come out of the solution. When the flask nears room temperature, it is cooled off some more with some cold water. This will cause even more methylamine hydrochloride to come out of the solution.

The chemist filters out the crystals and puts them in a Mason jar. The crystals look different from the crystals of ammonium chloride, so he should have no trouble telling the two apart. These crystals soak up water from the air and melt, so he does not waste time getting them in the Mason jar after they are filtered.

He pours the filtered reaction mixture into a 1000 ml round bottom flask and again sets up the glassware as shown in Figure 11. He reattaches the vacuum and continues boiling off the water and formic acid under a vacuum. When the volume of the mixture reaches 500 ml, he removes the flask from the hot oil and places it in cool water. As it cools off, more crystals of methylamine hydrochloride appear. He filters the cold reaction mixture to obtain these crystals. He transfers them to a beaker and adds 200 ml of cold chloroform to the beaker. He stirs the crystals around in the chloroform for a few minutes, breaking up any chunks. This dissolves any dimethylamine hydrochloride in the product. He filters the crystals in the beaker, then puts them in the Mason jar along with his first crop of methylamine hydrochloride crystals. He throws away the chloroform and returns the reaction mixture to the 1000 ml flask.

He boils the reaction mixture under a vacuum again. When its volume reaches about 150-170 ml, he turns off the vacuum and removes the flask from the hot oil. He pours the reaction into a beaker and stirs it as it cools down, to prevent it from turning into a solid block. Once it has cooled down, he adds 200 ml of cold chloroform to the slush. He stirs it around with a glass rod for a couple of minutes, being sure to break up any chunks. The mixture is then filtered. The crystals of crude methylamine hydrochloride are kind of gooey, so it may not be possible to filter out all the chloroform.

This batch of crystals is added to the Mason jar along with the rest of the crude product. The yield of crude product is around 425 grams. It absorbs water easily from the air, and melts. Its smell has been described as "like old woman's pussy." The main contaminant of the crude product is ammonium chloride, along with some dimethylamine hydrochloride, and some of the reaction mixture. The 425 gram yield of crude product is therefore deceivingly high.

Purification would best start with drying under a vacuum. This could be conveniently done by placing the crude crystals into a large vacuum flask, stoppering the top of the flask, and applying aspirator vacuum for about half an hour. Gentle heating of the flask with warm water during the vacuum drying helps speed along the process, as does some shaking around of the contents of the vacuum flask. If one has an aspirator that likes to spit water back into flasks under vacuum, then one should use a vacuum pump.

Now to get nice and pure crystals of methylamine hydrochloride, we leave those crude crystals in the filtering flask, and add around ¾ of a quart of 190-proof vodka to the crystals. One-hundred-ninety-proof vodka won't dissolve ammonium chloride, but it will dissolve methylamine hydrochloride when it is hot. Leave the top of the filtering flask stoppered to prevent steam from getting into the flask, then warm up the flask using hot water. Water fresh off the stove, almost boiling hot, would be best. Swirl around the flask as it warms to get the methylamine hydrochloride dissolved.

Once the alcohol solution gets hot, stop swirling to let suspended crystals settle out. Then decant off the alcohol solution, taking care to keep the crystals inside the flask. Filtering is necessary. Then put the alcohol which has been decanted from the flask in the freezer. As it gets cold, methylamine hydrochloride crystals will come out of solution. When the alcohol is good and cold, filter to collect these pure crystals of methylamine hydrochloride. Store them in a Mason jar with a lid.

Return the filtered cold alcohol to the filtering flask containing the crude product. Once again heat the alcohol with swirling to dissolve some more methylamine hydrochloride. Then let the suspended crystals settle once again, and decant the alcohol as before, and cool that down in the freezer to get another crop of pure methylamine hydrochloride. A few cycles through this process will get all the methylamine hydrochloride soaked out of the crude product and recovered as pure recrystallized methylamine hydrochloride. The yield of pure methylamine hydrochloride will be around 350 grams or so.

Sometimes, the methylamine hydrochloride is used directly as such in the reaction, such as, for example, in reductive alkylation using aluminum foil as the reducer. More generally, the free base is used. To obtain a strong solution of methylamine in water, 100 grams of methylamine hydrochloride is placed in a flask with 50 ml water. This is chilled in an ice-salt bath to a temperature nearing 0° F. Then a cold solution of 60 grams of NaOH in 100 ml water is slowly added with stirring. The addition must be slow enough, and the cooling strong enough, to avoid losing the free base as a gas. Methylamine solution produced in this way is roughly comparable to the commercial 40% methylamine, except that it also contains salt, and maybe a little NaOH if too much was added.

This solution should either be used immediately, or stored in a tightly stoppered bottle. Refrigeration of the solution is optional, but desirable.

The reader should be aware that chloroform is a poison for Raney nickel catalyst, so if that particular method is going to be used in meth production, the crystals must be vacuum-dried. Also, it is possible that the excess NaOH may interfere with methods using catalytic hydrogenation. I can't say. If it does, an apparatus like that in Figure 18 can be used to boil out the methylamine free base into a stirred, chilled solution of alcohol.

Other methods of making methylamine exist, but they are not well-liked by the pioneers mentioned at the beginning of the chapter. Presented here is their preferred method. For example, it can be made in 71% yield by reacting methyl iodide with hexamine, also known as hexamethylene tetramine. Good directions for making this substance from ammonia and formaldehyde can be found in *Home Workshop Explosives* by yours truly. The production details for methylamine are found in the *Journal of the American Chemical Society,* Volume 61, page 3585 (1939). The authors are Galat and Elion.

It can also be made by degrading acetamide with Clorox. See *Journal of the American Chemical Society*, Volume 63, page 1118 (1939). The authors are Whitmore and Thorpe, and the yield is 78%.

It can also be made via the Curtius reaction in a yield of 60%. See *Helvetica Chimica Acta*, Volume 12, page 227 (1929). The authors are Naegeli, Gruntuch and Lendorff.

References

Journal of the American Chemical Society, Volume 40, page 1411 (1918).

Chapter Fourteen
The Ritter Reaction:
Amphetamines Directly From Allylbenzene

A most interesting sidelight appears in an article by Ritter and Kalish found in the *Journal of the American Chemical Society*, Volume 70, pages 4045-50 (1948). This sidelight was a bit of research done by a grad student as part of his master's thesis. The grad student just happened to work out the experimental details for converting allylbenzene directly into amphetamine.

The main thrust of the article was the good Dr. Ritter telling of his new method for converting double bonds into amines. The method which he pioneered has since come to be known as the Ritter reaction. This versatile reaction can well serve the underground operator as an alternative pathway to the amphetamines.

The Ritter reaction in general is a reaction whereby amides are made by adding an alkene to a mixture of a nitrile in sulfuric acid. After the amide is made, it is then boiled in hydrochloric acid solution to give the corresponding amine.

The particular variation on this theme in which we are interested deals with the case in which the alkene is the now familiar and highly useful allylbenzene. When it is added to a solution of acetonitrile in sulfuric acid, the following reaction takes place:

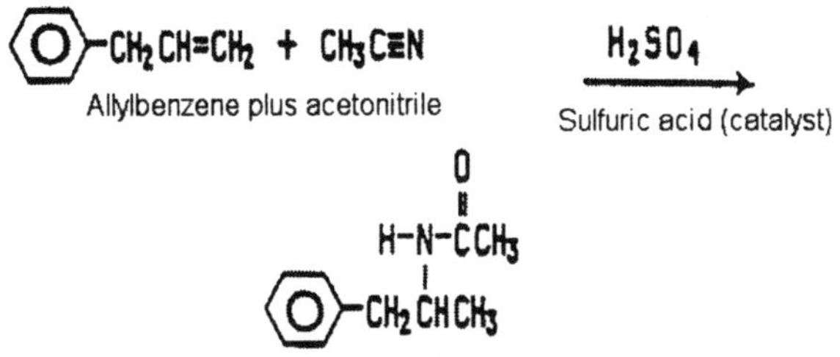

Acetyl amide of amphetamine

The acetyl amide thusly produced is not isolated and purified. Rather, it is added in the crude state to hydrochloric acid, and boiled for several hours. A hydrolysis reaction almost identical to the one seen in Chapter Five takes place producing the prototype amphetamine, benzedrine.

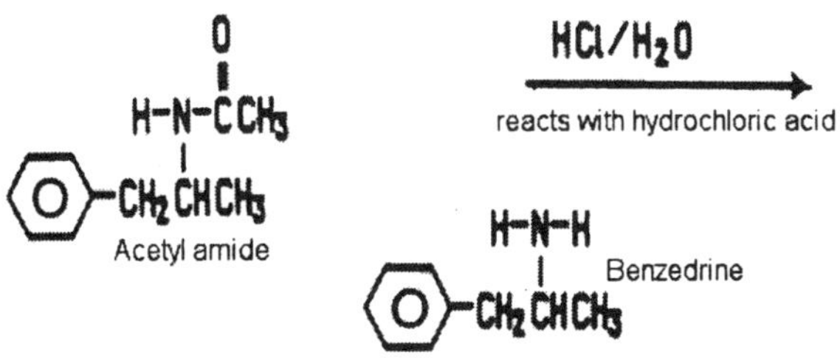

The acetyl amide of amphetamine is very similar to the formyl amide of methamphetamine produced by the Leuckardt-Wallach reaction. Its main difference is that it is more difficult to hydrolyze to amphetamine by the action of boiling hydrochloric acid. It must therefore be boiled with the acid for a longer period of time than the formyl amide. The manufacturer may well find it to his advantage to boil the tar left over at the end of the process once more with fresh hydrochloric acid. This will likely yield an additional measure of amphetamine from the stubbornly unreactive amide.

This small hassle with the hydrolysis process could be avoided if HCN were used as the nitrile in sulfuric acid solution. However, the extreme danger of dealing with hydrogen cyanide more than outweighs the additional work needed when using acetonitrile.

To do the reaction, a solution of 450 grams (243 ml) of concentrated sulfuric acid in 400 grams (530 ml) acetonitrile is made by slowly adding the acid to the acetonitrile. Both ingredients are cold when they are mixed together, and the temperature of the mixture is kept in the 5-10° C range during the mixing by setting the reaction container in ice. An admirable reaction vessel is a glass beer pitcher.

When the addition of the acid to the nitrile is complete, the pitcher is taken out of the ice, and 236 grams (262 ml) of allylbenzene is slowly added to it with stirring. The mixture quickly turns an orange color, and begins to warm up.

Stirring is continued and the temperature of the mixture followed. It slowly climbs to 50° C, and then more rapidly to 80° C, as the color of the mixture darkens.

This is a tenfold scale up of the original recipe, so be watchful and protected in case the reaction gets out of control. One wouldn't want this mixture to go postal on you. Once the 80° C temperature is reached, pour the mixture out of the pitcher, and onto a few pounds of ice cubes. Smaller batches can be cooled just by immersing the reaction vessel in ice, but on this scale, go right onto ice.

Once the reaction mixture has cooled down, the acid should be neutralized by slowly pouring it into a 15% solution of lye dissolved in water. About a pound of lye will be required to neutralize all the sulfuric acid and produce an alkaline solution. Most of the unreacted acetonitrile will end up in the water layer, but some will evaporate during the neutralization. Stay upwind!

The neutralization of the acid by the lye solution produces a great deal of heat. The lye solution is gently stirred during the addition, and then stirred more vigorously during the following minutes. After a few minutes of stirring, the mixture is allowed to sit for a few minutes. A yellow oily layer floats on the top of the solution. This yellow oil is the crude amide. If the oil were to be allowed to sit for a while longer, it would begin to form crystals of crude amide. There is no need for this, however, so the processing continues immediately.

The top yellow layer is poured off into a sep funnel, and any water carried along is drained off. Then the yellow oil is poured into a 2000 ml round bottom flask. It is now ready for hydrolysis with hydrochloric acid solution to make amphetamine. The approximate volume of the crude amide is determined, and five times that volume of 15% hydrochloric acid solution is added to it. Fifteen percent hydrochloric acid solution is easily made by starting with the 28% hardware store hydrochloric acid, and adding just about an equal volume of water to it. A wise move here is to rinse the inside of the sep funnel with acid. This rinses off the amide clinging to the glass insides of the sep funnel.

When the acid has been added to the amide, the mixture is swirled. They usually mix together well. If they don't, stronger acid is used. Adding some full strength acid to the mix should do the job. Then a few boiling chips are added to the flask, a condenser attached to the flask, and heat applied to boil the mixture at reflux.

The reflux boiling is continued for 10 hours. During this time the mixture will turn black. At the end of the boiling period, the mixture is allowed to cool down. When it is cool, 200 ml of toluene is added to the flask. The mixture is shaken well for a couple of minutes, then allowed to sit. The toluene floats up to the top, and has dissolved in it most of the unreacted amide, and other unwanted garbage.

The toluene layer is then poured off into a sep funnel, and any water layer carried along drained back into the flask. The toluene layer is poured off into another container for future processing. It may be difficult to tell exactly where the toluene layer ends and the water starts because of their similar color. A sharp eye and good lighting help to spot the interface of the two fluids.

The acid solution of the amphetamine is now made alkaline to liberate the free base for distilling. To do this, lye is added to the acid solution in the 2000 ml flask. Assuming the use of about 1200 ml of 15% hydrochloric acid solution, one 12 ounce can of lye does the job. The mixture is first swirled to release heat, then shaken vigorously for five minutes. I cannot emphasize enough the importance of vigorous and prolonged shaking here because the amphetamine base initially formed tends to dissolve unneutralized amphetamine hydrochloride. The oily droplets protect the hydrochloride from contact with the lye solution unless the shaking is strong and prolonged.

When the shaking is completed, the mixture is allowed to cool down. Then 300 ml of toluene is added to the flask, and shaking continued for a minute or two. After sitting for a couple of minutes, a toluene-amphetamine layer floats above the water layer. This is poured off into a sep funnel, and the toluene-amphetamine layer poured into a 1000 ml round bottom flask.

The amphetamine-toluene mixture is distilled in exactly the same manner as described in Chapter Five. The boiling point of benzedrine is 10° to 20° C lower than meth. The yield of benzedrine is in the range of 100 to 150 ml.

The benzedrine produced by this reaction is either used and removed as is, or it is converted to methamphetamine. A very good and simple process for doing this can be found in the *Journal of the American Chemical Society*, Volume 62, pages 922-4. The author is Woodruff. The yield for this process is over 90%, so a greater volume of methamphetamine comes out of the reaction than the benzedrine input. This is because the gain in molecular weight achieved by adding the methyl group outweighs the small shortfall from 100% yield.

For those who have difficulty reading the Woodruff article, meth is described as B-phenylisopropylmethylamine. The amine is benzedrine.

If the benzedrine product is used as is, the producer makes it as the hydrochloride salt. This is made the same way as methamphetamine hydrochloride. An alternative to the hydrochloride salt is the sulfate salt. This more hasslesome procedure calls for the use of cooled solutions of amphetamine base in alcohol and cooled solutions of sulfuric acid in alcohol. Furthermore, a recrystallization from alcohol-ether is required because trapped excess sulfuric acid in the crystals causes them to turn to mush or worse. By using HCl gas, the excess acid floats off as gas.

An excellent review of this reaction can be found in *Organic Reactions*, Volume 17. Nearly double these yields should be obtained if the underground chemist is willing to risk using hydrogen cyanide instead of acetonitrile. The hydrogen cyanide is made inside the reaction flask from sodium cyanide and sulfuric acid. For complete directions, see *Organic Syntheses*, Collective Volume 5, pages 471 to 473. The name of the compound is alpha, alpha, Dimethyl beta phenethylamine.

My opinion is that anyone attempting this variation with hydrogen cyanide in any place other than a well ventilated shed, well upwind from the batch, is just nuts. This variation isn't recommended, nor do I know if it has been specifically tested for efficacy with allylbenzene. It sure as hell is worth checking out, if the required precautions are taken for dealing with hydrogen cyanide solution. *This is not for beginners!*

PILL EXTRACTION UPDATE

It seems the only practical way to extract the ephedrine-guaifenesin pills now on sale is by use of steam. An easy way to do this is to grind the pills and mix in ¼ volume of washing soda. This mixture is put into a flask and put a two hole stopper on the flask. Run a glass or plastic tube down one hole just about to the bottom of the flask, and attach this to plastic tubing which is clamped to the steam vent of a pressure cooker. Put some glass tubing into the second hole of the two hole stopper, and attach some plastic tubing to this tube. It will be the steam outlet, and the tubing should run to a 40 oz beer bottle nestled in ice in a minnow bucket.

Wet the pill-washing soda with some water to free base the ephedrine, and tighten all tubing connections with wire to prevent steam leaks. Then turn the heat on the pressure cooker to get it to start producing a stream of steam. Blow this live steam through the based pill mass until roughly a quart of steam condensate has been collected for each 100 pills used. This steam condensate now contains the ephedrine from the pills. To collect it, acidify the steam condensate with muriatic acid. Then boil it down in a pan until half the original volume of water has been lost. Once it cools, base the ephedrine-water solution with lye and extract out the ephedrine with toluene. Let the toluene extract sit for a couple of hours to shed water, then pour it into a dry beaker and bubble dry HCl through it to get the product ephedrine hydrochloride crystals. Filter and rinse and dry. Yield 60% or so.

Pseudoephedrine pills can be steamed in a similar manner, To do them, the pills are ground up and poured into an ehrlenmeyer flask or similar container. About 2/3 ml of water is then added for each 30 mg pill. Double the amount of water for 60 mg pills, etc. Let this mixture sit with some swirling for about an hour to draw the active ingredient into the water. Then add about one-fourth volume of washing soda (sodium carbonate) relative to the volume of the ground up pills. After that, add at least 50 gr of Lite Salt (a 50-50 mix of salt and potassium chloride) for each 100 ml water added to the pills. This is very important as the salt mixture raises the temperature of the steaming. Without it pseudoephedrine will not steam distill.

The same steam distilling procedure is then used as with the ephedrine pills. Care must be taken during the steaming that the water level in the ehrlenmeyer flask does not rise. Diluted salt mixture will not be hot enough to force out the pseudoephedrine. Adjusting the temperature of the stove burner on which the flask sits during the distillation is called for so that the water level remains constant. It should not be overheated either as a dried out mixture is likely to burn.

When roughly a quart of water per 100 of the 30 mg pills has been collected in the ice chilled 40 oz beer bottle, the distillation will be complete. The water distillate in the beer bottle will look cloudy and milky. If you taste it, you will notice the pseudoephedrine in it. It also contains some of the detergent package from the pills. This is removed by making the steam distillate in the beer bottle acid by adding some hydrochloric acid. About half a shot glass full should be enough for a quart of steam product. Pour the water out of the beer bottle into a beaker or glass baking dish. Boil down the water until it is about one fifth its original volume. Most of the unwanted material will go off in the steam. Once it has cooled down, this water can be extracted with a bit of toluene to remove any left over unwanted material. Finally, the boiled down steam distillate should be made strong basic by adding lye. This will free base the pseudoephedrine. It can then be extracted out of the water with a couple portions of toluene. After the separated extracts have sit for a while to shed any entrained water, they can be bubbled with dry HCl gas to get the crystals of pseudoephedine hydrochloride. These are then filtered out and dried.

Ephedra can be steamed also. The procedure is the same as with pseudoephedrine pills except that the water added to the ground up ephedra should be heated to near boiling to make a tea from the herb. The steam should be kept in! Then washing soda and the same 50 grams plus of lite Salt per 100 ml water are added and steaming commenced.

Chapter Fifteen
Methamphetamine From Ephedrine or Pseudoephedrine; Amphetamine From PPA

Ephedrine and Pseudoephedrine

Ephedrine and pseudoephedrine are structurally mirror images of each other. This is possible because they have a chiral center, the carbon atom attached to the alcohol group of these two substances. Theoretically, reduction of these two materials should both give the same product, the "d" isomer of meth because when the alcohol group is reduced, the chiral center disappears.

Theory and reality aren't quite in agreement. Pseudoephedrine is less willing to undergo reduction in a couple of the methods given in this chapter than is ephedrine. This gives a lower octane product which had myself and several of my correspondents from the pen convinced that pseudoephedrine was giving the low octane "l" isomer of meth. The reality was that this low octane product was a mixture of meth and unreduced pseudoephedrine.

The most popular clandestine methods for producing meth work almost equally well with either ephedrine or pseudoephedrine. These two most popular methods are the lithium in anhydrous ammonia reduction, and the hydroiodic acid and red phosphorus reduction Some methods that work just fine for ephedrine fail miserably when used on pseudoephedrine. The prime example of this situation is the "Cat" recipe given in the next chapter. Also, catalytic hydrogenation to meth works better with ephedrine than it does with pseudoephedrine.

So long as the meth chef is planning to do the cooking with either lithium metal or the hydroiodic acid methods, little difference will be noted between the products obtained using either ephedrine or pseudoephedrine. In this case, production is done using the pills most easily available to him. In general, this will be pseudoephedrine (Sudafed) pills.

Procedure for Obtaining Pure Ephedrine, Pseudoephedrine or PPA From Stimulant and/or Cold Pills

No aspect of methamphetamine manufacture has changed so radically in the past few years as the composition and availability of the OTC "stimulant" pills which are useful as raw-material feed-stock for methamphetamine production. Back in early '92, when I penned the third edition to this book, ephedrine pills were available by mail order. These pills were quite well-suited for a very simple water extraction to get out the active ingredients, because they were about 30-40% active ingredient, and the fillers were mostly non-water soluble. Sudafed pills were almost that easy to extract, except that they had a red colored coat that had to be soaked off in solvent before commencing a water extraction of them.

Since that time, the mail-order outfits have been heavily leaned upon. Several of their heads are now in the slammer. Other such companies have been taken over and are now agents of the enemy. I wouldn't trust most of them as far as I could spit. All ephedrine orders now must be accompanied with photocopies of driver's licenses, etc. Ephedrine is now on the chemical diversion list with no minimum threshold quantity. To quote one of the DEA's top dogs on this matter: "We're keeping track of where they are going."

Pseudoephedrine, aka Sudafed, and PPA, aka phenylpropanolamine or Dexatrim, are now subject to close sales scrutiny as well due to the Meth Act of 1996, passed in response to the 4th edition of this book. A single sale limit of 24 grams of base, which corresponds to a little under 500 of the 60 mg pseudoephedrine pills or a little over 300 of the 75 mg phenylpropanolamine pills, has been established. Mail-order pill companies are now required to turn over, on a regular basis, a complete customer list with names and addresses and amounts purchased.

As a result, it must be emphasized that the procedures given in this chapter are most suitable for making "stash" amounts of meth or dexedrine. Large pill purchases will attract the heat like blood in the water does sharks. The recipes given in this chapter can easily be scaled down to whatever amount of pill feedstock one is able to obtain. Retail store sales aren't federally regulated, but the prices there are very high compared to the mail-order prices.

Starting in 1998, Wal-Mart began limiting sales of all cold medicines (read: ephedrine, pseudoephedrine and phenylpropanolamine) to three packages per customer. Try to buy more and the cash register throws a fit. Some localities in areas such as California have passed ordinances decreeing similar retail sale limits. State laws may be coming imposing similar limits. Some states have gone so far as to declare pseudoephedrine a Schedule 5 drug which must be dispensed one package at a time by a pharmacist. Unless one is willing to spend all day shopping for pills, it may be difficult to accumulate thousands of them.

An alternative to the use of ephedrine pills is what are called "herbal extract" pills. These don't at present come under reporting requirements because they wear that "herbal" label. These pills are loaded with ephedrine, so expect laws to be passed shortly bringing them under reporting requirements. They are at present available by mail order at a reasonable price. Ads can be found in our favorite sleazy magazines. I would suggest using a fake name and a rented mailbox when ordering them.

Another source of ephedrine is the herb Ma Huang, which is the source of those herbal extract pills, and the natural source of ephedrine. It contains around 8% ephedrine, along with some pseudoephedrine. This material too isn't under reporting requirements as I write this, but stay tuned. That could change at any moment. I've looked for this herb at my local health food stores with no success, but your area may be better stocked.

The "doctoring" of the pills over the past several years has been similarly dramatic. In the case of ephedrine pills, the first thing which was done was to add more filler to make them less suitable to a simple water extraction. The more filler, the more water required to extract the active ingredient, and the more inert pill-gunk co-extracted. This gunk had a bad effect on the meth-production reactions, which follow the GIGO principle.

Then the pill doctoring became more scientific. The insoluble filler was replaced by some type of water-soluble fiber which played much greater havoc with the ensuing reaction if no purification beyond just water extraction was done. Even in the case of making cat, it would screw up the reaction by causing the entire reaction mixture to take on a milkshake consistency upon neutralization with NaOH. The gunk equally filled the water and toluene layer (which, by the way, were pretty hard to spot because of the floating gunk). The floating gunk could be filtered out of the toluene layer, and the hydrochloride then precipitated, but the octane numbers were greatly reduced below normal.

The next version of ephedrine pills contained 25 milligrams of ephedrine along with 100 milligrams of guaifenesin. They were available at gas stations and other stores because they came under the definition of "mixture" in the Chemical Diversion Act, and so were not regulated.

These 100 milligram guaifenesin pills were followed by 200 milligram guaifenesin pills, which is what are on the market now. The definition of "mixture" has been changed so that ephedrine pills containing guaifenesin are just as reportable as the others.

The best extraction procedure to use has changed a lot with the passage of time, and the introduction of new pill formulations. I'll go through the various pill extraction techniques which have been used over the past decade so that the reader can have a sense of history on this subject. It's also wise to cover the old methods because many of the old techniques can be made useful again with some tweaking to counteract the effects of new polluting ingredients in pills. Let's start with the basic pill extraction procedure. It's an extension of the old standard water extraction procedure that was so successful with the old mini-thin ephedrine pills. It also used to work fine with the old types of pseudoephedrine pills. By substituting 1-2% HCl solution for the water in this extraction, it can also be used to extract Ma Huang, and the "herbal extract" pills.

The first step is to grind the pills. A mortar and pestle gives the best grind size, as overly fine grinding makes the subsequent filtering steps more difficult. With herbal extract pills and Ma Huang, the initial grind should be done in a blender, mixing the substance to be ground with its initial grind charge of 1-2% hydrochloric acid (hardware store muriatic acid diluted 15-30 fold) and blenderizing at medium high speed until small particles are obtained.

The next step after the grind is to determine whether these pills need to be degummed and desplooged.

The ephedrine-guaifenesin pills are really loaded with gum. They must be degummed and desplooged by soaking the ground up pills in toluene, then filtering. Other types of pills I've come across aren't so loaded with this ingredient, although it may be more prevalent in the future. Failure to degum and desplooge these pills results in a milkshake later when the water extract is made basic and extracted with solvent. It also really slows up the filtering of the water extract.

I think this emulsion-forming ingredient is some type of fatty acid which forms a soap when sodium hydroxide is added later on to free base the ephedrine, or whatever. Toluene is also quite good at removing guaifenesin from pills. Colored pills should be tested with solvent. If toluene is going to be used as the extractant at the end of this procedure, check to see if toluene dissolves the coloring matter. If it does, then soak the ground up pills in toluene and filter to remove the color. Ditto if Coleman camper fuel is going to be used as the final extractant. Allow the ground up pills to dry after desplooging so that the solvent is removed from them.

Then water extraction is done. Mix 1,000 ground up pills with 350 ml of water, and stir for about an hour. Another variation is to just mix 1,000 pills with 350 ml water, and after the pills have softened, mush them up and stir for an additional hour.

Now the pill mush should be filtered. Vacuum filtration through a Buchner funnel is greatly preferred, because it will suck the filter cake dry, giving better extraction with less use of water. The need to keep the amount of water used to a minimum arises from the fact that the "pill extraction deterrents" are less soluble in water than the desired ingredient, so the more water used, the more effective they are. It may be difficult to get the mush to filter easily through filter paper, so a preliminary filtering through clean white cotton cloth cut like a filter paper will be helpful in these cases.

The filtrate should be clear, and very bitter tasting, as it contains the active ingredient. Hopefully, most of the pill fillers didn't dissolve, and they are sitting in a filter cake in the Buchner funnel.

Now take this filter cake of pill sludge, remove it from the filter, and mix it with an additional 300 ml of water. Stir this around for an hour, then filter this. If a Buchner funnel was used, this is enough water to extract the pills. If only gravity was used to aid filtration, then the pill sludge should be soaked with a final 100 ml portion of water, and filtered.

To the combined filtrates, add a dash of hydrochloric acid to suppress steam distillation, and boil its volume down to about 200 ml. With pseudoephedrine, this isn't so important because it isn't as water-soluble as ephedrine or PPA free bases, but the volume should be reduced some for it, too.

Now let the solution cool, and then add 20% NaOH or lye solution with stirring or shaking until the solution is strongly alkaline to litmus paper. Indicating pH paper should say 12+. A pH meter may not be as useful as paper for this reading. The solution at this point should smell strongly of the kind of fishy free bases.

Extract the water solution with about 100 ml of toluene. This solvent can be found in the paint-thinner section of the hardware store or paint-supply outlet. If you can't find this solvent, Coleman camper fuel will work almost as well. The water layer should remain a liquid, and the toluene layer should be clear and transparent. If the particular "deterrent" formulation results in a milkshake consistency, just estimate how much is that top 100 ml of solvent, separate it off, and filter it. Rinse the filtered out gunk with solvent. Repeat this extraction with two additional portions of toluene. With the ephedrine-guaifenesin pills, extract with petroleum ether, hexane or Coleman camper fuel.

The combined toluene extracts should be placed in a 400 ml beaker and allowed to sit for a few hours. This serves two purposes: first, entrained water will settle to the bottom of the beaker and stick to the glass. When it is poured into a fresh beaker, the water will be removed. The second reason involves an observation I made some time ago with one particular "deterrent" formulation. In that case the water layer became almost solid after the second toluene extraction, because the solvating action of ephedrine free base was lost for these fillers. The toluene extract in this case, upon standing, grew a mat of white solid about ¼- to ½-inch thick on the bottom of the beaker. After letting this mat grow, and pouring the solution off of it, all proceeded well from that point.

Once the toluene has been poured into a fresh beaker, dry HCl gas should be bubbled through it to precipitate pure ephedrine, pseudoephedrine or PPA hydrochloride. This is done just like the bubbling to get meth hydrochloride in Chapter Five. The yield from 1,000 25 mg ephedrine pills is about 20 grams, from 1,000 60 mg pseudoephedrine pills about 50 grams, and from 1,000 75 mg PPA pills is about 65 grams.

With Ma Huang and herbal extract pills, the dilute acid solution used as extractant should be boiled down to concentrate it, just as with pills. Then take some toluene and extract the acid solution concentrate. This will remove coloring and other plant material, but not the desired ephedrine. One, of course, has to wait for the solution to cool before doing a solvent extraction or the solvent will boil and fume and make a mess on you.

Then after extracting the acid concentrate, this concentrate should be made strongly alkaline by adding lye solution and shaking. We now have free base as in the pill example. It can be extracted out with toluene, just as with pills, and the hydrochloride collected by bubbling with HCl, just as with pills.

Another pill extraction procedure which was briefly touched upon in the fourth edition of this book has proven quite useful when extracting those 200 milligram guaifenesin pills. Through the fifth edition of this book and partway through the life of the sixth ed until roughly the beginning of 2004, this method also was very effective in extracting the various brands of pseudoephedrine pills.

First the pills are finely ground in a blender. Shaking the blender some while it is running will help to get large pill chunks off the bottom of the blender and into the blades. Next one can pour the powdered pills into a beaker, and desplooge with roughly one ml of toluene for each pill used. Stir it around for about half an hour, then filter. Spread the pill mass out to air dry. I'm not really certain if this step is absolutely necessary. Feel free to skip the toluene desplooge step, and see if it makes any difference.

Return the dried pill mass to the beaker, and add about 4 grams of lye for each 100 of the 25 mg ephedrine pills used. Stir this in. Then slowly add 91% isopropyl rubbing alcohol, or hardware store denatured alcohol or 190-proof vodka, with stirring, until a moderately runny paste is achieved. Too much alcohol could make it difficult to precipitate the hydrochloride crystals at the end of this process. Using water instead of alcohol can result in a regrettable mess, especially if too much water is used. That extraction deterrent formulation really kicks in with water, and a horrendous milkshake emulsion easily forms. Stick to alcohol.

Stir this fairly light paste for about half an hour. The lye dissolves, and produces the free base of the ephedrine. We now extract out the ephedrine free base.

Add 50-75 ml of Coleman camper fuel for each 100 pills used, and stir this mixture for about half an hour. Then filter the mixture. Doubled up coffee filters or lab filter paper will be fine enough to catch the pill particles. A clear blue filtrate should result. The blue color is from the camper fuel; it causes no problems.

Return the pill mass to the beaker, and add another 50-75 ml of Coleman camper fuel for each 100 pills used. Stir this for about half an hour, then filter.

The combined clear blue filtrate is now ready for bubbling with dry HCl. This is done just like in all the other examples where we bubble dry HCl to get the crystalline hydrochloride product. The blue color of the camper fuel doesn't color the crystals at all, so long as it is sucked away using a Buchner funnel and vacuum flask. If you don't have such equipment, a final rinse of the crystals with toluene will wash off the camper fuel.

Camper fuel evaporates quickly, and doesn't leave a lingering smell on the crystals of ephedrine hydrochloride. One can expect to get close to 100% extraction of the pills by this method, so long as the pills were finely ground in the first place. Your Uncle has tried and likes this method!

Others have also used this method, and offered their variations on the techniques. Suggested refinement number one is to replace the Coleman camper fuel with the naphtha, a common solvent which one can pick up at the hardware store. People have found that naphtha evaporates away faster than Coleman camper fuel, and that it is also a considerably cleaner solvent than camper fuel.

Suggested refinement number two is to do away with the isopropyl alcohol/lye mixture for the free basing of the pills. Instead what they suggest is to take the ground up and toluene-desplooged pills, and wet the pill mass with a 5% solution of lye in water. After a bit of mixing, the wetted pill mass is spread out on wax paper to dry. Once the water has evaporated away, the free based pill mass is then extracted with naphtha or camper fuel, just as given above.

All of these methods use a lot of solvent, and so produce a lot of waste solvent. The question naturally arises, "What do I do with my waste solvent after it has been used?" There is an easy answer to this. One should never pour waste solvent down the drain, or dump it into the ground, or otherwise dispose of it haphazardly. What one should do is store the used solvent in champagne bottles or other containers that can be sealed up. Then at the earliest convenient moment, pay a visit to the waste oil recycling drop off point. These can be found in most any town. Just pour your waste solvent into the waste oil container. The waste oil will be recycled as a fuel, and your solvent will do no harm there.

The years 1999 through 2001 brought with them a series of unfortunate events for meth cookers. Bad news item number one was the pulling from the shelves of products containing phenylpropanolamine. This unfortunate occurrence was due to a number of strokes traced back to taking too much Dexatrim and related products. The resulting lawsuits and FDA restrictions meant the end for OTC (Over-the-Counter) phenylpropanolamine.

Of even greater significance for meth cookers has been the gradual appearance of and now market domination by a series of "doctored" OTC pseudoephedrine pills. These pills first showed up in Australia in late 1999. The parent company distributing these "gak" pills chose Australia as their test ground because if people keeled over from ingesting the mile long list of "inert ingredients" the publicity in the US would be much easier to kill. One must also consider that the price that a lawyer can extract for a dead Aussie is much less than an American lawyer could get for dead Americans. Add to that the fact that the Aussies had a nice and growing clandestine meth "problem" based upon pseudoephedrine pills. The Aussies were the perfect test market for these new pills.

Once Warner Lambert demonstrated that their new pill formulations didn't seem to kill people, and also showed that the new pill formulations confused the hell out of clandestine chemists trying to extract them to use as raw material for meth cooking, the pills moved to the US. Then, as of Spring 2001, was very hard, regardless of brand, to find a pseudoephedrine pill which one could extract by the old method of grinding the pill, and extracting with water or alcohol. These "simple extractions" should now be considered to be completely worthless for the pills now on the market.

The new pseudoephedrine pills can be recognized by looking at the list of inert ingredients on the pill package. The list of ingredients will be a mile long, and will include such things as povidone, crospovidone, lactose, carnauba wax, acacia gum, Xanthane gum, soaps like stearic acid or magnesium strearate, polymers with many names like Polyox or Pluronic. They may also choose to say "may contain one or more of the following" or they may just say, for ingredients, see US Patent 6,136,864 or World Patent 00/15261. These patents make for great reading. I'm mentioned over and over. Check them out.

The two Fester-obsessed jackasses at Warner Lambert Pharmaceuticals who have devoted their working life to me will go nameless here. What won't go unnoticed is the fact that they and their parent company are making a killing off these new pills, charging $15 per hundred for them. Generic versions also using their formula charge almost as much while paying royalties to them. They didn't bother to cut me in on the gravy train while getting rich off me, so it was time to send them back to work.

The new pseudoephedrine pills were roughly modeled after the gas station ephedrine pills. This should come as no surprise, since the same crew was responsible for them It is also possible to extract them by a roughly similar method to that used with the gas station ephedrine pills. The difference at present is that the new pseudoephedrine (Sudafed) pills are much more heavily loaded with waxes like carnauba and gums like acacia. Dragging these waxes and gums into a meth production batch will kill the yield, and make isolating any product very hard. The whole idea behind these patents, which claim to make "illicit drug product impossible," is that they are formulated so that the "gak" which has been added gets brought over through the extraction process into the batch, thereby ruining any batch.

A method which works on these new pseudoephedrine pills is to first grind them in a blender. It is hard to get a fine grind because they are so gummy. Then soak the ground-up pill mass in at least 3 ml of toluene for each pill used. Stir the mixture from time to time, and after a couple hours of soaking, filter off the toluene, and allow the pill mass to dry. Then free basing using isopropyl alcohol/lye or lye solution in water can be done just as given previously. This is fooled by extraction of the free base using toluene rather than naphtha or camper fuel. There is no need to use the naphtha to extract these pills since they don't contain guaifenesin. After a couple extractions with toluene, the pooled toluene extracts can be bubbled with dry HC1 to get nice looking crystals of fairly pure pseudoephedrine. With the present formulations on the market, this procedure will give a clean enough pseudoephedrine to make meth. Further cleaning of the pseudoephedrine crystals could be done by taking the crystals, and for each gram of crystals, dissolving it in 10 ml of water. Add lye with strong shaking until the pH of the water is 12+ to pH paper. Now extract the water with some toluene, and bubble the toluene extract with dry HC1 to get really pure pseudoephedrine.

This procedure worked well until late 2003, although towards the end of that period, new pill ingredients were added that required the product pseudoephedrine to be soaked in a little cold methyl ethyl ketone (MEK) and then filtered to get a clean product. Then a new generation of gas station ephedrine pills appeared which were completely unextractable by the old method. Within six months, the same formulation had spread to the pseudoephedrine pills as well, and clandestine chemists everywhere were stumped as to how to defeat this new menace.

I had foreseen the emergence of this threat, and in the sixth edition of this book suggested using steam distillation when they made their way to the market. It turned out that steam distillation worked OK with Ma Huang and herbal ephedrine pills. The result with gas station ephedrine pills was bad because of a combination of foam producing ingredients they contained. Pseudoephedrine pills proved to be impossible to steam distill successfully using standard methods.

The culprit behind this plague for meth cooks was US Patent 6,359,011, again by those two jackasses Nichols and Bess at Warner Lambert. This Patent relied heavily upon polymers to mess up the extraction and isolation of ephedrine or pseudoephedrine from pills, and included polymers which mimic the solubility of ephedrine and pseudoephedrine so that chemical methods of isolation of the product was next to impossible. The polymers would be pulled into whatever product could be isolated from the pills, and they would proceed to kill any reaction used to convert the ephedrine or pseudophedrine into meth.

I love a good challenge like this from "the man." It's always been my belief that playing offense is much easier than playing defense, so I knew the Patent holders were in a losing game. The question was just how to crack this new and heavily polymer dependent formulation.

After about a month of thinking about the subject in my spare time (oh, there is so little of that!) and a couple of weeks of experimentation again in that very rare spare time, I hit upon the Achille's Heel of all formulations dependent upon polymers.

Polymers are built up of smaller subunits linked together into long chains. The most common linkage used in the pill additives is the ester link. When I was a kid, I used to watch Granny Clampett cook lye soap by the Cement Pond. She was cleaving ester links in fat to make soap. If Granny can cleave esters, so can I to get something more interesting than lye soap. I knew that once the links were broken in the polymers, they would no longer have the solubility characteristics which make them such a problem.

There are two general methods for cleaving esters, be they in polymers or anywhere else. They are alkaline hydrolysis and acid hydrolysis. Granny was doing an alkaline hydrolysis of the ester links in fat by the Cement Pond using lye, which is NaOH. For the cleavage of the pill polymers, the more general chemical method is called for. That method is hydrolysis using potassium hydroxide (KOH) in alcohol solvent.

To put this method into use, I revived the old alcohol extraction method. To hydrolyse, the alcohol should be the azeotropic mixture. This gives maximum yields, as less or more added water cuts the yield of the desired product. There are two commonly available azeotropic alcohols. They are 95% ethyl alcohol, exemplified by 190-proof vodka and hardware store denatured alcohol, and 91% isopropyl alcohol found in the bandage section of your local drug store. I chose the latter for my work because it doesn't have as great a smell as the denatured alcohol found in the hardware store.

The extraction method I discovered doesn't require the pills to be pre-soaked in solvent to remove their copious supply of gums and waxes prior to extraction. The KOH hydrolysis removes them as well as the new breed of polymer additives. Simply add the Sudafed pills to a blender, and grind them up. When the dust settles inside the blender, empty it onto a plate and then use a spoon to crush any pill bits and pieces which escaped grinding in the blender. When finished with that job, just put the ground up pill mass into a beaker or measuring cup with a pour spout.

An example batch size of 100 Sudafed pills containing 30 mg each of pseudoephedrine will require about 250 to 300 ml of alcohol to extract completely. Measure out about 100 ml of alcohol and add it to the pill mass, then stir. An orange red solution will start to form immediately as the pills extract. Swirl or stir around this mixture from time to time for about an hour, then it is time to collect the first extract.

Generic Sudafed pills obtained from Walgreen's are very closely formulated according to the previously mentioned Patent. Obtaining the first extract from them is very simple. Just pour the mix through a coffee filter and collect the filtrate. Brand name Sudafed pills and Wal-Mart store brand pills are of a bit more advanced formulation which plugs filters. For them, just let the pill sludge in the beaker settle for about an hour, then pour off the alcohol solution from the pill sludge. Try to keep as much of the pill mass in the original container as possible.

If filtering was done, then return the filtered pill mass to the original beaker or cup. Then add another roughly 100 ml portion of alcohol to the pill mass for another extraction. If pouring off of the first soak was done, then just add another 100 ml portion of alcohol to the pill mass. Let the next pill soak proceed for an hour or so like the first one with some swirling or stirring. Then filter or decant off the second alcohol soak just like the first one.

Finally, do a third alcohol soak of the pill mass with another roughly 100 ml of alcohol just as before. By now, the alcohol extracts are becoming pale in color, indicating that the pills are nearing complete extraction. Filter or decant this soak just like the previous ones.

If the pills being used allowed one to filter, then the next step can be done immediately. That step is hydrolysis with KOH. If filtering was skipped due to filter plugging pills, then let the extracts settle overnight to shed floating crap. I am told that putting some sand into the filter one is using defeats the filter plugging qualities of these new pills, so do give this a try. It is far more preferable to filter the pill extracts, as this allows more complete extraction.

Now for the big part of the show, the KOH hydrolysis. The combined alcohol extracts are now poured into a Pyrex beaker. A Mr. Coffee pot will do fine if you don't have a beaker. Put it on a stove top at medium heat, and then add 10 grams of KOH pellets for each 100 of the 30 mg Sudafed pills extracted. If one has picked up the new 60 mg Sudafed pills, then one would use 20 grams of KOH per one hundred pills, and of course, the amount of alcohol required to extract them would be double as well. Do avoid the new 120 mg Sudafed time release pills, as they are of a gooey formulation you don't want to mess with. The goo is their time release mechanism, and is unlikely to spread to the lower dose pills.

The question which is likely to pop into one's head at this point is "Where do I get KOH?" This close chemical cousin of lye, NaOH, is pretty easy and safe to get from those mail order chemical outlets that advertise in the classifieds sections of some magazines. Keep the order to KOH, and there will be little chance of bad things happening as a result. An alternative source can be found at everybody's favorite department store which is open 24/7. Head to the plumbing section, and you will find a product which is roughly 50% KOH and 50% NaOH. This can be used in a pinch, just by upping the amount used to about 15 grams per hundred pills. One can also do some Internet shopping. Search under "soap making." KOH is used to make soft soap. One can also search under "hide tanning." KOH is used in that craft as well. Avoid solutions of KOH in water, as the alcohol already has the best amount of water for the reaction in it. If you can only get a KOH solution in water, boil it down. In no case just use NaOH, as it doesn't work. Also avoid getting hardware store drain openers which contain KOH plus bleach. The bleach will simply destroy the pseudoephedrine or ephedrine in the extract and leave you with nothing. KOH pellets can also be found at pool supply stores as pH Up (caustic potash).

As the KOH pellets dissolve into the alcohol extracts, they begin to chew up the polymers, gum and waxes in the pills. This only happens when the alcohol solution is at or near boiling, so a gentle boiling of the alcohol is needed. Adjust the heat setting on the stove top accordingly. Within a few minutes of boiling with the KOH, you will note that the original red orange color of the alcohol extract is fading rapidly, and that an oily layer is forming at the bottom of the beaker, or Mr. Coffee pot, whichever you are using. Continue to gently boil for half an hour, then set aside to cool. With the variety of pills exemplified by the Walgreen's store brand which closely follows the Patent formulation, you will have a coffee colored oil layer at the bottom of the beaker, and a weakly colored alcohol solution containing the pseudoephedrine you want floating above it. Just pour this mix into a sep funnel and let the oil settle to the bottom and drain it off to get rid of the crap you just destroyed. We can now move on to the evaporation and getting product portion of the process.

With the brand name Sudafed pills and the Wal-Mart store brand pills, the ones which plugged up the filters, the KOH boil takes a bit of a different course. With these pills, as the KOH dissolves in and the alcohol gets boiling, the red orange color fades, and the solution turns milky. Don't be alarmed by this, as the milky appearance is caused by little white flakes of crap which you have destroyed. Boil in this case for half an hour just as with the previous example. The only difference is after it is done boiling. In this instance, let the solution cool and settle overnight, or at least for several hours. The white flakes will settle out leaving a clear alcohol solution containing the pseudoephedrine, and a mat of gooey white flakes overlaying a layer of gak oil on the bottom of the beaker. When fresh, this oil will most times be yellow colored, but in a few hours it too will turn coffee colored. The white flake mat is gooey, and given a few hours it sticks to itself. Then in this case, one can just pour off the alcohol solution from the white flake mat and oil layer to get a clean alcohol extract.

Now to get the product! Pour the cleaned alcohol extract into a clean beaker, or a Teflon-coated pan. Then simmer down the alcohol extract. There are two points here to be wary of. Point number one is that alcohol is flammable, and mildly toxic. Be sure to use a good draft of air to clear the vapors away from the boiling spot. Alcohol isn't nearly as flammable as naphtha or other solvents, but fire precautions need to be observed.

Point number two concerns the last phases of the evaporation. It is VERY important that one not boil down the solution to dryness. This will result in a yellow colored product that is crap. As the solution gets to nearly all evaporated down, switch to boiling water heat, or just letting it evaporate with some aid of heat and blowing off the alcohol vapors. It is far better to let a bit of alcohol remain than burn the product!

Now the remaining product in the bottom of the flask will consist of left over KOH, pseudoephedine free base, and assorted crap. Start by adding about 50 ml of water for the 100 pill batch example to the beaker. Swirl it around, and let it work for a few minutes. Then pour it into a sep funnel. Homemade substitutes for sep funnels are easily constructed. Check out Jack B. Nimbles' book for starters to get some ideas. Then add 50 ml of toluene or xylene to the beaker to dissolve your product. Swirl and let it work for a few minutes, then pour that too into the sep funnel. Chase the residue in the beaker with a little bit (about 15 ml) more toluene (that means add a bit more!) and pour it too into the sep funnel.

Now we are on our way home. Shake that sep funnel for about half a minute, and let it set to settle the layers. In the case of the close Patent formulation pills, the water layer will look like Pepto-Bismol. Using the pills which give a milky look upon boiling, this color will not be seen in the water layer. Drain off the water layer, and add about 50 ml of water for this 100 pill example. Shake the toluene or xylene layer again with this fresh water. Now let things settle in the sep funnel. You should have a clear toluene solution floating above a reasonably clear water layer.

Drain off the water. Now check the toluene layer. It should be just clear solution. If there is floating crap in it, pour it through a coffee filter. This will give you a clear toluene or xylene solution containing the pseudoephedrine free base. Let it set in this beaker for a couple of hours to settle any water you dragged in. Then pour it into a clean beaker, and bubble the solution with dry HCl gas. You will get roughly 70% yield of the possible pseudoephedrine available from the pills after you filter out the pseudephedrine hydrochloride, and rinse the product with fresh toluene or xylene. This is the same HCl bubbling procedure which has been used throughout this book, and for the last 20 years. Become familiar with it!

Gas station ephedrine pills extract and behave in a very similar fashion to the pseudoephedrine pills, except that they don't give the color change while they are boiling in the alcohol and KOH. They aren't colored in the first place, so that is to be expected. The same procedure can be used on them to get similar yields of good product. The complication with them is that they contain guaifenesin. To remove most of the guaifenesin from the pills, they should first be ground up in a blender, then add 2 or 3 ml of toluene or xylene for each pill used to the ground up pill mass and with stirring let it soak out the guaifenesin for a few hours. Next filter the pill mass to remove most of the toluene, then spread the pill mass out to dry on a plate. Once dry, the extraction with 91% isopropyl alcohol followed by boiling with KOH is done just like in the procedure for Sudafed pills. In this instance, naphtha should be used as the solvent to extract the residue from the beaker, and to bubble HCl gas with. This solvent is used instead of toluene or xylene because it doesn't dissolve the guaifenesin still left in the pill extract.

I know what you are going to say at this point: "That seems like a complicated procedure." Actually, it's really simple. It's just that I put in all the possible details for you. It's just an extension of a method which was used to extract the pills from the mid 90's. One could even add the old method of blowing into the bottom of the beaker to get rid of the last of the alcohol to reveal big crystals of pseudoephedrine. Now add to that KOH boil, and destroyed crap removal, and you have the exact same method which was used roughly 10 years ago with the pills that I prescribed a solvent pre-soak and water extraction for. Extraction methods do turn a circle. If this method seems complicated to you, check out my website www.unclefesterbooks.com. The Cookin' Crank with Uncle Fester video will make it all just as plain as it really is.

The next reasonable question is.. "I just can't get KOH!" I worked out a method just for people like you, but the yield is closer to 50% rather than the 70% gotten from my original method. Dig harder for KOH, but here it is:

The pill extraction is done exactly like before, except when it is time to boil the pills add instead 20 grams of KCl. This is salt substitute found at your grocery store. Read the label, and get the brand that just says KCl with a couple of other minor ingredients. Then as the pot is warming up to a boil, slowly, with stirring add 10 grams of NaOH (lye) per one hundred pills used. The best stirring tool is a rubber spatula, and stir well because the alcohol can only dissolve around one gram per 250 ml alcohol, and you want that bit of KCl to react to make KOH plus NaCl. This is making KOH on the sly. The addition of NaOH should take about 20 minutes, with lots of stirring. A big snow fest of NaCl crystals in the solution will be seen during this process.

About the time that the last of the NaOH has gone into solution, it's time to boil this mix. During the course of a half hour boil, the initial red orange color of the alcohol will fade to a shade which is best described as melon. Then set aside the beaker, and allow the mixture to cool and settle.

The next thing to do is to pour the alcohol solution off of the settled layer of salt on the bottom of the beaker. Using a coffee filter makes this separation much more efficient. Then the alcohol solution is carefully boiled down, just as in the previous example. When it gets close to being evaporated down to the bottom of the beaker, its best to let the evaporation finish by itself at room temperature, or with mild heating.

Now add about 50 ml of water to dissolve the salt, KOH and NaOH in the bottom of the beaker. Let it work for about 10 minutes, and pour the water solution into a sep funnel. Then put about 50 ml of toluene or xylene into the beaker to dissolve the residues of pseudoephedrine in the bottom of the beaker. Let that soak work for about 10 minutes, then add this also to the sep funnel. Finally, chase the last bits of product out of the beaker with a rinse using about 10 or 15 ml of toluene or xylene. Pour this into the sep funnel as well.

The sep funnel is then shaken hard for about 20 seconds, and set aside to settle. What is left of the orange color will go into the water layer at the bottom of the sep funnel. A lot of floating crap will be seen, as well as a fairly clear toluene layer at the top.

Now drain off the water layer. You will note that the floating crap is in both the water layer and the toluene layer. You don't want to be draining off and throwing away the toluene, because that is where your product is. It's better to be leaving some of the water rinse in the sep funnel at this stage than throwing away your toluene product layer.

When the water layer has been drained away, add 50 ml of clean water to the sep funnel, and shake again. When the mix inside the sep funnel settles in a few minutes, you will see a nicely clear toluene layer at the top mixed with floating flakes of crap, and a lightly colored water layer at the bottom mixed with floating flakes of crap. Drain off the water layer, and pour the toluene layer through a coffee filter to remove the floating crap.

Rinse out the sep funnel with water to remove clinging flakes of crap from the glassware, then pour the filtered toluene back into the sep funnel and drain off any water which carried through the filtration. Then pour the toluene layer into a clean beaker and let it set for an hour or so to shed any water still entrained in the toluene. Finally, pour the toluene into a clean beaker and bubble dry HCl gas through it to get crystals of pseudoephedrine hydrochloride. Filter them out with a coffee filter, rinse them with a fresh portion of toluene, and spread them out on a plate to dry. The yield in this case will be around 50% of the maximum possible. In the case of 100 of the 30 mg pseudoephedrine pills, that will be around a gram and a half of product.

This variation isn't as good as the one using only KOH pellets, but it does avoid any problems associated with getting them as it uses instead KCl salt substitute and lye. I tried one more variation on this procedure. That variation was to use water to extract the pills, and then to the filtered water extract I added 3-5% by weight of sulfuric acid and boiled for about 45 minutes. I obtained the product by making the solution basic by adding lye to the solution when it was cool, and then extracted the product with toluene, washed it a couple of times with water, filtered the toluene extract, and then bubbled dry HCl gas through it. In that instance I got less than 50% yield of a product which was still dirty and would need to be recrystallized to be used successfully in making meth.

The alcohol extraction followed by boiling KOH treatment procedure should once and for all put the pill formulators out of the polymer additive game. It has been my greatest pleasure to wreck their Patents, and fill their work lives with consternation and dread. As they lay in bed trying to sleep while muttering my name, I have only one suggestion for them. If you had cut me in on the gravy train from those Patents, I might have been persuaded to keep quiet about this.

The pill formulators have one more card to play, and we may see it sometime soon. That card is to replace the natural "d" isomer of pseudoephedrine with the synthetic "l" isomer. Reduction of this "l" isomer would then give only the very weak "l" isomer of meth. A Patent has already been published claiming that the "l" isomer of pseudoephedrine works just as well as the real McCoy for cold relief. If this new product should come to replace the pseudoephedrine presently on sale, the following procedures will prove very helpful. One can racemize "l" pseudoephedrine to "d,l" pseudoephedrine. Reduction of this product would then give "d,l" meth, which is a very nice buzz indeed. In fact it is much better than the strictly "d" isomer meth one generally gets from ephedrine or pseudoephedrine as the starting material.

Racemization of Pseudoephedrine

This procedure is taken from *Chemical Abstracts*, Volume 23, pages 3452-4 (1929). It yields racemic ephedrine or racephedrine from pseudoephedrine, thereby allowing the use of pseudoephedrine to get d,l-meth.

Pseudoephedrine hydrochloride prepared as described above is dissolved in 25% hydrochloric acid solution. Stronger acid must be avoided, as the use of this stronger acid would produce a significant amount of chloroephedrine. One hundred ml of 35% lab-grade HCl can be diluted to 25% by adding 40 ml of water. In the example given in the *Chemical Abstracts*, a fairly dilute solution of pseudoephedrine was used, but I can't think of any reason why one can't mix this solution much stronger. Adding more pseudoephedrine to a given volume of HCl solution allows much more material to be processed at once, and also makes recovery by extraction at the end of the process much easier.

This solution of pseudoephedrine in HCl is then heated at 100° C for at least one day, preferably two. Heating the solution to reflux is to be avoided, as correspondents have informed me that reflux temperatures lead to the burning of the product. Simply heat the flask in an oil bath whose temperature is kept at about 100° C. A reflux condenser must be attached to the flask to keep the acid from evaporating away. In the example from *Chemical Abstracts*, the acid was heated inside a sealed tube, but I can't see why this simpler procedure won't work just as well.

At the end of the heating period the solution is cooled, and then sodium carbonate is added to the acid solution a bit at a time until all of the acid is neutralized. This point can be spotted because the carbonate will stop fizzing once all the acid is gone. Now shake the solution strongly for a few minutes to ensure that all of the racephedrine hydrochloride has been converted to the free base. Then extract a couple of times with toluene. The pooled toluene extracts can then be bubbled with HCl gas to precipitate the product as the hydrochloride.

Method Two

The hot hydrochloric acid isomerization has the advantage of using easily available hardware store muriatic acid to do the job. The drawback is the tendency this method has of burning the product. A much more refined procedure can be found in US Patent 2,214,034. To use this isomerization method, one will have to get a chemical named sodamide, aka sodium amide ($NaNH^2$). This material is reasonably cheap and isn't on anyone's "watch list" at the present time. It's also a dangerous chemical which can suck up water and CO_2 from the air and become explosive. Keep bottles of this powder tightly sealed. Weigh and use portions of this substance quickly, and on dry days. Don't leave it laying around in the open air! Wipe residue of the powder off the threads of the bottle before resealing it. This white to greenish substance turns yellow or brown when it has reached a dangerous state. Small amounts of contaminated sodamide can be just flushed down a toilet, but if you have more than a gram or so of dangerous material, the way to neutralize it is to pour in enough xylene or kerosene to easily cover all of it, then add a 10% solution of alcohol in toluene or kerosene slowly with agitation to the contaminated sodamide until most of it has reacted. Then it can be flushed.

To use sodamide to isomerize "l" pseudoephedrine, one starts with a solution of pseudoephedrine free base in solvent. This is pretty convenient, as all the pill extraction methods given here at one point end up with a solution of the free base in solvent just before bubbling dry HCl to get the hydrochloride crystals. The best solvents for this isomerization procedure are high boiling point liquids such as xylene or kerosene. Both of these materials can be picked up at the hardware store.

The examples given in the Patent are for 50 gram batches of pseudoephedrine, but the method can of course be scaled up or down as desired. 50 grams of pseudoephedrine free base dissolved in roughly 500 ml of xylene or kerosene is placed into a 1000 ml flask. The glassware is then rigged for simple distillation as shown in Figure 11, and roughly 10% of the solution is distilled off to make sure that all water is gone from the mixture. If kerosene is being used as the solvent, the oil bath will smoke a lot and may burn before it gets hot enough to boil the kerosene, so in that case heat the flask directly on the buffet range.

Once about 10% of the solvent has been distilled off, let the solution cool down to about 100° C. Then add roughly 7.5 grams of powdered sodamide in small portions to the flask with some swirling between adds. Ammonia will be fumed off of the mixture as the sodamide goes in.

When the sodamide has been added, rig the glassware for reflux as shown in Fig. 10. It is important that the mixture be protected from atmospheric moisture, so don't skip the drying tube, or at least use a balloon over the top of the condenser.

Now gently boil the solution at reflux for at least a couple of hours. Two hours is about right using kerosene. Xylene boils at a lower temperature, so stretching the reflux out to three hours or so is probably a good idea using that solvent.

When the reflux boil is finished, let the solution cool down. Then add around 200 ml of 5% hydrochloric acid solution to the flask and shake. This will pull the isomerized pseudoephedrine into the acid solution and out of the solvent.

Pour the liquid into a sep funnel, and drip out a little bit of the lower layer, which is the acid layer onto a pH paper. The paper should tell you that this water is still acidic. If not, add a bit more acid until the pH paper says that the water is acid. This is important, as the pseudoephedrine will not be extracted from the solvent unless the water layer is acid.

Once the water has been confirmed as being acidic, drain off this acid layer into a beaker. The solvent can then be thrown away.

Finally, slowly add lye to the acid water in the beaker with stirring until pH paper tells you it is pH 13+. Once it has cooled down, you can pour the water into a sep funnel. Add a few hundred ml of toluene or other convenient solvent, and bubble dry HCl gas through it to get crystals of isomerized (d,l) pseudoephedrine HCl which are then filtered out and dried as usual. If one is planning to use the lithium metal in the anhydrous ammonia reduction method for making meth, the bubbling with dry HCl can be skipped, and the free base solution used as is. In that case, using ether starting fluid as the extraction solvent would be the best choice.

This method should also be useful to isomerize the "l" meth extracted from Vick's inhalers to "d,l"-meth.

Indirect Reduction

A popular alternative method for making methamphetamine uses ephedrine as the starting material. This method was not covered in the original edition of this book. It is now presented in all its glory for the education of the reader.

The reasons for the popularity of this method are twofold. Firstly, this method does not require the use of methylamine because the methylamino group is already incorporated in the ephedrine molecule.

The utility of this method is not limited solely to ephedrine. Pseudoephedrine and phenylpropanolamine can also be used as starting materials. This means that Sudafed and Dexatrim, and their generic equivalents, can be used as raw materials for clandestine amphetamine manufacture. The active ingredient is easily separated from the diluents in the pills by the method given in this book.

The bad thing about this method is that foul impurities generated during the manufacturing process are easily carried into the final product. Due care must be practiced by the chemist during the purification to exclude this filth. Unscrupulous and/or unskilled manufacturers turn out large volumes of crank containing this abomination. The impurities not only ruin the finer aspects of the meth high, but they also have a pronounced deleterious effect on male sexual function.

Study the compounds pictured below, and compare them to the meth molecule:

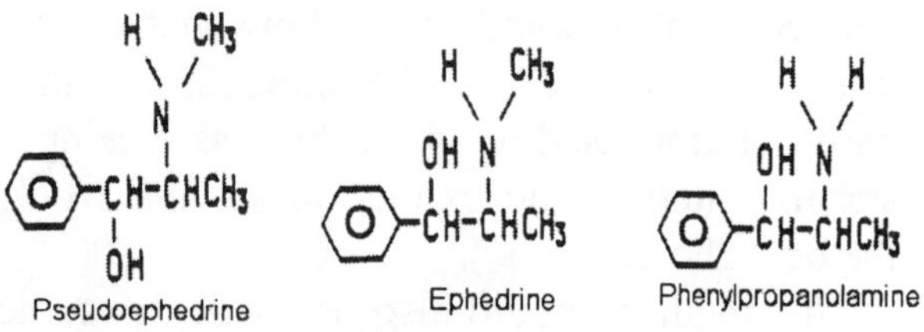

One can quickly see that all a chemist needs to do to turn ephedrine into meth is to replace the alcohol OH grouping with a hydrogen atom. This is not done directly. Instead, a two-step process is used whereby the OH is first replaced by a chlorine atom, and then this chlorine is removed by one of several reductive processes, to be replaced with a hydrogen atom. To illustrate:

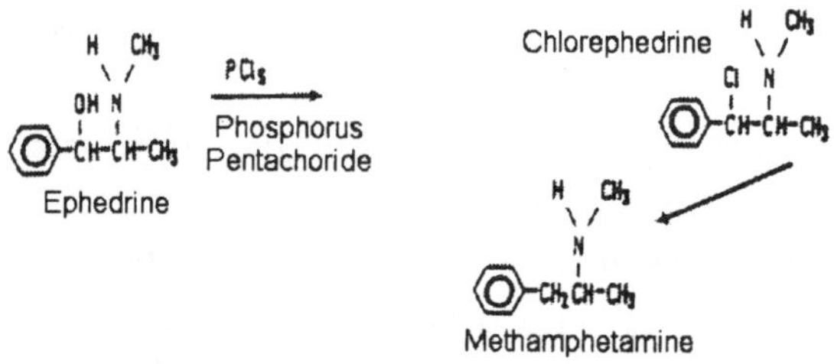

There are several general methods for converting an alcohol group into a chlorine atom. Substances such as thionyl chloride ($SOCl_2$), phosphorus pentachloride (PCl_5), phosphorus trichloride (PCl_3), phosphorus pentabromide (PBr_5) and phosphorus tribromide (PBr_3) can all be used to convert the alcohol group to either a chloride or bromide. Essentially the same reaction conditions are followed when using any of the above listed substances. The only difference is how much ephedrine or PPA (phenylpropanolamine) the substance can chlorinate or brominate. See the table below:

Substance	Molecular Weight	Reacts with this many moles of ephedrine
$SOCl_2$	119	1
PCl_3	137	2
PBr_3	271	2
PCl_5	208	3
PBr_5	430	3

molecular weight of ephedrine HCl=202, PPA-HCl = 188

Using the above table, a person can quickly calculate how much ephedrine or PPA will react with a given amount of chlorinating agent. Use of excess chlorinating agent will result in a higher percentage yield based on the ephedrine used, but after a point, this is wasteful. The following example takes this largess to an extreme, but achieves 100% conversion of ephedrine to chloroephedrine. This procedure can be followed with all the chlorinating agents. The reaction is fairly easy to do. The main precautions are to make sure that the glassware is free of water, and taking one's time to be sure the mixture stays sufficiently cold. It is also wise to avoid doing this reaction in very humid conditions.

The following procedure for the conversion of ephedrine, racephedrine or PPA to the chloro compound and then its reduction can be found in *Chemical Abstracts*, Volume 23, page 3453. It results in a little racemization, but mostly keeps the structure of the starting material.

To do this reaction, a 2000 ml Erlenmeyer flask is filled with 360 ml of chloroform and 360 grams of PCl_5. This mixture is then cooled down in ice water, and once it has cooled down, 240 grams of ephedrine HCl is added in little portions, with shaking of the slushy reaction mixture after each add of ephedrine, racephedrine or PPA. The addition should be completed in about ½ hour. Then, for an additional two hours, the reaction mixture should be shaken to mix around the contents. Cooling in ice must be continued throughout the reaction time to keep the contents from overheating.

When two hours of reaction time has passed, let the contents settle in the flask. After about 45 minutes, when all has settled inside the flask, the mixture is carefully decanted off into a one-gallon glass jug. Great care is taken during this decanting to make sure that all of the settled PCl_5 remains behind. If any of it were mixed in with the product chloroephedrine it would be reduced in the succeeding hydrogenation to phosphine, PH_3, an exceedingly deadly gas. If it appears any is being carried along, the mixture is filtered.

The PCl_5 left in the flask should be rinsed with 150 ml of chloroform to get the trapped product out of it. This is done by adding the chloroform, shaking the sludge to mix, allowing it to settle, then decanting off the chloroform as before into the glass jug.

There will be a lot of unused PCl_5 in the flask, and it would be a shame to just trash it. The obvious thing to do is to save it by stoppering the flask, and try using this material to run another batch.

Next, the product is precipitated from the chloroform solution in the gallon jug. This is done by slowly adding ether or, better still, mineral spirits (cheap and easily available in large amounts) to the gallon jug until it is nearly full. For best results, the mixture in the gallon jug is continuously stirred during the addition of the ether or mineral spirits. Chloroephedrine does not dissolve in ether or mineral spirits, so as the solution changes from chloroform to predominantly ether, the product is thrown out of solution in the form of crystals. If an oily layer forms at the bottom of the jug, this means a dirty batch. The oil may eventually crystallize, but more likely it must be separated, dissolved in an equal volume of chloroform, and precipitated once again by adding ether or mineral spirits.

After the addition of the ether or mineral spirits, a large mass of crystals fills the jug. This is the product. The jug is stoppered, and put into the freezer overnight to let the crystals fully grow. The crystals are then filtered out and rinsed down with a little bit of cold acetone. Then the crystals are spread out to dry on china plates or glass baking dishes. The yield of chloroephedrine hydrochloride is in the neighborhood of 250 grams.

A similar recipe can be found in US Patent 6,399,828. In this example they use thionyl chloride to make chloroephedrine, and they did it as follows:

Into a 500 ml round bottom flask they placed 150 ml of thionyl chloride and 55 grams of ephedrine hydrochloride. Pseudoephedrine hydrochloride would probably work just as well in this reaction. One should also be forewarned of the noxious fuming properties of thionyl chloride, and if it gets on your skin it will eat holes through it unless promptly rinsed off.

They next rigged the flask for reflux, and boiled the solution for about an hour. Once the reaction mixture had cooled down, they then evaporated away the remaining thoinyl chloride under a vacuum to get a crystalline mass of crude chloroephedrine in the bottom of their flask.

They next added a few hundred ml of ether to the flask, and stirred up the crystals to allow the ether to wash off unwanted residues from their surfaces. They then filtered out these washed crystals, and once they had collected and dried them, they recrystallized them. In this instance, they used methanol to dissolve the chloroephedrine, and then precipitated the pure chloroephedrine by slowly adding ether with stirring until no more crystals came out of solution. They then spread them out to dry as in the previous example.

The discerning reader will note that in the Patent example, they used no chloroform as the solvent for their reaction. One must understand that patents are also sales pitches, and they were less than artful in their demonstrations of "Prior Art," to try to show the superiority of their new method. By using no chloroform solvent in this recipe for chloroephedrine, they got 50-60% yield of chloroephedrine. If one had used a hundred ml or so of chloroform along with the 55 grams of ephedrine and 150 ml of thionyl chloride, the expected yield would be around 90%. Making the "Prior Art" look bad is just one of the things you have to be aware of when reading a Patent.

Production of Meth

To make meth from chloroephedrine, the chlorine atom is replaced with hydrogen. This reduction is accomplished by any of several methods. Lithium aluminum hydride does the best job of completely converting the chloroephedrine into meth, but it is very expensive, and a watched chemical. Zinc dust, on the other hand, is cheap and easily available, but it leaves a large proportion of the chloroephedrine trashed. The most practical and effective way to turn out large volumes of meth is by catalytic hydrogenation. It is possible to use Raney nickel as the catalyst for this hydrogenation, but it has to be used in quite large amounts to do a good job. Ammonia, amine or some other base also has to be added to the bomb in an amount equal to the chlorine given off by the chloroephedrine, (i.e., one mole of chloroephedrine would require one mole of ammonia, amine or other base added). Platinum can also be used to reduce the chloroephedrine, but it too has to be used in large amounts to get good results. Furthermore, it is rapidly poisoned by the hydrochloric acid generated by the removal of the chlorine atom from chloroephedrine unless one mole of base per mole of chloroephedrine is included in the hydrogenation mixture. But I have seen one example in which good yields were obtained with Pt and chloroephedrine without addition of any base.

The best catalyst to use for this reduction is palladium, in the form of palladium black on charcoal, or palladium on barium sulfate. The palladium stands up well to the chlorine, and can be used to run many batches before it needs to be recycled. Palladium works fine at low pressures of hydrogen, and can be used with the champagne bottle hydrogenation system pictured in Chapter Eleven.

Important: *The valve on the hydrogen tank is only opened when adding more hydrogen to the bomb. Otherwise, it's kept closed. Failure to do this may result in explosive accidents!*

"The Poor Man's Hydrogenation Device" is a good deal more resilient than a champagne bottle, and it will only accept a feed of hydrogen when its valve is opened. It also is much easier to seal against leaks. So long as the inside surface is coated either with Teflon or high phosphorous electroless nickel, this container for hydrogenation is superior to the champagne bottle used in the following example.

To do the reaction, a champagne bottle of at least 1.5 liter volume is filled with 50 grams sodium acetate (anhydrous) and 700 ml of distilled water. The pH of this solution is then made neutral (pH 7) by dripping in diluted acetic acid. This forms an acetic buffer which prevents the solution from becoming acidic when chloroephedrine hydrochloride is added to it. It also neutralizes the hydrochloric acid formed when the chlorine atom is removed from the chloroephedrine molecule.

Then 40 grams of 5% palladium black on charcoal (palladium content 2 grams) is added, and finally 125 grams of chloroephedrine hydrochloride is added.

Palladium on $BaSO_4$ catalyst gives a faster reduction, but barium compounds aren't as easily available as activated C. The choice is up to the reader.

Sodium acetate is now on California's list of less restricted chemicals, so it is wise to avoid using sodium acetate as such. This is not the least bit troublesome, and shows just how stupid the people are who put it on the restricted list. To avoid the need for sodium acetate purchases, acetic buffer is made from vinegar and sodium hydroxide. To do this, 700 ml of vinegar is used instead of distilled water. It should be the cheapest grade of white distilled vinegar, because this is likely to be made just by diluting glacial acetic acid with water down to a 5% strength. Then to this 700 ml of vinegar, sodium hydroxide pellets are slowly added until the pH of the solution is around 7. This takes about 22-23 grams of NaOH.

The champagne bottle is then attached to the hydrogen line pictured in Figure 32 in Chapter Eleven, and the air is sucked out and replaced with hydrogen as described in that chapter. Then the pressure of hydrogen is increased to 30 pounds, and magnetic stirring is begun. The solution soaks up hydrogen for several hours, during which time the pressure is maintained around 30 pounds by letting more hydrogen into the bottle.

When absorption of hydrogen ceases after several hours (up to one day for Pd/C), the reaction is complete. The hydrogen valve is turned off at the cylinder, and hydrogen inside the bottle released outside through a line of tubing as described in Chapter Eleven. Stirring is stopped, and the palladium on charcoal catalyst is allowed to settle in the bottle. When it has settled, the solution is carefully poured out of the bottle into a beaker, taking care to try to leave all the catalyst behind in the bottle. The solution is then filtered to remove suspended Pd on charcoal catalyst. The catalyst is returned to the bottle, which is then refilled with a fresh batch, or filled with hydrogen to protect the catalyst. If another batch isn't going to be done soon, rinse the catalyst with some water.

Before proceeding further with the processing of the filtered batch, it is wise to look more closely at the nature of the byproducts produced by this method of making meth. There are twin villains to be dealt with here:

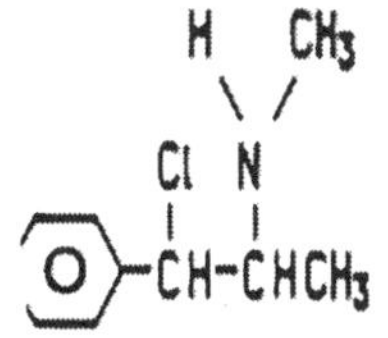
Chlorephedrine

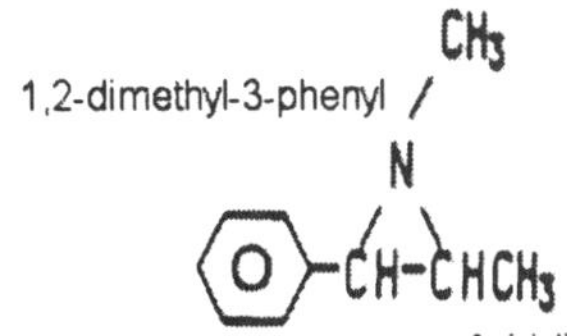

Aziridine

These substances, or closely related ones, will always be formed when making meth by this method. The chloroephedrine is the result of incomplete reduction to meth, and the aziridine the result of an intermolecular reaction between the chlorine atom and the nitrogen atom of the chloroephedrine. It is likely that the aziridine byproduct is more easily formed when the bromoephedrine variation of this synthetic route is chosen. There are two things which aid in the formation of the aziridine. They are exposure to strong bases such as lye, and heat. To minimize formation of the aziridine, one first of all aims for as complete a reduction as possible of the chloroephedrine to meth. Next, during processing, one backs off on the heavy duty use of lye, using bicarb instead to neutralize the last of the acid. Finally, the distillation is done as quickly as feasible under vacuum to get the least heat exposure to the unreduced chloroephedrine. Obviously, the first point is the most important.

To proceed, the filtered batch is reacted with lye with strong shaking until litmus paper says that the pH is around 7. Then bicarb is added to finally make the solution basic. One needs only go over pH 7 here. Neutralization is complete when fizzing stops when adding bicarb. The fizzing and venting of CO_2 gas is a hassle at this point, but it is worth it to avoid the formation of the aziridine. A 2000 ml flask is a good vessel in which to do the neutralization procedure. One must periodically vent off the built up CO_2 gas after bicarb has been added.

Upon standing after the shaking, a layer of meth floats on top of the water layer. Then 200 ml of benzene or toluene is added, and the jug is shaken again. After standing for a couple of minutes, the benzene-meth layer floats nicely upon the water. This is carefully poured off into a sep funnel, and the benzene-meth layer is poured into a 500 ml round bottom flask. The water layer is discarded.

Next, the product is distilled as described in Chapter Five. Here also is a point at which lazy or under-equipped operators err and thereby leave their product polluted with chloroephedrine. You see, it is next to impossible to completely convert the chloroephedrine into meth. The conversion can be encouraged by using plenty of catalyst, sufficient pressure, and ample reaction time in the bomb, but there will still be some left unreacted. As the catalyst wears out from doing repeated batches, the proportion of chloroephedrine in the product will increase. Only by doing careful fractional distillation, can the chloroephedrine be removed. Chloroephedrine's solubility characteristics are so similar to meth's that it can't be removed by crystallization or rinsing the crystals. When doing the distillation, the meth distills at the usual temperature range. The next fraction which distills is chloroephedrine. Since this chloroephedrine can then be cycled back into the hydrogenation step, it makes both economic and ethical sense to remove it from the product. By skipping the fractional distillation, lazy operators cost themselves an added measure of meth yield from their raw material inputs.

The chloroephedrine free base thusly obtained is too unstable to keep as such. It must immediately be reacted with HCl to form the hydrochloride.

It has become kind of obvious that you wonderful readers out there have been having trouble using the table presented earlier in this chapter, so some examples of the use of other chlorinating agents other than PCl_5 are called for. See *Chemical Abstracts*, Volume 23, page 3453. For example, with thionyl chloride ($SOCl_2$), one puts into a flask 100 ml of chloroform, 100 ml of thionyl chloride and a magnetic stirring bar. The contents are chilled in an ice bath, then 50 grams of ephedrine or racephedrine is slowly added. Stirring in the ice bath is then continued for a few hours, as the reaction of $SOCl_2$ is slower than that of PCl_5. After the reaction time is up, about 500 ml of ether or mineral spirits is slowly added with stirring to precipitate the chloroephedrine hydrochloride. This is filtered out, rinsed with a little cold ether, and spread out to dry as in the previous example.

In the above example, the 100 ml of $SOCl_2$ could have been replaced with 60 ml of PCl_3, or 65 ml of PBr_3.

Along a similar line, a correspondent named Yehuda has written to tell of his experience with the use of trichlorethane as a chloroform substitute. He was not pleased with the results, and wrote with his homebrew method for making your own chloroform. It's interesting and I'll pass it along. In a sep funnel, he puts 35 ml of acetone (hardware store) and either 500 ml of Clorox bleach or 170 ml of 15% sodium hypochlorite solution. This 15%-strength bleach is easily available from swimming-pool suppliers. The sep funnel is shaken vigorously with frequent breaks to vent the gas from the sep funnel. The solution gets pretty warm. The shaking is continued until it stops producing gas. Then let the solution sit for a few minutes for the chloroform produced to settle to the bottom. Drain it off. Then shake the sep funnel again to get a little more chloroform. Total yield: about 15 ml of chloroform. The crude product should be distilled. Then preserve the distilled chloroform by adding .75 ml of ethyl alcohol to each 100 ml of chloroform. The boiling point of chloroform is 61° C. Yehuda also writes to remind the readers that all of the chlorinating agents in this section produce noxious fumes, and should be handled with extreme care. Good ventilation, gloves, protective clothing and eye protection are highly recommended.

Palladium Black on Carbon Catalyst

Since palladium black on carbon catalyst is on the narcoswine's watch list of chemicals, it is wise for the operator to make his own supply. Luckily, this is not too difficult, and gives a catalyst that is fresher and more active than off-the-shelf catalysts.

To make the catalyst, the chemist first obtains Norit or Darco brand activated charcoal, and washes it with nitric acid. This is done by measuring out about 100 grams of the charcoal, and then putting it into a beaker along with 10% nitric acid. They are mixed together into a watery slurry, and heated on a steam bath or in a boiling water bath for 2 or 3 hours. After the heating, the carbon is filtered and rinsed liberally with distilled water until the last traces of acid are rinsed from it. This requires about a gallon of water.

The acid washed carbon is then transferred to a 4000 ml beaker. A few grams of the carbon sticks to the filter paper and is otherwise lost, but this is OK since the idea is to get about 93-95 grams of carbon into the beaker. 1200 ml of distilled water is added to the beaker, and it is heated with stirring to 80° C. When this temperature is reached, a solution of 8.2 grams of palladium chloride in 20 ml of concentrated hydrochloric acid and 50 ml of water is added. This acid solution of palladium chloride is heated for a couple of hours before it is added, because $PdCl_2$ dissolves slowly in the acid solution. It is not added until all the $PdCl_2$ is dissolved. If $PdCl_2$ dihydrate is used, the amount used is increased to 10 grams.

When the $PdCl_2$ solution has been added and stirred in, 8 ml of 37% formaldehyde solution is added and mixed in. Next, the solution is made slightly alkaline to litmus by adding 30% sodium hydroxide solution to the beaker dropwise with constant stirring. Once the solution has become slightly alkaline to litmus paper, the stirring is continued for another five minutes.

Next, the solution is filtered to collect the palladium black on charcoal catalyst. It is rinsed ten times with 250 ml portions of distilled water. Then after removing as much water as possible by filtration, the catalyst is spread out to dry in a glass baking dish. It is not heated during the drying process since it could burst into flames. When it has dried it is stored in a tightly stoppered bottle and used as soon as possible. This process gives about 95 grams of 5% palladium black on charcoal catalyst.

An alternative to the filtration and rinse is the settle, rinse, and decant procedure used in the $Pd/BaSO_4$ procedure which follows.

Palladium on Barium Sulfate Catalyst

As mentioned earlier, $Pd/BaSO_4$ catalyst will reduce the chloroephedrine to meth a good deal faster than Pd/C. It is useful in other reduction methods in this chapter, so its preparation will be covered here.

$PdCl_2$ is used to make this catalyst, just like the Pd/C catalyst, so some more mention should be given to sources of supply for this very useful material. As I mentioned before, $PdCl_2$ and H_2PtCl_6 are both used in the plating industry. $PdCl_2$ is used to activate plastics so that they can be electrolessly plated. It is also used to electroplate palladium. The typical bath formulation is 50 gr/1 $PdCl_2$, 30 gr/1 NH_4Cl and HCl to adjust the pH to 0.1 to 0.5. Similarly, a platinum plating bath is mixed up with 10 gr/1 chloroplatinic acid and 300 ml/1 HCl. Companies which supply platers carry these materials. See the *Metal Finishing Guidebook and Directory*. Your library can get it by interlibrary loan if they don't carry it. Turn to the back of the book to the "product, process, and service directory" and look under palladium and platinum to get a list of suppliers. This is far better than dealing with a scientific supply house loaded with snitches, and their prices are much better. For example, 10 grams of $PdCl_2$ was going for just under $60 in 1995. By naming yourself XYZ Plating instead of Joe Blow, easy access to these materials is assured.

To make about 45 grams of 5% Pd/$BaSO_4$, a solution of about 5 grams of $PdCl_2$ dihydrate (the usual form) in 10 ml concentrated hydrochloric acid and 25 ml water is made. The $PdCl_2$ will take a while to dissolve, and heating the solution to about 80° C speeds the process. It can take about 2 hours. Once it is dissolved, set this solution aside.

Then in a 2000 ml beaker, a solution of 600 ml distilled water and 63 grams barium hydroxide octahydrate is made and then heated with stirring to 80° C. When this temperature is reached, 60 ml of 6N sulfuric acid (3M; 160 ml concentrated H_2SO_4 diluted to one liter with distilled water is 3M or 6N) is added all at once to the barium hydroxide solution with rapid stirring. Then some more 6N sulfuric acid solution is slowly added to the barium suspension until it is just acid to litmus.

Now add to this suspension of barium sulfate the $PdCl_2$ solution prepared earlier. Stir it in, then follow it with 4 ml of 37% formaldehyde.

After the formaldehyde has been stirred into solution, make the barium sulfate suspension slightly alkaline to litmus by cautiously adding 30% NaOH solution with constant stirring. Stir for an additional 5 minutes, then let the Pd/$BaSO_4$ catalyst settle to the bottom of the beaker. Decant off the water (it should be clear) and pour fresh distilled water into the beaker. Stir up the settled catalyst to rinse it off. Then let the catalyst settle again, and decant off the clear rinse water. This rinsing procedure is repeated about ten times to get clean catalyst. It can then be poured into the champagne bottle hydrogenation bomb.

Reference

See *Organic Syntheses,* Collective Volume 3.

Direct Reductions

This section deals with the direct conversion of ephedrine, pseudoephedrine, or phenylpropanolamine to meth or dexedrine respectively. This conversion can be accomplished by one of five methods. These conversions are all possible because ephedrine, pseudoephedrine, and phenylpropanolamine are all benzyl alcohols, and benzyl alcohols are the easiest of all alcohols to reduce to the corresponding hydrocarbon.

These methods all have the advantage of being quick and simple, but they also have their unique disadvantages, along with the general shared disadvantage that the starting material must be gathered bits at a time from bottles of pills.

Method 1: Lithium Metal in Liquid Ammonia Reduction

This is a new method, and is the best one I've seen come down the pike in ages. This procedure was pioneered by a clandestine operator in California. Unfortunately, he was busted because he bought a jug of ephedrine to use as his starting material. Had he been more cautious, and isolated the ephedrine from legal pills, he may well have gone undetected. This method is ideally suited for the rapid production of truly massive amounts of crank. It suffers from the need to use liquid anhydrous ammonia. This is very smelly stuff, especially in the quantities needed to make large amounts of meth. The smell problem means that this method can only be used in countryside locations, preferably in a large shed with a strong breeze passing through it. In this way, the production masters can position the reaction so that they are upwind from the fumes.

The countryside location has the further advantage that tanks of anhydrous ammonia are not at all out of place in such a location. In every agricultural area, tanks of anhydrous ammonia ply the roads all through the growing season. Farmers use it for nitrogen fertilizer on their crops, especially corn. The local co-op hauls out the tank to the farmer, who then applies it to his crops at his leisure. The implication of this is obvious. A well thought out large-scale meth production scheme would center upon renting some nondescript piece of land, planting some corn on it, and then getting a tank of "anhydrous" to fertilize the crop. The resulting product will pay much better than corn. A less well thought out plan would involve getting a tank of anhydrous ammonia from a chemical or welding supplier and taking it to a countryside location for further use. In either case, the ammonia is of the same grade.

Farmers often leave tankers of anhydrous ammonia parked out in their fields overnight. Clandestine cookers have noticed this, and often walk out into a field to tap a little anhydrous out of the tanker. An emptied propane cylinder is a favorite container for the formerly dull-lived ammonia. This container is convenient because it has the need valves and the ability to withstand pressure already built in. Some states have made it a felony just to be caught stealing anhydrous. Check the local laws, and beware! Some areas send police patrols wherever anhydrous tankers are parked. Anhydrous ammonia is some god-awfully fuming and dangerous stuff to deal with. Always stay upwind from it, and have a gas mask which covers the eyes as well!

Locks of various types are sometimes placed on the valves of these tankers when they are parked overnight. They have the obvious function of frustrating people wandering out to them to tap off a few gallons of anhydrous. A patent has just been filed for a new locking mechanism, and this patent is so recent that it has only been granted an application number as of the Summer of 2004. The application number is 20030234043, and you can read the hilarious hate filled text just by going to the Patent Office website at www.uspto.gov and typing in the application number. The guy is such a dope he calls methamphetamine "methylene" in the intro to his patent. He does, however, provide a public service with his drawings of typical valve systems on tankers and how he would like to place his locks. A couple of his drawings are reproduced here for your enlightenment.

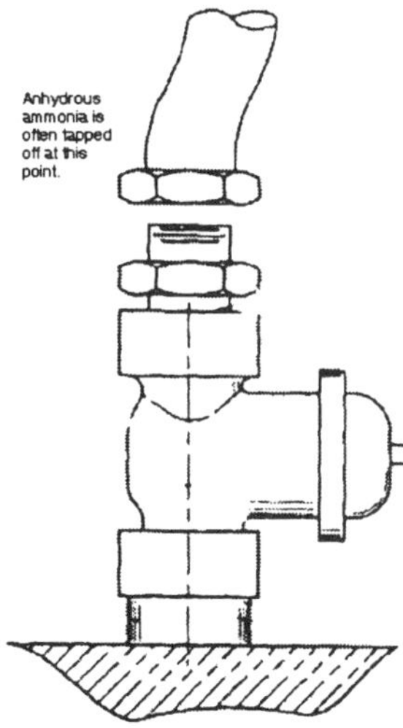

Figure 36a

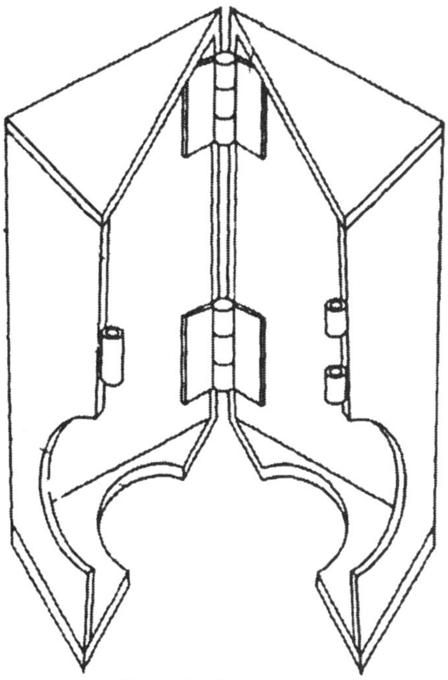

The valve locking cage

Figure 36b

This method of making crank is based on the research of Gary Small and Arlene Minnella as published in the *Journal of Organic Chemistry*, Volume 40, pages 3151 to 3152 (1975). The article is titled "Lithium-Ammonia Reduction of Benzyl Alcohols to Aromatic Hydrocarbons. An Improved Procedure." It results in the 100% conversion of ephedrine, pseudoephedrine or PPA in a reaction time of 10 minutes or so.

This method requires the use of the free base rather than the hydrochloride salt. Both the hydrochloride salt and water act as quenchers to the so-called dissolved electrons formed by either lithium or sodium metal in ammonia. This quenching activity could probably be overcome by using more lithium or sodium metal in the reaction, but this is wasteful of a very valuable commodity.

Lithium ribbon currently sells for about $33 per 25 grams, and sodium sticks sell for about $70 per pound. These prices won't break a cooker, but purchasing large amounts of them may result in unwanted attention. Sodium metal can be easily made at home from lye, using DC current in a version of the Down's cell. How to do this will be covered at the end of this section.

A detailed procedure for disassembling lithium batteries to get the lithium metal contained within them is given in *Advanced Techniques of Clandestine Psychedelic & Amphetamine Manufacture.*

This method works equally well with ephedrine, pseudoephedrine, or phenylpropanolamine as the raw material. A high-quality product is obtained that doesn't cause the dreadful hangovers that one will experience from crude undistilled meth made from the HI/red phosphorous method.

The need for free base in this reaction is no real problem. The toluene or mineral-spirits extract of the pills contains the free base in solution. This can just be added directly to the reaction mixture. This in fact saves the added work of bubbling HCl through the toluene or mineral spirits solution to precipitate the hydrochloride. Correspondents also indicate that the ether or THF used as co-solvent in the reaction mixture with ammonia can be entirely replaced with toluene or mineral spirits. Of these two, mineral spirits is preferred, such as, for instance, Coleman camper fuel or naphtha. One could also consider using ether starting fluid as the extractant for the free base ephedrine or pseudoephedrine obtained from the pill extraction methods given in this book. Then by drying any traces of water out of the ether extract with sodium sulfate or other drying agents, one would have a water free ether extract containing ephedrine or pseudoephedrine free base ready for conversion to meth.

With a supply of free base in hand, it is now time to consider the lithium metal in ammonia reduction method. A very good review of this procedure can be found in the book *Reduction: Techniques and Applications in Organic Syntheses,* by Augustine, pages 98 to 105. At the heart of this method is the fact that lithium metal, or sodium metal, or even potassium metal can dissolve in liquid ammonia to form blue-colored solutions that have powerful reducing properties. Such solutions are often referred to as "dissolved electrons." These solutions are stable unless water gets in them, or unless they are contaminated with iron from the ammonia tank. When the free bases of ephedrine or PPA are added to these "dissolved electrons," they are quickly and easily reduced to meth or dexedrine respectively. To do the reaction, a 3000 ml round bottom 3-necked flask is set inside a Styrofoam tub. The purpose of the tub is to provide insulation, because once liquid ammonia gets out of the cylinder it starts to rapidly boil away until the liquid is lowered to its boiling point of -33° C. This boiling can be kept under control by adding dry ice to the tub. If a cylinder of ammonia is being used, it is a good idea to cool it down before use by putting it in a freezer. With a tank from the co-op, this is not practical. To get the liquid ammonia out of the tank or cylinder, either clear plastic tubing or rubber tubing is placed over the exit valve of the tank or cylinder, and run into the 3-necked flask. Use of metal, and especially copper, is to be avoided. Then the cylinder is tipped upside down, so that the valve is at the bottom of the cylinder. This assures that liquid comes out rather than gas. Next the valve is cautiously cracked open, and liquid ammonia is run into the flask until it is about ½ full. It will quickly boil away until the volume of the ammonia is down to about 1000 ml, and then more slowly because the ammonia has cooled to its boiling point. Then, wearing rubber gloves and eye protection to keep the fumes out of the eyes, a magnetic stirring bar is placed in the flask, and the tub is put on a magnetic stirrer, and stirring is begun. Now 7 grams of lithium metal is put into the flask. Lithium usually comes in the form of turnings inside a sealed glass ampule under inert atmosphere. It can be used directly as such. If lithium wire is being used, it should be cut into short lengths, and rinsed off with petroleum ether prior to use. The lithium metal quickly dissolves, forming a blue solution. Next, 500 ml of tetrahydrofuran is added to this solution. The purpose of the THF is to aid in the dissolution of the ephedrine or PPA which is to be added next. I can see no reason why anhydrous ether can't be used instead of THF, if it is easier to obtain. Many Backwoods Cookers get by with just using Coleman's Camper fuel or naphtha or toluene as the co-solvent. Next 55 grams of ephedrine (or 50 grams of PPA) is dissolved in 500 ml of THF or ether or naphtha or toluene, and this solution is added to the lithium in ammonia solution over a period of 10 minutes.

Some care should be taken to make sure that this solution of the free base of ephedrine in solvent is pretty much free of water, which would quench the dissolved electrons. If one has, for example, pill extract dissolved in camper fuel, the simplest way to assure dryness of this solution would be to add a couple of grams of drying agent such as magnesium sulfate to the extracted solution, stirring it around for a few minutes, then decanting the extract solution off the settled drying agent. Magnesium sulfate drying agent can be made by pouring a layer about ¼ inch thick of Epsom salts into the bottom of a glass baking dish, and baking at 450° F in an electric oven for about an hour. Allow the dish to cool to the point where you can handle it with oven mitts, then pour the baked Epsom salts into a glass jar and seal the top. Another way to dry the water out of such a solution would be to boil it for a few minutes to drive off water.

After allowing the reaction to proceed for an additional 10 minutes, the reaction is quenched by slowly adding water to the ammonia. This is done dropwise at first, and then more rapidly until the blue color disappears from the ammonia solution. The flask is then taken out of the Styrofoam tub, and the ammonia is allowed to evaporate overnight. Back Woods Cookers generally omit the water add, and just let the ammonia evaporate overnight. They find this to be safer and to give a more pleasing product. When the ammonia is gone, some more water is added to the remaining ether (or THF) solution to dissolve the salts of lithium in the bottom of the flask. After separating the water layer, the ether layer is dried using anhydrous sodium sulfate, and the meth or benzedrine is obtained as the hydrochloride salt by bubbling HCl gas through the ether solution as described back in Chapter Five. Distillation is unnecessary because of the lack of formation of byproducts in this reduction. It would just be a colossal waste of ether. When using naphtha or toluene as the solvent, drying before bubbling with HCl isn't necessary. Simply let the extract sit for a couple of hours to shed entrained water, then pour it into a clean, dry glass container before bubbling with dry HCl.

An alternative procedure has become popular among clandestine cookers. In this alternative procedure, liquid ammonia is first added to the reaction vessel, followed by free base dissolved in solvent. Some go so far as to add solvent and then ground-up pseudoephedrine or phenylpropanolamine pills. That this works is a testimony to the power of this alternative procedure. When simply using ground up pills, they generally have to stir the reaction using a wooden stick. Otherwise the surfaces of the pill particles gunk up, and complete reaction isn't achieved. A plastic 5 gallon pail is the standard reaction vessel used by these Backwoods Chefs. Simply grinding up the pills and tossing them in can no longer be used with the gunked up pills now available. The info is given solely for historical reference.

Then to the liquid ammonia containing solvent and ephedrine, pseudoephedrine or phenylpropanolamine, they add lithium metal obtained by taking apart lithium batteries. It takes roughly five minutes to dissect a lithium battery and pull out the lithium metal. See *Advanced Techniques of Clandestine Psychedelic & Amphetamine Manufacture* for the details. The extra-long-life lithium battery contains a bit under .4 grams of lithium metal. They just take apart the batteries one at a time, and add the lithium metal to the pot. When using ground up pills rather than pure pill extract, stirring with a wooden stick is done between adds of lithium. They continue adding the lithium until the solution takes on a blue color, rather than the blue color from the dissolving lithium being rapidly sucked up by the solution. Once the blue color persists for a bit, the addition is complete. Then the anhydrous ammonia is allowed to evaporate away overnight and processing is done as usual. A smoky campfire can help to cover the smell of the evaporating anhydrous ammonia.

This variation probably works better because one isn't relying upon the stability of a pot full of dissolved electrons. As they form, they can just go on to react with ephedrine. In this way, the reaction becomes more tolerant to the presence of potential quenching agents. People like this variation.

One may justifiably ask now, "How is this such a great mass production method, when one is only getting 50 grams of product out of each batch?" The answer is that the work can easily be organized so that one batch after another is quickly turned out by this method. Each individual batch only requires a few minutes of attention. After one flask is filled with ammonia, another may be set up and filled, resulting in a virtual assembly line procedure.

Before moving on, there is a possible complication which must be addressed. This is the possibility that a tank of ammonia may only be putting out ammonia gas, rather than spewing liquid. This is no great hassle. In that case, the 3000 ml 3-necked flask is well packed in dry ice, and rubbing alcohol poured on the dry ice to create a very cold bath. When the ammonia gas hits the very cold flask, it will be condensed to a liquid. This may actually be a better procedure because it will assure that the ammonia does not have dissolved iron in it from the tank. Iron interferes with lithium in ammonia reductions.

Clandestine cookers should also be aware of another new Patent which is also so recently filed that it doesn't have a Patent number, only an application number. In this case the application number is 20040049079. The evildoers behind this Patent are more malicious than the patent application mentioned previously, and they are also better educated. The text of their Patent, however, is equally hateful, so do read it for fun. They propose to purposely incorporate a little bit of various substances into the anhydrous ammonia used to fertilize crops, so that the dissolved electrons we want to do the reaction to make meth are not stable in solution. The best substance of all they tested was an iron compound called ferrocene, and since iron can be considered a trace plant nutrient, they are really hot to get their Patent to market and make a killing. If in the future, anhydrous from tankers just refuses to produce that "royal blue" color of dissolved electrons, one might consider the possibility that it has been laced with ferrocene. Luckily, ferrocene is a solid that will not distill with anhydrous ammonia. The purification procedure given in the previous paragraph will remove it, and life will be good again.

Similarly, in some areas, a pink dye is being added to the anhydrous. This pink dye gets on the cooker, and takes days to wash off. It also gets into the product and colors it. This dye can also be removed from the anhydrous ammonia by using the distillation purification procedure.

Sodium metal can be used with just as good of results as lithium. The higher atomic weight of sodium requires that 23 grams of sodium metal be used instead of the 7 grams of lithium used in the preceding example. Potassium metal probably works too, but is not so common as sodium metal.

Before leaving this topic, a couple of issues should be addressed. Issue number one is the very bad effect that the presence of water in the reaction mixture has upon the success of the reaction. Lower levels of water will quench the dissolved electrons, preventing that beautiful blue color from forming in the reaction mixture. This low level of water contamination can probably be overcome by using more lithium metal.

At higher levels of water contamination, the lithium metal just seems to fizz away in the liquid ammonia, forming hydrogen gas and lithium hydroxide from the water contaminant. At this level of water contamination the batch is guaranteed to be a failure.

Where is this water coming from? Assuming that the pill extract has been dried of water, either by use of a drying agent or distilling the water off the extract as the water-solvent azeotrope, then the water can only be coming from the "anhydrous ammonia" used. It must not be so anhydrous, so get another source of anhydrous. Low levels of water contamination can also come from absorption of water from the air. Liquid ammonia will pull water out of the air. This method shouldn't be done on very humid days.

Issue number two is the smell produced by this reaction. If this is going to be tried in any place other than a remote country location, the ammonia fumes will alert anyone nearby to the goings on. They'll also knock you over without really good ventilation. The fumes can be held down during the course of the reaction by cooling in a dry ice bath, but after the reaction, the liquid ammonia must be allowed to evaporate away. There comes a cloud of ammonia fumes! On a small scale, these can be sucked up with an aspirator, and flushed down the drain. Simply run some clear plastic tubing from the aspirator to the top of the reaction vessel, and turn on water flow to the aspirator. This will control the smell from small-scale batches.

Another way to address the strong smell and actual physical assaultiveness of liquid ammonia is to replace it as the material in which the dissolved electrons are generated. The most clandestine-suitable substitute for liquid ammonia is ethylenediamine. This substance was mentioned earlier in the Knoevenagel reaction section. It can be obtained from laboratory chemical suppliers for around \$20 per liter. It is also used in the electroplating industry as a complexor in nickel stripping solutions. Component A of these strips is m-nitrobenzene-sulfonic acid, component B is ethylenediamine. This industrial grade of ethylenediamine will cost about \$20 per gallon. See the *Metal Finishing Guidebook and Directory* under "stripping solutions," for suppliers.

The industrial grade of ethylenediamine should first be fractionally distilled (boiling point 116° C) to see if there is a significant amount of water in the material. We do this by looking for the toluene-water azeotrope, which boils at 85° C. Mix two parts ethylenediamine with one part toluene, and fractionally distill at atmospheric pressure through a good fractionating column, such as a claisen adapter packed with broken glass. If an insignificant amount distills at the azeotrope temperature, there isn't much water. If a fair amount does distill, the solution can be dried by distilling off the azeotrope. The toluene left in the mixture can just be left there as co-solvent for the reduction.

If only a small amount of water was found in the industrial grade ethylenediamine, then in the next run drying can be done by letting the ethylenediamine sit in contact with KOH pellets for about half a day, then distilling the dried ethylenediamine. The amine has to be dry, because water really fucks with this reaction! Once distilled, the ethylenediamine should be stored in a tightly stoppered bottle to prevent absorption of water from the air.

To do a reduction using ethylenediamine in place of ammonia, I would follow the method given in the *Journal of Organic Chemistry*, Volume 22, pages 891-4. To get the solid lithium metal, such as would be found by taking apart a lithium battery, to dissolve into the ethylenediamine and form the blue-colored dissolved electron solution, one must heat the ethylenediamine. Lithium powder will dissolve at room temperature, but the battery anodes now on the market will not.

Into a 2000 ml flask, I would put about one quart of ethylenediamine. I would then flush out all the air using argon or nitrogen gas from a cylinder. Argon is commonly used for TIG welding. I would then stopper that flask to keep air from getting back in. The *Journal* article tells us that lithium metal will not dissolve into ethylenediamine until the ethylenediamine is heated to around 70° or 80° C. I would add about 3½ grams of lithium metal to the flask (roughly 10 battery anodes). I would then heat the flask using boiling hot water. As they dissolve, some fizzing occurs. A cork would be a good stopper because it will allow the gas to escape. Once the lithium metal has dissolved to produce the blue solution we need, I would then cool down the solution in a cold water bath.

Once roughly room temperature has been reached, I would then add, with stirring, pseudoephedrine or ephedrine to the mixture until enough has been added to make the blue color disappear. That should be around 25 grams of pseudoephedrine or ephedrine. Care should be taken to maintain the atmosphere of argon or nitrogen during the adding of the ephedrine because air reacts with the dissolved electrons.

In what form would one add the ephedrine? The best way to my knowledge of the matter would be to add the ephedrine or pseudoephedrine as a free base in solution with toluene. Toluene is one of the few solvents that dissolve into anhydrous ethylenediamine. Ether will not, nor will naphtha/camper fuel. If one doesn't get mixing of the two solutions, one will not get the reaction we want. One could also add the ephedrine in the form of the hydrochloride crystals without any solvent added to the mixture.

Once the blue color has been consumed, one would be left with the problem of how to recover the meth from the mixture. The best way to my thinking would be to pour the mixture into about 5 quarts of water. Then after some stirring, add some toluene to the mixture and extract out the meth. If the ephedrine was added in toluene solvent, one wouldn't have to add so much. Ethylenediamine doesn't dissolve in toluene when water has been added. I would then wash the toluene extract with some fresh water, and finally bubble the toluene extract with dry HC1 to get the meth hydrochloride product. Others have suggested the possibility of distilling off the ethylenediamine, preferably under a vacuum, and then taking up the residue with toluene, washing the extract with water, and then bubbling dry HC1 through it. This method should work, and not stink to high heaven.

If the clandestine cooker wants to substitute sodium metal for lithium, he may find it difficult or risky to buy sodium metal through standard channels, so a procedure to manufacture sodium metal from household items is of definite value. A process called the Downs cell has been in use since the early days of this century to make sodium on an industrial scale. In its early versions, sodium hydroxide (lye) was the raw material used, but in later versions this was replaced with salt. By using salt instead of lye, another valuable material — chlorine gas — was also obtained. In this procedure the earlier version will be used because we don't want the poisonous chlorine, and because lye has a much lower melting point.

Some basic precautions must be followed with this cell. The last thing someone wants on their skin or in their eyes is molten lye or sodium metal, so gloves, protective clothing, and a face shield must be worn. Also, molten lye gives off fumes, especially when a current is being passed through it, so good ventilation is mandatory.

To make sodium metal, a steel or copper pan about 1 pint in capacity is placed on a stove burner. A gas stove shouldn't be used here because of its open flame. Also, we want to take advantage of the insulation on the electric heating element. An aluminum pan can't be used, nor is Teflon-coating acceptable.

Now, into this pan put lye until it's ¾ full. Begin heating this lye at medium to medium-high heat, and melt it. More lye can then be added to get the volume of lye back to about ¾ full.

When the lye has melted, the standard DC electrochemical cell wiring is done. A graphite anode is inserted into the molten lye. It should be about 1 inch away from the side of the pan, and it must not touch bottom, or a short will occur. Clamp it in place with a ringstand and insulated clamp. Make an electric contact with the graphite anode by wrapping some bare copper wire around the top of the graphite anode, and twisting it down tight. Run a wire from the positive pole of the DC source, through an amp meter, and to the graphite anode. Then run wiring from the negative pole of the DC source to the pan near where the graphite anode is located.

Now turn on the DC power source. The best source would be a 0-110 volt transformer. Also workable would be a DC arc welder with variable voltage control. One could also hook a couple of car batteries in series to raise the applied voltage to 24 V. I have heard from reliable sources that 12 volts isn't enough to push current through the molten lye.

Now turn up the applied voltage from the DC source until a current of one or two amps flows. This is enough current to give a good rate of sodium production. How much current can be passed is largely limited by how big the graphite anode is. As in so many other things, bigger is preferable to smaller. The molten lye obeys Ohm's law, so increasing voltage gives higher current flow.

As the current flows, little globules of shiny metallic sodium should start floating up to the top of the molten lye. They will mostly show up along the edge of the pan near the handle. Sodium is very light, so it floats. As they appear, scoop them up with a stainless-steel spoon. Separate them from the molten lye also scooped up by rolling them around in the spoon, and once separated, pour them into a jar of mineral spirits. Sodium metal reacts rapidly with air to make sodium peroxide, so this should be done as fast as they appear. The mineral spirits will protect the product. It's important to keep the lye out of the product, as this would mess up its intended use in the reaction. It's probably quite possible to separate the two later by melting the sodium (it melts below the boiling point of water) and pouring it off the entrained lye, but this hazardous procedure is best avoided in the first place.

Current flow is continued until one has as much sodium metal as needed. The serious experimenter will want to try using iron or steel as the anode if graphite can't easily be found. The serious experimenter will also note that the amount of current that he can pass through the molten lye is related to the size of the anode used. After a certain current density, the electricity gets wasted by fizzing at the anode, which produces water and hydrogen.

Before leaving this topic, another warning should be included. When scooping up the sodium metal with the spoon, don't make contact with both the anode and the pan at the same time! You will short the cell, and make sparks that will startle you to no end! This could lead to spills and messes and burns! Be careful!

Method 2: Wolff-Kishner Reduction

This method of directly reducing ephedrine, pseudoephedrine, or phenylpropanolamine to meth or dexedrine uses hydrazine hydrate as the reducing agent. The Wolff-Kishner reduction is generally used to deoxygenate ketones to the corresponding hydrocarbon, but in this case, it can be used on these particular substances to reduce them. No doubt this is because the benzyl alcohol grouping has a ketone nature due to tautomerism.

I came across the reference for this reduction several years ago, but have been unable to find it since. I've also received mail from the pen from cookers telling me that they have had some success with this reduction. This method should be considered a minor pathway to meth, as there are better and more convenient routes.

The Wolff-Kishner reduction has the advantage of not producing great plumes of stink. It could likely be done in an urban setting without arousing the suspicions of nosey neighbors. Further, the reactants are only moderately expensive, and not tightly controlled at present. Fair amounts of product can be turned out at a rate of one batch per day.

The disadvantages of this method are twofold. First, hydrazine is a carcinogen. The chemist must wear gloves while doing the reaction, and do a careful clean-up when finished. If any should be spilled on the skin, a serious, prolonged, and immediate shower is called for. Care must further be taken that the fumes of hydrazine are not breathed in, as this could cause the same problem. Ever try giving your lungs a shower? Fumes in contact with the eyes can so burn the surfaces of them that blindness results! To make matters worse, hydrazine is on the FBI explosives scanning list. This is because mixtures of ammonium nitrate and hydrazine are very powerful explosives. The other disadvantage to using this method is that the free bases must be used. This necessitates the free basing and distillation procedure described in Method 1.

The mechanism by which this procedure works involves first the formation of a hydrazone by reaction between the ephedrine and hydrazine. Then at the high temperatures at which this reaction is done, the hydrazone loses nitrogen (N_2) to form meth. This is illustrated below.

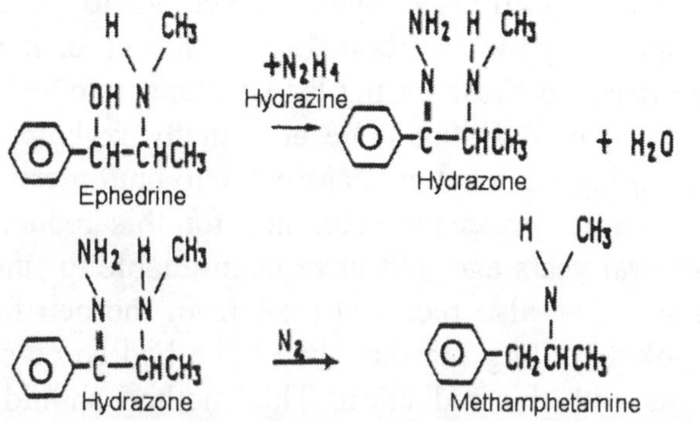

To do the reaction, a 3000 ml round bottom flask is placed on a buffet range, and then 1500 ml of diethylene glycol and 336 grams of KOH (potassium hydroxide) pellets are put in the flask. Next a condenser is attached to the flask, and water flow is begun through it. Gentle heating of the flask is now begun, with occasional swirling of the flask to try to dissolve the KOH pellets. The operator must be ready here to quickly remove the buffet range, because once the solution warms up, and the KOH pellets start to dissolve, a great amount of heat is released which could cause the solution to boil wildly and squirt out the top of the condenser. Since diethylene glycol has a boiling point of 245° C, this would definitely not be good stuff to be splashed with. Eye protection is, of course, necessary. The heat source is periodically removed, and then reapplied until the dissolution of the KOH pellets is complete.

Once the KOH pellets have dissolved, the heat is removed, and the temperature of the solution is allowed to fall to about 80° C. Then 300 ml of hydrazine hydrate (85% to 100% pure material is OK) and either 303 grams of PPA free base or 332 grams of ephedrine free base is added to the flask. The condenser is then immediately replaced, and the mixture is heated with great caution until any exothermic (i.e., heat-generating) reaction has passed. Then stronger heat is applied to maintain gentle boiling for one hour.

Now heating is stopped, and as soon as boiling ceases, the condenser is removed, and the flask is rigged for simple distillation as shown in Figure 11 in Chapter Three. The stillhead should have a thermometer in it reaching down into the middle of the liquid mass in the flask. A cork or rubber holder for this thermometer is unacceptable because hydrazine attacks these materials. The holder must be made of all glass only.

Now the heat is reapplied, and distillation is commenced sufficiently slowly that the froth does not rise out of the flask. Froth can be broken up by occasional applications of weak vacuum, as mentioned back in Chapter Five. When the temperature of the liquid has reached 200° C or so (around 200 ml of distillate will have been collected by that point), the heating is stopped. Once boiling ceases, the stillhead is removed, and the condenser is reinserted into the flask. Now heat is reapplied, and the mixture is boiled gently for 3 additional hours.

The reaction is now complete, and it is time to get the product. The heating is stopped on the flask, and once it has cooled down, the contents of the flask are poured into 2000 ml of water. The 200 ml of distillate obtained earlier is also poured into the water. This mixture is stirred to get the hydrazine out of the meth layer which floats on the top, and into the water. The solution of KOH in water makes the water fairly hot. Once it has cooled down, 500 ml of toluene is added, and the mixture is shaken. A one gallon glass jug is a good vessel to do this in. The top layer of meth dissolved in toluene is then separated and distilled as described earlier. The yield is 250 to 275 ml of meth. If a careful fractional distillation is not done, the product may be contaminated with a small amount of hydrazine. This is definitely not good, and may be avoided by shaking the separated meth dissolved in toluene layer with a fresh portion of water.

Method 3: Direct Reduction of Ephedrine With Palladium

These methods are pretty similar to the indirect reduction of ephedrine, racephedrine, pseudoephedrine and PPA presented earlier. The difference here is that these variations on that theme are one-pot methods. For example, chlorination and reduction can be done simultaneously in a solution containing dry HCl gas and palladium in a hydrogenation bomb. Other variants use sulfuric, perchloric or phosphoric acid to either first form an ester with the ephedrine, or whatever, which is then reduced to meth in the hydrogenation bomb, or these same substances act as promoters to cause the direct reduction of the benzyl alcohol (ephedrine, or whatever) to the meth or benzedrine. The exact mechanism of how this actually works is a matter for debate in the patent literature.

These direct routes have the advantage of using very common materials as feed stocks. The various chlorinating agents given in the indirect-reduction section aren't particularly common lab chemicals. Also, chloroform is becoming less commonly used in industry with the new ozone-depletion rules.

There are a few drawbacks to this method. First and foremost, in one of these procedures the contents of the bomb must be heated to about 80-90° C during the reaction. This leads to the danger that the champagne bottle hydrogenation bomb may crack and burst due to heat stress. This is a possibility even if the outside of the bottle is coated with fiberglass resin. Another problem is the high cost and suspiciousness of purchases of the various palladium catalysts used in these methods. This can be avoided by getting one's palladium chloride from the plater's supply outfits mentioned earlier in the palladium-catalyst section. One note on this source of palladium. Platers tend to use archaic or weird technical language. For example, a company may offer what he calls 60% $PdCl_2$. This refers to the Pd content of the $PdCl_2$, and indicates that it is actually quite pure.

A big improvement to this procedure is to use "The Poor Man's Hydrogenation Device" detailed in the second edition of *Practical LSD Manufacture* and also in *Advanced Techniques of Clandestine Psychedelic & Amphetamine Manufacture*. With this improvement, the danger of breaking champagne bottles under heat is removed. A genuine Parr hydrogenation device from a scientific supply outfit is even better.

For this particular application, I would use a fire extinguisher having a bottle made out of a non-magnetic stainless steel. The technical sheet that comes with the extinguisher will state what the bottle is made out of, and a magnet will tell you if it is a non-magnetic alloy. A non-magnetic stainless steel will pass a magnetic field, so magnetic stirring will be possible.

The reason for choosing stainless steel over aluminum in this application is the corrosive nature of the hydrogenation mixtures used in these direct reductions. Acetic acid is the solvent used in these procedures, and it will dissolve aluminum. The promoters used in these procedures, either HCl, sulfuric or perchloric acid, all eat away at aluminum as well. If the coating should fail inside the bottle, then a hole could be rapidly eaten through the pressure bottle if it were made of aluminum. The light weight of the bottle will quickly tell you if it is made of aluminum rather than stainless, although a magnetic field passes much better through aluminum than stainless steel.

To coat the inside of an aluminum pressure bottle, one has the choice of Teflon and Teflon-based paint, or high phosphorous electroless nickel. To pursue this latter alternative, one goes to the Yellow Pages and looks under electroplaters. Ask them if they do electroless nickel, if they have a high phosphorous (phosphorus content in alloy of about 12%) plating bath, and if they are set up to plate over aluminum. To plate aluminum, one must be prepared to enter the fumes of hell.

If they answer yes to the above questions, you are set to go. Say you want the plating to make the extinguisher match your techno décor or some such cock and bull story. I work in this business and I've heard it all by now. Make sure that the inside of the bottle gets plated, as that is actually the priority item, not the outside. Also ask that threads be masked off so that you can screw the top back on the extinguisher after plating. Also ask for a plating thickness of about one thousandth of an inch for good protection.

Let's look at variation number one of this procedure.

To do this reaction, the chemist first prepares palladium black catalyst. This is done as follows: In a 2000 ml beaker, 50 grams of palladium chloride is dissolved in 300 ml of concentrated hydrochloric acid (laboratory grade, 35-37%). Once it has all dissolved, it is diluted with 800 ml of distilled water. Next, the beaker is nestled in a bed of ice that has been salted down. This is an ice-salt bath. The contents of the beaker are stirred occasionally, and once it is cold, 300 ml of 40% formaldehyde solution is added with stirring. After a few minutes, a cold solution of 350 grams of KOH in 350 ml distilled water is added slowly over a period of 30 minutes. The palladium solution must be vigorously stirred during the addition. Now the beaker is removed from the ice and warmed up to 60° C for 30 minutes with occasional stirring during the heating.

When the heating is complete the beaker is set aside to cool and to let the catalyst settle. Once the catalyst has settled, the chemist pours off as much of the water solution as possible, without losing any catalyst. Then fresh distilled water is added to the beaker, the catalyst is stirred up to wash it off, then the chemist lets it settle again and pours off the water. This washing is repeated a total of six times. Finally, the catalyst is suspended in a bit of fresh distilled water, and filtered, preferably through sintered glass to be sure of catching all the catalyst. Any catalyst still clinging to the sides of the beaker are rinsed down with water and poured in with the main body of catalyst. It is wise to rinse off the catalyst again with still another large portion of water while it is in the filtering funnel. This process yields 31 grams of palladium black catalyst, once it has dried. It is important that the catalyst be allowed to dry completely, because the presence of water in the reaction mixture is to be avoided.

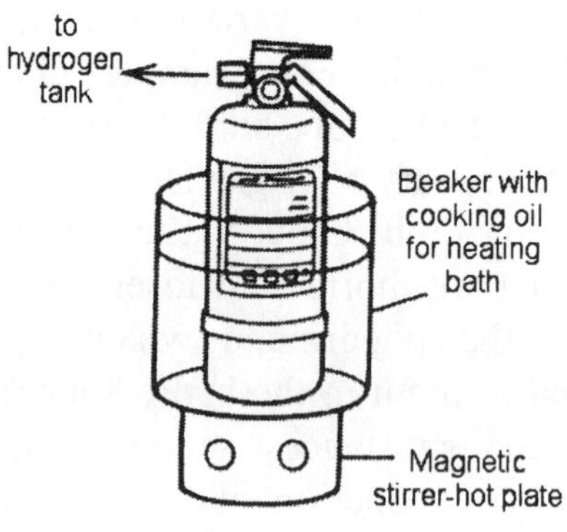

Figure 37

With a supply of catalyst on hand, the chemist can move on to production. To begin, 600 ml of glacial acetic acid is poured into a 1000 ml beaker. Now the glassware is set up as shown in Figure 20 back in Chapter Five. The glass tubing is led into the acetic acid, and bubbling of dry HCl gas into the acetic acid is begun as described in that chapter. It is a good idea here to magnetically stir the acetic acid solution during the bubbling. The whirlpool formed will help the bubbles of HCl gas to dissolve in the acetic acid, rather than escape and waft away on the breezes. This bubbling is continued until the acetic acid solution has gained 30 grams in weight.

Next, this acetic acid-HCl mix is poured into the 1.5 liter hydrogenation device along with 60 grams of either ephedrine, pseudoephedrine or PPA (sulfate or HCl salt is okay for any of these), and 50 grams of palladium catalyst. Since the mixture is going to be magnetically stirred, a magnetic stirring bar, of course, is put in the bottle. Now the apparatus is set up as shown in Figure 32 in Chapter Eleven. The air is sucked out of the bottle as described in that chapter, and then hydrogen is put into the bottle at a pressure of 30-40 pounds per square inch. Stirring is then begun, and the oil heating bath warmed to the 80-90° C range. Hydrogen absorption begins, and fresh hydrogen is put into the bottle to keep the pressure in the prescribed range. In an hour or two, hydrogen absorption stops, and the reduction is complete.

The heating is then stopped, and the stirring is halted. The hydrogen is vented outside as described back in Chapter Eleven, and the product solution is carefully poured out of the bottle, taking care not to pour out the palladium catalyst. If any comes out, it is filtered, and the palladium returned to the bottle for the next run.

The product mixture is poured into a 1000 ml round bottom flask along with a few pumice chips, and the glassware is set up as shown in Figure 11. The chemist distills off 500 ml of acetic acid (b.p. 118° C). This acetic acid can be used over and over again just by bubbling some more dry HCl through it to dry the solution and to recharge its HCl content.

The residue left in the distilling flask has the product. Once it has cooled down, lye water is added to it, and it is shaken vigorously. The solution should be strongly basic. Now toluene is added, the top layer separated off, and this top layer is distilled as described so often in this book to yield a little over 50 grams of meth (or dexedrine if PPA was used). This is about 95% yield.

If there is a high boiling residue in the flask, it's likely to be the acetyl amide of meth or dexedrine formed by heating with the acetic acid. This can be hydrolyzed by boiling with concentrated hydrochloric acid, then making the solution basic with lye, and extracting out the product as described in Chapter Five.

You thought that method was a hassle? You ain't seen nothing yet. Check out the method given in *Journal of Medicinal Chemistry,* Volume 9, pages 966-67. In this variation, the phosphoric acid ester of ephedrine, pseudoephedrine is reduced to meth or phenylpropanolamine is reduced to dexedrine by hydrogenation with Pd on carbon catalyst. The phosphoric acid ester isn't made just by reacting the ephedrine with phosphoric acid. No, that would be too easy. Instead, phosphorus oxychloride ($POCl_3$) must be used. The free base of the ephedrine, or whatever, must be used to make the ester. Here's how they did it:

165 ml of freshly distilled ephedrine free base was put in a flask fitted with a drying tube. Then with stirring, 300 ml of triethylamine was added along with 5000 ml of benzene. The triethylamine absorbs HCl given off in the reaction. The benzene could be replaced with toluene. Water in the reaction mixture must be avoided. Then with vigorous stirring, 100 ml of $POCl_3$ dissolved in 500 ml of anhydrous benzene is added dropwise at such a rate that the temperature stays below 50° C. Stirring was then continued for 4-5 hours.

Next, the reaction mixture was filtered and the solvent was removed under a vacuum. The residue left in the flask was next extracted with boiling petroleum ether. Several portions of petroleum ether were used to do the extraction, a total of about 6000 ml.

The petroleum ether extracts were next combined and put in a freezer for 24 hours. The product, 2-chloro-3,4-dimethyl-1,3,2-oxazaphospholine 2-oxide precipitates out in the cold, and it is filtered out. It is stable only in a dry inert atmosphere. The yield is about 165 grams.

This product is then put in a flask, along with 5 ml of 1N hydrochloric acid for each gram of the 2-oxide. 1N hydrochloric acid is made by adding 11 volumes of water to one volume of concentrated hydrochloric acid. This reaction mixture is heated on a hot water bath for one hour. The reaction mixture becomes clear. It should be decolorized with some activated charcoal, filtered, and then the solution is evaporated away under a vacuum, with a warm water bath no hotter than 40-45° C to speed along the evaporation. The residue is the phosphate ester of ephedrine. This material will hydrogenate to meth.

The residue can be dissolved in 15 ml of ethanol for each gram of residue, and then add one tenth gram of 5% Pd on C for each gram of residue. It is hydrogenated at room temperature, and they claim no more than atmospheric pressure of hydrogen was required to get 85% yield of meth in about 2 hours.

The hydrogenation mixture was filtered to remove catalyst, and the filtrate was evaporated under a vacuum. The residue was dissolved in water, then the water solution was made strongly alkaline by adding sodium hydroxide solution. The free base was then extracted out using toluene, and this solution can either be distilled to yield the pure free base, or the toluene extract could be directly bubbled with dry HCl to get crystals of meth hydrochloride. That was some pain in the ass, huh?

It's time to move on to some good and convenient hydrogenation methods, isn't it?

See *Chemical Abstracts* from 1949, column 1025 to 1026. Here meth was made in 95% yield by placing 40 grams of ephedrine HCl in 900 ml of glacial acetic acid with 47 grams of 84% sulfuric acid and 10 grams of palladium. Concentrated sulfuric acid can't be used because it breaks down under hydrogenation to H_2S, which poisons the catalyst. 84% sulfuric acid is made by taking 84 grams of concentrated sulfuric acid (46 ml) and adding 16 ml of water. The palladium catalyst specified is Pd wool, but Pd black will work as well. This solution was hydrogenated at about 30 pounds per square inch at room temperature for a few hours until hydrogen absorption ceased to give meth with no change in its stereochemistry.

If the hydrogenation is done at room temperature, a champagne bottle would be good enough to do the reaction in. One has the option of heating the hydrogenation mixture to get a much faster reduction time. See *Chemical Abstracts*, Volume 38, column 1219, for this example. In this case, the heated hydrogenation should be done in "The Poor Man's Hydrogenation Device," preferably one with a non-magnetic stainless-steel bottle coated on the inside with Teflon or high phosphorus electroless nickel.

I'm pretty sure that the mechanism whereby this hydrogenation works is that the sulfuric acid acts as a catalyst for the formation of an ester between the acetic acid solvent and the alcohol group in the ephedrine, pseudoephedrine or PPA. This benzyl ester then undergoes what is termed hydrogenolysis to yield meth or dexedrine. See *Organic Reactions*, Volume 7, for an article on hydrogenolysis of benzyl esters. This reaction mechanism is the basis for the Fester Formula, revealed in *Advanced Techniques of Clandestine Psychedelic & Amphetamine Manufacture*, and also the Advanced Fester Formula, given later in this chapter.

Workup of the reaction mixture to isolate the meth product consists of first either letting the catalyst settle to the bottom, or filtering it out. It can be reused and when it wears out it can be reworked like platinum catalyst.

Now the acetic acid solution containing the meth is poured into a distilling flask, and with the aid of vacuum, the acetic acid is distilled off. If the vacuum isn't really high, and if the condenser and receiving flask are pretty cold, then most of the acetic acid should be recovered for reuse.

Then to the residue left in the flask after most of the acetic acid has distilled away, add a 20% solution of lye in water until the mixture is strongly alkaline. This would be best done in a sep funnel, with shaking between adds of lye solution. You now have meth free base. When the solution has cooled down a bit, add 200-300 ml of toluene and shake to extract the meth into the toluene, which will float on top of the water after the shaking stops. Separate this top layer of meth in toluene, and after any entrained water has settled out, bubble HCl gas through it to get meth hydrochloride crystals. Filter them out, and air dry as usual to get the product.

My commentary on this procedure? That sure is a lot of acetic acid used, a lot of expensive catalyst used, and heating the hydrogenation bomb is a hassle that prevents the use of glass hydrogenation vessels. Also that need for distilling off all that excess acetic acid at the end of the hydrogenation is a lot of work, and it requires that one have a set of distillation glassware to do the hydrogenation. Let's get around all that.

This hydrogenation works by way of the hydrogenolysis of the acetic acid ester of the benzyl alcohol of the ephedrine, or whatever. This hydrogenolysis is very easy, even easier than reducing a double bond by hydrogenation. It's the formation of the acetic acid ester that requires the heating of the hydrogenation mixture. Also, ephedrine hydrochloride isn't very soluble in cold acetic acid. That's why such a large amount of glacial acetic acid is used in the old standard recipes found in the patent literature. Let's get around all the practical cooking problems in one broad stroke just by performing the acetic acid ester prior to hydrogenating the solution.

The general method for producing an ester from an alcohol like ephedrine and an acid like acetic acid is to add a bit of sulfuric acid to the mixture of the alcohol and acid, and then heat the mixture. In spite of the claims of the preceding Patent recipe, almost everybody is in agreement that sulfuric acid in this recipe ruins the yield and potency of the product. The recipe given later in this chapter using perchloric acid instead of sulfuric acid works just fine, however.

The holders of a recent Patent, US Patent 6,399,828 tried sulfuric acid in this method and agree that the use of this substance ruins the yield of product. They ascribe the problem as stemming from bad conversion to the ester when sulfuric acid is in the mixture. It may also be that the very bulky sulfate anion attached to the nitrogen atom of the ephedrine has a steric effect which blocks the reduction of the neighboring alcohol group. This is my favored interpretation of the results.

In the recipes they give in their Patent, they formed the acetic acid ester of ephedrine or pseudoephedrine by reacting it with acetic anhydride. They did it as follows:

Into a 100 ml flask they put 18 grams of ephedrine hydrochloride or pseudoephedrine hydrochloride, 18 ml of glacial acetic acid and 13 ml of acetic anhydride. The mixture was stirred magnetically during the addition of the ingredients, and then with continued stirring, the mixture was heated to 80° C. The mixture cleared as the ephedrine went into solution, and they continued heating at 80° C for two hours.

When two hours had passed, they slowly with strong stirring added 36 ml of heptane to the hot reaction mixture. Heptane could be replaced with naphtha or even gasoline to make the method more clandestine. The addition of the heptane to the reaction mixture precipitated crystals of their product, the acetic acid ester of ephedrine, from the reaction mixture. They continued stirring as the reaction mixture cooled down. It was at this point a slurry. They kept the stirring going overnight at room temperature, and then filtered out the crystals of their product. After spreading the crystals out to dry on a plate, they got around 18 grams of product, which works out to a 90% yield. The batch size in the example can be scaled up or down as desired.

Then to make meth from the ester, they tried two variations of the palladium hydrogenation. Both of them worked great, so pick the method most convenient to you.

In their first example, they did catalytic transfer hydrogenation. They used ammonium formate to pump hydrogen to the palladium catalyst rather than using a pressurized vessel full of hydrogen gas. Into a 500 ml three-necked flask equipped with a condenser, stirrer, and addition funnel, they put 47 grams of the acetic acid ester of ephedrine or pseudoephedrine, 130 ml of water, and one gram of 10% Pd on carbon catalyst. Their catalyst is soaking wet with water, to the extent that half its weight was water. I'm pretty sure that one could also use Pd black catalyst made by adding some sodium borohydride to a water solution of $PdCl_2$, and that Pd on carbon made by the same method with about 5 times the weight of activated carbon in the water solution as compared to the $PdCl_2$ would work just fine.

They made a solution of 15.5 grams of ammonium formate in 20 ml of water, and warmed the contents of their reaction flask to 70° C using a hot water bath. When it had warmed up to 68° C, they took 6 ml of their ammonium formate solution and added it to the reaction flask in two ml portions at 5-minute intervals with good stirring. The mixture started fizzing, and as the fizzing slowed down, they dropwise added the remainder of their ammonium formate solution to the reaction. When the addition of the ammonium formate was complete, they continued stirring the mixture at 70° C in the hot water bath for another hour. Then they let the mixture cool down while continuing to stir.

Several hours later, they filtered out their palladium on carbon catalyst, and kept the catalyst wet before transferring it to a glass bottle for storage under water in their fridge for reuse in the next batch.

Then they took the filtered reaction mixture and put it into a sep funnel. Slowly, they added 20 grams of lye to the sep funnel to free base the meth. The mixture will get warm, and ammonia from the ammonium formate will also be free based. The addition of the lye should be done in roughly 5 gram portions, with hard shaking between each add.

When all the lye has been added, the meth free base should be floating atop the water in the sep funnel. For this size batch, it will be roughly 40 ml. When it has cooled down, one could add about 50 ml of toluene or xylene to the sep funnel and shake to get all the meth drawn together into the top floating toluene layer, and then drain off the water layer, and finally distill to get VERY pure meth free base which can then be mixed with toluene and bubbled with dry HCl gas to get crystals of meth hydrochloride.

Alternatively, one could just add 50 ml of toluene or xylene to the sep funnel that contains the newly free based meth, and shake to get all the meth collected into the toluene solvent. After letting the layers settle, the water layer could be drained away. Pour the toluene layer, which contains the meth into a beaker, and let it set for a while to drain entrained water and to fume off any ammonia which happened to be pulled into the "goods." A water wash of the toluene extract might be very helpful for removing ammonia. When the toluene no longer reeks of ammonia, it can be further diluted with a few hundred ml of toluene, and then bubbled with dry HCl gas to get high quality meth hydrochloride crystals. The yield will be around 80%.

In their second example, they put 12 grams of the acetic acid ester of ephedrine or pseudoephedrine into a Parr hydrogenation vessel similar to The Poor Man's hydrogenation Device. They then added 75 ml of water and a gram of the 10% Pd on Carbon catalyst. They flushed out the air with hydrogen, and then pressurized their hydrogenation device to about 50 pounds of pressure, and stirred or shook the device to get the catalyst into contact with the hydrogen. After about 4 hours, they released the hydrogen from the vessel, filtered out the catalyst, and poured the filtered reaction mixture into a sep funnel. They free based the meth with lye until the pH was 13+, and after cooling they extracted out the meth with toluene or xylene. The water layer was drained away, the toluene layer with the meth was poured into a beaker, and then after it had shed entrained water, it was poured into a clean beaker and bubbled with dry HCl to get crystals of meth hydrochloride. The yield on this run was 60%, so the ammonium formate method is much better, as well as being easier.

Since the fourth edition of this book came out in 1996, I've been hearing the narcoswine rant about some Nazi Meth recipe, and this Nazi Meth recipe has been a topic of speculation on the Internet for some time. These narcoswine aren't exactly clear as to what they mean when talking about this recipe. Could they be talking about the lithium metal reduction in liquid ammonia? That wouldn't fit because that reduction technique was invented long after the Nazis went defunct. The following recipe would fit, as it dates to the correct time period, and was invented by guys named Karl Kindler, Erwin Karg and Ernst Scharfe, and published in *Berichte der Deutschen Chemischen Geshellschaft*. That fits, but since when do narcoswine and reality have anything in common?

See *Chemical Abstracts*, Volume 38, column 1219. In this variation, the 84% sulfuric acid is replaced with 70% perchloric acid. The specified catalyst is palladium on barium sulfate, but I would think any palladium catalyst would work. The reaction mechanism would be the same as in the earlier example, with the perchloric acid acting as catalyst to form the ester between the acetic acid solvent and the ephedrine, and this ester then gets reduced to meth. Care must be taken not to use too much perchloric acid, as this over-hyped-up reaction mixture would also reduce the benzene ring to cyclohexane.

The hydrogenation using perchloric acid as the activator is complete in about ¾ of the time required when using sulfuric acid, but other than that, the reaction is pretty much identical to the earlier example. It can be done at room temperature, or heated to around 80° C to speed up the hydrogenation. Workup of the hydrogenation mixture to get the meth product is the same as in the earlier example given from the 1949 volumes of Chem Abstracts which used sulfuric acid as the "promoter" for the hydrogenation.

In my book, *Advanced Techniques of Clandestine Psychedelic & Amphetamine Manufacture,* I revealed the Fester Formula. This is an electrocatalytic hydrogenation where one takes an ingot of palladium metal, anodizes the surface of the metal in dilute sulfuric acid to coat the surface of the metal with black palladium, and then in an electric cell this ingot of palladium hydrogenates the acetic acid ester of ephedrine, pseudoephedrine or PPA by means of electrically generated hydrogen at the surface of the palladium ingot. This is a good and very low profile method for making stash quantities (a gram or two at a time) of meth. It does, however, have a few drawbacks.

The primary drawback to this method is that the small size of the ingot (submerged face surface area about 6 square cm) limits the batch size to a gram or two at a time. That is fine for stash, but not production of sizable amounts. Drawback number two is the kind of low specific catalytic activity of the anodized palladium metal ingot. These two roadblocks to production using this method are now addressed here with the Advanced Fester Formula.

Starting back in the middle '70s, it was discovered that a black palladium electroplated in a very thin layer upon graphite produced an electrode for electrocatalytic hydrogenations that was much more active than a simple piece of palladium metal. In the scientific literature, the solid piece of metal is termed a mass electrode. Handling and using the mass electrode (an ingot of Pd) is simpler than Pd electroplated onto graphite, but the higher activity and unlimited size of the plated graphite has almost completely displaced the mass electrode in truly modern methods. One has, in effect, Pd on carbon with the plated graphite electrode.

Electroplating graphite with a thin layer of black palladium is no great challenge. Your Uncle's day job is as a chemist in the electroplating industry, so count on me to walk you through the procedure with no hitches. Very small amounts of palladium are consumed in the plating of this electrode, so this is a very economical method as well as being stealthy and clean in both product and waste generated. The very small amount of Pd catalyst used requires that the ephedrine feed for the reduction be very pure to avoid fouling out the catalytic surface.

For some examples to read using this production method, see *Journal of Applied Electrochem-*

istry, Volume 5, pages 125-28 (1975), and *Electrochimica Acta*, Volume 21, pages 449-59 (1976), and for a 50 amp scale up, see *Transactions of the SAEST,* Volume 13, pages 161-67 (1979). The author of all these articles is Krishnan, a pioneer in the use of palladium plated graphite electrodes.

Let's take the case of a piece of graphite with a surface area of about 100 square cm on a side. We only consider the surface area of the graphite which will face the anode in the electric cell, because the current can't reach the back side. See the discussion earlier in Chapter Ten for sources of graphite bars, which can be cut as desired. Graphite rods could also be used; just hook a bunch of them together using copper wire to make electrical contact with the rods. The graphite must be clean to plate properly. Any grease on them should be removed by scrubbing with hot dish-soap water, then after rinsing and drying, a solvent rinse such as acetone would ensure grease and dirt removal. Have clean hands or wear rubber gloves. Foul and polluted graphite can be cleaned by soaking in hydrochloric acid (10%), rinsing and letting dry. This would remove metal contaminants, but good unused graphite will be pretty much free of these materials.

We're now ready to electroplate the graphite with that thin layer of black palladium. The plating solution is made up by starting with 350 ml of 3N hydrochloric acid. This is roughly 90 ml of concentrated (36%) HCl diluted to 350 ml with water. To this acid we add .35 grams of palladium chloride and 1.75 grams of ammonium chloride, and stir until the $PdCl_2$ has all dissolved to give a reddish-brown-colored solution. The palladium chloride can be purchased as such from those suppliers of precious metal salts to electroplaters mentioned earlier, or from photography shops or their suppliers. It can easily be made by the electrical method given in *Advanced Techniques of Clandestine Psychedelic & Amphetamine Manufacture* from an ingot of palladium. See the discussion there for calculating how much $PdCl_2$ is dissolved in the hydrochloric acid solution. One could also calculate how much $PdCl_2$ is in solution by means of current consumed. .35 gr $PdCl_2$ is .002 mole. It is a 2 electron oxidation to $PdCl_2$ from Pd ingot metal. This would require 386 seconds at a current of one amp, assuming 100% efficiency. Let's figure on getting 75% efficiency. Then at a current of one amp on the ingot, that would require 515 seconds (roughly 8½ minutes) to dissolve .35 grams of $PdCl_2$ into solution.

The ammonium chloride is very important to the plating solution, and shouldn't be left out! It acts as a complexor, preventing the hydrogen generated at the graphite during plating from sludging out the palladium solution. You want electroplate on the graphite, not sludge on the bottom of the beaker! If you don't have ammonium chloride right at hand, the same effect could be obtained by adding a couple ml of strong (28% NH_3) ammonia to the plating bath.

We're now ready to plate the graphite with palladium. Wire up your plating cell as shown in Figures 38 and 39.

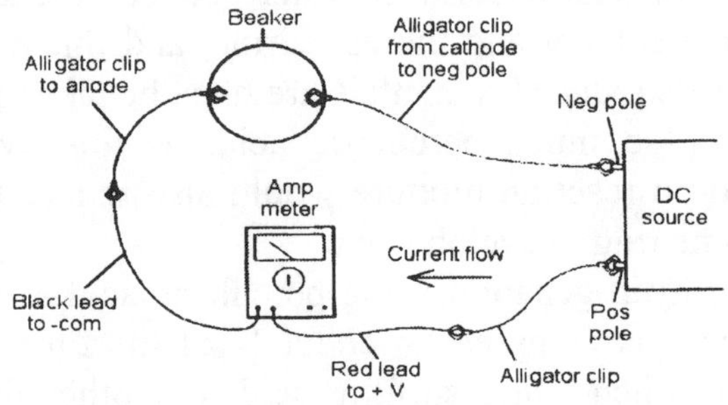

Figure 38

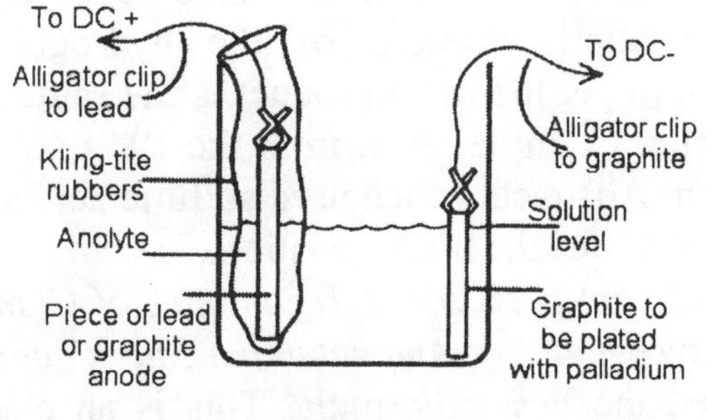

Figure 39

I'll make the assumption that the reader is already familiar with the basic Fester Formula given in *Advanced Techniques of Clandestine Psychedelic & Amphetamine Manufacture*, so I won't go through the complete explanation of all the terms and other details given there. A DC power source is needed to do the electroplating. A DC rectifier, such as a Hull Cell rectifier sold by suppliers to electroplaters for around $500, is perfect. See the *Metal Finishing Guidebook and Directory* for suppliers. A battery can also be used, such as a 1½ volt "D" cell battery. This will deliver about the right amount of voltage and current. This is very important! The Hull Cell rectifier has controls that easily adjust the applied current to the desired level, and meters which display both voltage and current.

Wiring is run from the positive pole of the DC source, to the amp meter. In Figure 38 I show the $50 Radio Shack multi-tester, which works fine for the purpose. Then from the amp meter, wiring is run to the anode. Lead, or preferably graphite, is acceptable to use as the anode. It functions solely as a current pump into the solution, and at the low currents used for plating, its size relative to the graphite to be plated isn't very important. The anode is placed inside two scrubbed Kling-Tite Naturalamb rubbers. One rubber is placed inside the other to give a double wall of separation of the anode and its associated anolyte from the plating solution (catholyte).

The anolyte is a 3-5% by volume solution of sulfuric acid in water. Pour it inside the Kling-Tite rubber, and put some of it in the space between the two layers of rubbers to carry current. The anolyte should fill the rubbers to about the same depth as the plating solution fills the beaker. A clothes pin can be used to hold things in place.

The piece of graphite to be plated should be almost completely immersed in the plating solution, with just enough sticking out of the solution to keep the alligator clip which makes electrical contact with the graphite out of the plating solution. The plating solution will dissolve the clip, and result in an alloy being plated rather than pure palladium, so this is important! The alligator clip from the graphite being plated then goes to DC negative to complete the circuit.

A 10 cm by 10 cm piece of graphite has the face area of 100 square cm mentioned in this example. A 1000 ml to 2000 ml beaker or similar size plastic measuring cup or pitcher are suitable for doing the plating in. If the 350 ml of plating solution doesn't fill up the container enough to completely submerge the graphite to the extent shown in the drawing, switch to a different container, or dilute the plating solution with water or dilute HCl.

A moderate rate of stirring is begun on the plating solution, preferably with a magnetic stirrer. Then the DC power is turned on. With a rectifier, you just turn on the power switch. With a battery, the circuit is completed by making contact with the negative pole of the battery. The current output is adjusted to 2 or 3 milliamps per square centimeter of graphite actually facing the anode. With the 100 square cm graphite example, the output is 200-300 milliamps (.2 to .3 amp).

Output on a DC rectifier is adjusted just by turning up or down the voltage output. E=IR. Output from a battery is adjusted by choosing a battery with the right voltage output.

After a few minutes of plating at this rate, enough palladium has deposited on the graphite surface to begin the formation of hydrogen gas on the surface. It bubbles off the surface at a moderate rate. As the palladium plates out of solution, the initially reddish-brown colored plating solution begins to lighten in color. Continue plating at the prescribed current density until the plating solution becomes clear, indicating that all the palladium has plated out of solution onto the graphite.

How long does this take? In my experiments at the prescribed current density, plating was complete within three hours. I started plating at the beginning of a Packers game, went to my favorite pub to guzzle beers and watch the game, and when I got back home after three hours, plating was complete.

The palladium plates from this bath at the prescribed current density of 2-3 milliamps per square cm, is a black form of palladium. It's next to impossible to distinguish it from the graphite itself. Mark the side of the graphite facing the anode, as this is the side you will want facing the anode when doing the electrocatalytic hydrogenation.

When plating is complete, remove the plated graphite piece from the plating solution, rinse it off with water, and stand it up to dry on a piece of wax paper. Don't put it in contact with paper, as it will discolor the paper.

In this plating cell, I used dilute sulfuric acid as the anolyte, and a double layer of Kling-Tite rubber as the cell divider. The reason is that one doesn't want to generate chlorine gas at the anode. It would work its way through a single layer of rubber into the plating solution and make an unplatable mess out of it. HCl working its way into the anolyte would also start to dissolve the lead anode, which would be bad. It might get into the plating solution, and do bad things to the catalytic activity of the plated graphite. It would also contribute to pollution in a minor way.

Now that the graphite has been plated with palladium, it can be used as a catalytic cathode in exactly the same manner as in the basic Fester Formula. Once again, I refer you to *Advanced Techniques of Clandestine Psychedelic & Amphetamine Manufacture*. The sole difference is that the palladium plated graphite electrode doesn't need to be anodized in dilute sulfuric acid, because the palladium has been electroplated in the form of palladium black.

Let's do a 10 gram batch example. First the acetic acid ester of ephedrine, pseudoephedrine or phenylpropanolamine must be made. We do this by taking 10 grams of ephedrine hydrochloride, or whatever, isolated from pills by the extraction procedures given earlier in this chapter, and we place it in a glass reaction vessel such as an Erlenmeyer flask or a volumetric flask. Then for each gram of ephedrine hydrochloride, we add 5-7 ml of glacial acetic acid. Swirl to mix, then stopper the opening of the flask with a cork. Heat the mixture with water heated to just about boiling, and swirl the contents of the flask to dissolve the ephedrine hydrochloride. When it has just about all dissolved, add two or three drops of concentrated (70%) nitric acid for each gram of ephedrine hydrochloride. One could also use perchloric acid instead of nitric. Swirl this in to mix, then continue the heating in the nearly boiling hot water for about four hours. Keep the opening of the flask stoppered with the cork during the heating to keep steam from getting into the ester reaction mixture. This would lower the yield of product. After the heating period, remove the flask from the hot water, and allow it to cool. The mixture should be perfectly clear.

The electrocatalytic hydrogenation cell is then wired up. It is identical to the cell used in the basic Fester Formula, and also, by the way, identical to the plating cell given earlier in this section. The anolyte inside the Kling-Tite rubber(s) is 3-5% by volume sulfuric acid diluted with water. The catholyte, which is the solution in contact with the catalytic cathode, is either 3-5% nitric acid solution, or probably more preferably 2N to 3N perchloric acid. This is concentrated perchloric acid diluted to four to six times its original volume with water. If nitric acid is used as catholyte, then a single layer of Kling-Tite rubber is OK as a cell divider, and either lead or graphite or platinum can be used as anode. If perchloric acid is used as catholyte, then a double layer of Kling-Tite rubber must be used as a cell divider, and only graphite or platinum can be used as anode.

The most convenient container to do this reduction in is a rectangular shaped plastic measuring cup. Stand your catalytic cathode up on one side of the measuring cup, hold it in place with a clothes pin, and make contact with it using an alligator clip. The anode inside the rubber(s) goes on the other side of the measuring cup, and one can hold it in place as well with a clothes pin, and make electrical contact with it. Put your stir bar in the cup to stir the catholyte, and turn on the DC power source. Apply 50 to 100 milliamps for each square centimeter of the cathode surface facing the anode which is actually immersed in the catholyte. If you have about 75 square cm immersed, then flow 3.75 amps through the cell. The cathode will fizz off hydrogen, and you will know that you are making good contact with anode and cathode. If you don't see fizzing, or read current flow on the amp meter, check the surfaces on which electrical contact has been made for dirt.

Now that you have current flow, pour the cooled ester reaction mixture into the catholyte, and continue moderate stirring. Adjust current flow to 30-50 milliamps per square cm of cathode immersed by the catholyte. Only count the area on the side facing the anode! For instance, if 90 cm2 of cathode is now immersed, flow 2.7 to 4.5 amps through the cell. When doing this reduction, one should choose the catholyte volume so that the ester reaction mixture is diluted at least five fold when it is poured into the cell, but no more than 10 fold. For this example, with an ester reaction mixture having a volume of about 80 ml, one would be using 400 to 800 ml of dilute nitric or perchloric acid as catholyte. Obviously, one wants as much of the catalytic cathode immersed into the catholyte as possible, without getting the alligator clip contact down into the solution. Choosing the proper size and shape of plastic cup makes this possible.

Continue current flow through the cell at this rate until at least 3000 coulombs per gram of ephedrine has passed. This will take roughly 2½ to 3 hours. Only a DC rectifier or car battery will be able to supply this much power for the required time period. The reduction mixture shouldn't be allowed to heat up, as that would encourage the hydrolysis of the ester of ephedrine back to ephedrine. If necessary, cool the reaction in an ice bath.

After the reduction is complete, the catholyte, which now contains the product, should be poured into a large sep funnel. The solution is made strongly alkaline by adding a 20% solution of lye in water. The lye solution should be added slowly, with shaking between adds of lye. When the solution is strongly alkaline (pH 13+ to pH papers) the meth will come out of solution as free base, which floats on top of the water. When the solution has cooled down, extract out the meth with 200-300 ml of toluene. Separate the toluene-meth layer from the water, and then bubble dry HCl through the toluene to get crystals of meth as usual.

The catalytic cathode can be reused at least a few times before it loses its catalytic activity. When this happens, the palladium black can be sanded off, and the graphite replated with new palladium black. Alternatively, one could just toss away the whole electrode, and start over by plating a new one.

This procedure can be scaled up as much as desired. One could conceivably line the walls of a fish tank with palladium plated graphite cathodes, and just cook away! One would need to use considerably heavier wiring than alligator clips to carry that much current. The 3 amps used in this example is near the upper limit for those flimsy little things. One might be advised to double up the alligator clips so that each clip is only carrying 1½ amps in this example.

The cooker will note that the product produced by the Advanced Fester Formula has higher octane numbers than the product made by the basic Fester Formula. That is because of the higher catalytic activity of the Pd on C used here versus blackened palladium metal. Passing more current through the basic Fester Formula should raise the octane rating to a similar level, but there is that competing reaction mentioned before. That reaction is the hydrolysis of the acetic acid ester of ephedrine back to ephedrine. In that form, it's not going to be easily reducible, and will act as a filler in the product.

As mentioned before, eventually the catalytic cathode will lose activity. In the early phases of this process, one can just use more current to get the same results. Eventually, one will just want to recover the product mixture in the usual way, and react it with fresh acetic acid and sulfuric to reform the ester, and run it through the electrocatalytic hydrogenation using newly plated Pd on graphite. Quality control is one of the joys of cooking. There's just nothing like the thrill of sampling batches, is there? Dirty ephedrine feed will quickly kill the catalytic activity of the cathode.

Someone who calls himself WizardX has suggested an improvement to this procedure. His suggestion is to add a little bit of palladium on carbon catalyst to the electrocatalytic hydrogenation mixture. About ½ gram of at least 10% Pd on C catalyst added to the reduction mixture would be plenty to increase the efficiency of the process. The directions for making this catalyst are found in this chapter. One could also vary the standard directions to make 20 or even 30% Pd on C for added power.

The addition of the catalyst to the reduction mixture results in much quicker and more efficient hydrogenation of the acetic acid ester of the ephedrine. The hydrogen which would be fizzed off by the palladium cathode can be caught by the catalyst and put to use doing the reduction rather than being wasted. This improvement would have the most striking beneficial effect on the basic Fester Formula procedure, as this basic procedure is more wasteful in its use of hydrogen than the Advanced Fester Formula given here.

One would simply add the catalyst to the reduction mixture during the charging of the cathode with hydrogen. This will give time for the Pd on C catalyst to also absorb hydrogen. At the end of the electrocatalytic hydrogenation, one would then just filter out the Pd on C catalyst for reuse in the next batch. Work-up of the reaction mixture then proceeds as usual. Your Uncle thinks this is a really good idea, and it comes with my highest recommendation.

Before leaving this topic, I have to report that some guys on the Internet have found that the two makeshift alternative power sources for doing electrocatalytic hydrogenation are completely unsuitable for use. A toy train transformer usually puts out a set voltage, around 10 or 12, and turning the knob on the transformer just causes the current output to be raised or lowered. This is way too much voltage. As a result, the reaction medium gets fried, and the catalyst surface gets poisoned.

Using a 12 volt battery is just as bad, as the dashboard dimmer knob similarly just turns the current up and down, with voltage staying at 12 volts.

The correct way is to start at zero applied volts, and then crank up the voltage a notch at a time until the desired current flow is achieved. In this way the resistance of the cell and the electrolyte determines the voltage of the reaction. In this Fester Formula, applying around 3 volts to the cell will give the desired current flow. The cathode will then measure around -.6 to -.8 volts versus standard calomel electrode. This is the right range. One only wants generation of hydrogen at the cathode, not a bunch of other electric reactions caused by excessive voltage.

Batteries can be used as power sources if the voltage they produce is in the right range. For example, a six-volt lantern battery can be used to anodize the ingot surface, and two "D" cell batteries hooked in series to give three volts can be used to generate hydrogen at the ingot surface.

I always use an electroplating rectifier. Just by turning the knob on the front of the rectifier, the voltage output is raised or lowered, and the amperage which flows through the solution is correspondingly increased or decreased. Used "hull cell rectifiers" or their generic equivalents are available cheap. See the *Metal Finishing Guidebook and Directory*. Look under Testing Equipment — plating cells, and also under Rectifiers. If your library can't get you a copy of the directory, call the publisher at 914-333-2500.

Considering the ease with which hydrogenation is done according to US Patent 6,399,828 just by adding some ammonium formate solution to a mixture of the acetic acid ester of ephedrine plus some small amount of palladium on carbon catalyst, that method may well be the best of all. The only problem with their method is the use of acetic anhydride to make the ester of ephedrine, and that would be easily gotten around by using ephedrine plus acetic acid plus a small amount of perchloric or nitric acid to give the acetic acid ester of ephedrine when the mixture is heated in boiling hot water. Then precipitate out the ester from the reaction mix by adding some naphtha, and the patent hydrogenation procedure can then be followed.

Method 4: Reduction With Hydroiodic Acid and Red Phosphorus

In this procedure, the alcohol grouping of ephedrine, pseudoephedrine, or PPA is reduced by boiling one of these compounds in a mixture of hydroiodic acid and red phosphorus. Hydroiodic acid works as a reducing agent because it dissociates at higher temperatures to iodine and hydrogen, which does the reducing. This dissociation is reversible. The equilibrium is shifted in favor of dissociation by adding red phosphorus to the mixture. The red phosphorus reacts with the iodine to produce PI_3, which then further reacts with water to form phosphorus acid and more hydroiodic acid. Since the hydrogen atom of the HI is being absorbed by the ephedrine, the red phosphorus acts as a recycler.

In clandestine cooking, the need for HI is dispensed with just by mixing red phosphorus and iodine crystals in a water solution. The red phosphorus then goes on to make HI by the above-mentioned process. With a small amount of due care, this is an excellent alternative to either purchasing or stealing pure hydroiodic acid. This substance is on the List 1 of restricted chemicals, so buying it is just asking for trouble.

This method has the advantage of being simple to do. It is generally considered the most popular method of making meth from ephedrine or pseudoephedrine. Now red phosphorus is on the List One of restricted chemicals, so an increased level of subterfuge is called for to obtain significant amounts. One might think that this is easily gotten around by making your own red phosphorus, but this is a process I would not want to undertake. Ever hear of phosphorus shells? I would much rather face the danger of exploding champagne bottles. Those who insist upon finding out for themselves, will see *Journal of the American Chemical Society,* Volume 68, page 2305. As I recall, *The Poor Man's James Bond* also has a formula for making red phosphorus. Those with a knack for scrounging from industrial sources will profit from knowing that red phosphorus is used in large quantities in the fireworks and match-making industries. The striking pad on books of matches is about 50% red phosphorus.

The determined experimenter could obtain a pile of red phosphorus by scraping off the striking pad with a sharp knife. A typical composition of the striking pad is about 40% red phosphorus, along with about 30% antimony sulfide, and lesser amounts of glue, iron oxide, MnO_2, and glass powder. These contaminants can be removed from the red phosphorus by soaking the mixture in a 10% solution of hydrochloric acid for about an hour with stirring, then filtering out the red phosphorus and rinsing it with some water. This procedure is very stinky as the antimony sulfide gets converted to hydrogen sulfide and antimony chloride. Ventilation is mandatory. Various correspondents have written to tell that the glue holding the red phosphorus striking pads to the paper is softened by soaking in rubbing alcohol, or acetone, or even hot water, and can then be easily scraped off.

The red phosphorus can be replaced with a solution of hypophosphorous acid. It is generally sold as either 30% or 50% solutions in water. The later is preferable, although concentrated solutions of hypophosphorous acid can spontaneously burst into flames in contact with air. It reacts in the same way towards iodine as red P does, and so functions the same as red P in both making the HI solution, and in recycling the iodine formed during the course of the reaction with ephedrine or pseudoephedrine.

Hypophosphorous acid isn't very commonly used industrially or in the lab. As a result, it too has become an item under intense suspicion. It has joined red phosphorus on the List One of restricted chemicals. No need to get discouraged at this point. The acid, H_3PO_2, is easily made from its sodium salt, sodium hypophosphite, NaH_2PO_2. This substance finds wide use industrially as the reducer in electroless nickel-plating solutions.

See *Advanced Techniques of Clandestine Psychedelic & Amphetamine Manufacture* for a fuller discussion of electroless nickel-plating solutions, but suffice it here to say that they are generally sold as a three-component package. The A solution contains nickel sulfate, the B solution contains complexors to prevent the nickel from sludging out during operation of the bath, and the C component contains anywhere from 300 to 400 grams per liter of sodium hypophosphite in water solution. One can also get nickel bath "annihilator," which is used to plate the nickel out of spent electroless nickel-plating solutions. It will be a solution of about 50% by weight of sodium hypophosphite in water. This solution is also sometimes called "plate out" solution by the suppliers. One could also get pure sodium hypophosphite crystals from the same suppliers who sell electroless nickel-plating solutions, but sodium hypophosphite is also on the List One of restricted chemicals. It is when it is in solution in these electroless nickel-plating products that it is free from restrictions. It is surprisingly easy to go "shopping" for these electroless nickel products if one has a bit of resourcefulness. I won't list dealers here, because that would make life miserable for them. Several cater to the "hobbyist" plater, so get hunting!

Before leaving the topic of electroless nickel-plating products, there is one complication which must be addressed. The "C" or hypophosphite replenisher component of many electroless nickel products will also contain a lot of ammonia. These products are termed "self pH regulating" baths because ammonia has to be added to the bath as it runs to keep the pH of the bath in operating range. The "self regulating" products have the specified amount of ammonia already incorporated into the hypophosphite replenisher. The plate out component will hardly ever have ammonia as part of its formulation, and the smell of the solution will instantly tell you if ammonia is an ingredient in the mix. Before making hypophosphorous acid from an electroless nickel product which contains ammonia, the ammonia should be removed. This is easily done by warming the solution up to around 110° to 120° F, and blowing a stream of air through it until the reek of ammonia is gone from the solution.

The best way to get hypophosphorous acid from the sodium salt is to pass a 10% solution of sodium hypophosphite in water through a column filled with strong cation exchange resin in H+ cycle. The resin is charged to the H+ cycle by slowly pouring a solution of 10% by volume sulfuric acid through the resin column. The cation exchange resin is then rinsed with water to remove the acid from the column and ready to go. Each ml of resin will react with about one or two milliequivalents of sodium hypophosphite, so depending upon column size, one can calculate how much solution of sodium hypophosphite one can run through before the resin is exhausted. One can also titrate the hypophosphorous acid solution made with standard NaOH solution to determine completeness of conversion. The resin is then recharged by passing more sulfuric acid solution down through it and rinsing. Resin can be used over and over thousands of times before it wears out. Cation exchange resin is easy to get. It is used in deionized water systems. Call your Culligan man. Insist on cation exchange resin, not a mixed bed of cation and anion exchangers.

The dilute solution of hypophoshourous acid in water which comes out of the ion exchange bed must next be concentrated to at least 30% strength. There are two methods for doing this. One can evaporate away the water, preferably under a vacuum, until the desired strength is reached. The other method is to extract out the hypophosphorous acid from the water solution using ether starting fluid, and then allow the ether to evaporate away. This is again best done using a vacuum. In the latter case, the hypophosphorous acid produced will be VERY strong, probably 80-90% hypo. In this state it is quite prone to catch on fire, and should be stoppered and kept in a freezer.

Clandestine chemists have come up with a more convenient procedure for making hypophosphorous acid from sodium hypophosphite. The method is to mix sodium hypophosphite with hydrochloric acid. This produces hypophosphorous acid plus salt. If one is starting with crystals of sodium hypophosphite, the preferred procedure is to add a couple of ml of lab strength 35% hydrochloric acid for each gram of sodium hypophosphite in the batch. They then with stirring warm up this mixture to get complete reaction. After allowing it to cool, the salt is filtered off.

Next, the filtered solution is poured into a glass-baking dish, and set onto a stove top. A fan is set up next to the dish to blow away the hydrochloric acid fumes, and the glass-baking dish containing the solution is warmed up to about 70° C. It is simmered down until about half of the solution has evaporated away, then allowed to cool.

If more salt has precipitated out of the cooled solution, it is then filtered off, and the filtered solution is returned to the glass-baking dish. Now to remove most of the remaining hydrochloric acid from the hypophosphorous acid solution, add approximately one half volume of fresh water to the hypophosphorous acid solution. Then repeat the simmering at 70° C until the solution has evaporated down again to the level it was before adding this last portion of water.

The result will be a solution of hypophosphorous acid in water somewhere between 30 and 50% strength, and it will work well for making meth. It should be carefully poured into a glass container, and stored in a freezer or refrigerator until ready for use.

When starting with sodium hypophosphite solutions in water, such as one would obtain from electroless nickel products, a very similar procedure is used. In this case, add one ml of hydrochloric acid to each ml of ammonia free electroless nickel product. For example, 100 ml of electroless nickel plate out solution is mixed with 100 ml of 35% hydrochloric acid. The 31% hardware store muriatic acid can be used instead and one still get a good product.

Now one pours this mixture into a glass-baking dish, and simmers it down just as in the previous example until half of the solution has evaporated away. When it has cooled, the salt is filtered off. The filtered hypophosphorous acid solution is then poured into a glass container and stored in a freezer until ready for use. It will be roughly 30% strength hypo, but may vary depending upon how much sodium hypophosphite was in the electroless nickel product.

The iodine one needs to do this reaction can be obtained at farmers' supply stores and similar outlets. Gone are the days when iodine crystals could be picked up straight off the shelves, but iodine tinctures and iodine disinfecting solutions are still widely used and available. When shopping for iodine, one needs to do some careful label reading. The solutions one wants contain iodine, one needs to do some careful label reading. One does not want povidone-iodine, nor does one want complexes which contain some iodine. The solutions one wants contain iodine, generally at about 7% strength (more would be better!), along with roughly three times that much potassium iodide, KI. The KI is in the mixture to keep the iodine in water solution. Without it, the iodine doesn't dissolve at all in water.

The first thing one wants to do to these iodine tinctures/disinfectants is to knock out the KI, and convert it to more iodine. To do that, one makes use of the following reaction:

$$\mathbf{2KI + CuSO_4 \rightarrow CuI + \tfrac{1}{2}I_2 + K_2SO_4}$$

In script, that reads potassium iodide plus copper sulfate react to make copper iodide and iodine. Copper sulfate crystals can be obtained for about $20 per five-pound container. It is found at the hardware store as root killer for clogged drains. The large blue crystals of copper sulfate are easy to recognize, even though the label should identity the contents. The other popular drain opener is sodium hydroxide.

Let's say one has a gallon jug of iodine disinfectant, containing 20% KI. At eight pounds per gallon, that's 1.6 pounds of KI per gallon. Half of that will be converted to iodine by reaction with copper sulfate, and we can add in the roughly half pound or so of iodine originally present in the disinfectant to give a total iodine yield of over a pound and a quarter from this very cheap jug of iodine disinfectant. More concentrated disinfectant would yield proportionally more.

Now to convert the KI in this hypothetical example jug containing 20% KI, one would slowly with stirring and cooling add 550 grams of copper sulfate crystals. After about a half hour of stirring, the KI present in the mixture has been converted to iodine and copper iodide, and the iodine is no longer soluble in the water.

To then use the iodine, one first puts the mixture into a large sep funnel. Add a couple hundred ml of toluene and shake. The toluene will dissolve about 40 grams of iodine. Separate off this toluene solution. In the example batch given in this section, we are using 300 ml of water and 30 grams red P. Add the toluene to the 300 ml of water and 30 grams of red P. Then shake to react the iodine with the phosphorous. Keep the mixture cool. Then separate off the toluene from the water/HI mixture. Wash it with a 300 ml portion of water which will be used in the next batch. This will remove HI from the toluene, and save it for the next batch. Finally, pour the toluene back into the sep funnel with the iodine and shake to get another load of iodine. One repeats this process until all the red P has been consumed. Then enough iodine will have been added to do the example batch given in this section. Finally add another 30 grams of red P to the mixture, and it is ready to cook ephedrine or pseudoephedrine to meth.

After a while, the iodine produced in our hypothetical jug of iodine disinfectant will begin to run out. The toluene extracts of the jog solution will no longer be so darkly violet in color. Then it is time to get the roughly one pound of iodine which is in the form of copper iodide in this jug. That is the tan colored solid which has been floating around in the solution.

To turn the copper iodide into iodine, first add hydrochloric acid to the jug solution until the pH is down to one. The tan solid will then dissolve. Next slowing with stirring add hydrogen peroxide solution to the jug solution. Make sure the pH of the solution stays in the region of one. It will tend to rise during the reaction. If it rises too much, a lot of gas bubbles will form from the peroxide, and it will be wasted. If the pH rises too much, just add more hydrochloric acid. Two hundred ml of 30% hydrogen peroxide should be enough to convert the copper iodide to iodine. Two thousand ml of 3% hydrogen peroxide would do the same thing. When enough has been added, a clear green-colored solution will result, and the iodine will fall out of solution in the form of big nuggets. To get these nuggets of iodine, just pour off the green-colored water. Then rinse the iodine nuggets with clean water. They are now ready to use just the same as iodine crystals one might buy.

Many people use a simpler and more direct method for getting iodine crystals from their tinctures containing iodine plus KI. They omit the copper sulfate, and go directly to the portion of the procedure using hydrochloric acid and hydrogen peroxide. They first add hydrochloric acid to the tincture until the pH is around 1. Then they add hydrogen peroxide with stirring to get crystals and nuggets of iodine to fall out of the solution.

When working with iodine, one must follow some precautions. Overexposure to the substance can poison you! Use good ventilation, and keep upwind. This shouldn't be done in a basement or closet. Rubber gloves will keep the fingers from becoming stained with the iodine. Such stains are not only incriminating, they also pose a health hazard. Store the iodine nuggets and crystals in a sealed glass container until ready for use.

One problem with the HI/red P method is that it can produce a pretty crude product if some simple precautions are not followed. From checking out typical samples of street meth, it seems basic precautions are routinely ignored. I believe that the byproducts in the garbage meth are iodoephedrine, and the previously mentioned aziridine. (See the earlier section concerning chloroephedrine.)

I don't think that even a good fractional distillation will remove these byproducts from the meth. The aziridine has to have a boiling point very near that of meth, and so is unremovable by that method. I also think that the heat of distillation would cause the iodoephedrine byproduct to form more aziridine, which would then distill with the meth. Steam distilling the meth produced by this method has become very popular, and is recognized as "the" way to get the contaminants removed from the product.

To some extent, these byproducts can be avoided in the first place, if when making hydroiodic acid from iodine and red phosphorus, the acid is prepared first, and allowed to come to complete reaction for 20 minutes before adding the ephedrine to it. This will be a hassle for some, because the obvious procedure to follow is to use the water extract of the ephedrine pills to make HI in. This should never be done, especially with the doctored pills now on the market. Pure ephedrine, racephedrine, pseudoephedrine or PPA hydrochloride made according to the directions at the beginning of this chapter must be used. Impure raw material leads to big reductions in yield, and isolation problems at the end of the reaction. Since the production of HI from iodine and red phosphorus gives off a good deal of heat, it is wise to chill the mixture in ice, and slowly add the iodine crystals to the red phosphorus-water mixture.

To do the reaction, a 1000 ml round bottom flask is filled with 150 grams of ephedrine hydrochloride (or psuedoephedrine-HCl). The use of the sulfate salt is unacceptable because HI reduces the sulfate ion, so this interferes with the reaction. Also added to the flask are 40 grams of red phosphorus, and 340 ml of 50%+ hydroiodic acid. This same acid, which by the way, is on the chemical diversion list and should never be purchased, can be made by adding 300 grams of iodine crystals to 50 grams of red phosphorus suspended in 300 ml water. After allowing the iodine and red phosphorous to react together for an additional 20 minutes after the last of the iodine has been added to the red phosphorus, the ephedrine or pseudophedrine can then be added to the mixture. A more refined procedure is to react 300 grams of iodine with 30 grams of red phosphorus in 300 ml water, and then distill this mixture, collecting the first ⅔ of it, and leaving the phosphorus-acid byproduct behind in the distilling flask. This home brew acid smells bad, but works really well. It loses its bad smell shortly after the beginning of the reflux in the reaction with ephedrine. When using this more refined procedure, remember that 20-30 grams of red phosphorus must be added to the reaction mixture. The 40 grams cited above is overkill, but the unused portion can be reused by filtering it out at the end of the reaction.

With the ingredients mixed together in the flask, a condenser is attached to the flask, and the mixture is boiled for one day. This length of time is needed for best yields and highest octane numbers on the product. While it is cooking, the mixture is quite red and messy looking from the red phosphorus floating around in it. It is best to take about an hour to warm up the reaction mixture to the boiling point, as this reduces the amount of byproducts made.

The progress of the reaction can be followed by watching the consumption of the red phosphorus. The majority of product is obtained in about 10 hours; after 16 hours over ¾ is obtained, and after 24 hours, the reaction is done.

When one day of boiling under reflux is up, the flask is allowed to cool; then it is diluted with an equal volume of water. Next, the red phosphorus is filtered out. A series of doubled-up coffee filters will work to get out all the red phosphorus, but real filter paper is better. The filtered solution should look a golden color. A red color may indicate that all the phosphorus is not out. If so, it is filtered again. The filtered-out phosphorus can be saved for use in the next batch. If filtering does not remove the red color, there may be iodine floating around the solution. It can be removed by adding a few dashes of sodium bisulfite or sodium thiosulfate.

Of these two, the thiosulfate is much preferred because it has the ability to destroy aziridine along with the iodine in the solution. You see, iodoephedrine makes aziridine (dimethylphenyl aziridine) by reaction between the iodine atom and the amino group. See earlier in this chapter for the drawing of the aziridine in question. The high temperature at which this reaction works encourages its formation.

The best way to add the thiosulfate or bisulfite is along with the sodium hydroxide or lye solution used to neutralize the reaction mixture. That is our next step. For the batch size given here, using 300 grams of iodine in 300 ml of water and about 150 grams of ephedrine hydrochloride, over 150 grams of lye or sodium hydroxide will be required to make the solution strongly alkaline (ph 13+). This amount of lye should be dissolved in about 600 ml of water and allowed to cool down.

Now add the few grams of sodium thiosulfate or sodium bisulfite to the cooled-down lye mixture and stir until it dissolves. Slowly and with periodic shaking or strong stirring, add the lye solution to the filtered reaction mixture. If one prefers, one could also do the opposite and add the reaction mixture to the lye water with shaking and stirring.

Some fizzing may be noted here as the reaction takes place. Then as the great heat produced by the neutralization reaction allows, the rest of the sodium hydroxide solution should be added with stirring or shaking to the reaction mixture. The meth free base which forms will float to the top of the water solution. Strong and prolonged shaking of the mixture is necessary to ensure that all the meth has been converted to free base.

Then check the pH of the water layer using pH papers. It should read strongly alkaline. If not, add more lye, and continue shaking.

With free base meth now obtained, the next step, as usual, is to form the crystalline hydrochloride salt of meth. To do this, a few hundred mls of toluene is added to the batch, and the meth free base extracted out as usual. If the chemist's cooking has been careful, the color of the toluene extract will be clear to pale yellow. If this is the case, the product is sufficiently pure to make nice white crystals just by bubbling dry HCl gas through the toluene extract as described in Chapter Five. If the toluene extract is darker colored, a distillation is called for to get pure meth free base. The procedure for that is also described in Chapter Five. The yield of pure meth hydrochloride should be from 100 grams to 130 grams.

If gummy binders from the stimulant pills are carried over into the reaction mixture, they produce a next-to-impossible-to-break emulsion of meth, gum, toluene, and water when the reaction is done and it is time to extract out the meth. It is absolutely necessary that the pill-extraction procedure given in this book be followed. If this emulsion is encountered, the only way to break it is to first let the emulsion sit in a sep funnel for a few hours. Water will slowly work its way out and settle to the bottom where it can be drained away. The stubborn residual emulsion should be transferred to a distilling flask, and the toluene slowly distilled off through a fractionating column. This removes water from the emulsion as the toluene-water azeotrope. It may be necessary to add additional toluene to the distilling flask to get all the water removed. The gum sticks to the glass flask, and causes no further problem. Once the emulsion is broken, distilling should be stopped. The toluene-meth solution should be poured from the distilling flask, and the meth precipitated as hydrochloride as per the usual dry HCl bubbling method.

When using hypophophorous acid instead of red phosphorus, a very similar procedure is followed. First, 50%+ hydroiodic acid solution is made. To do this, put 35 ml of the 30-50% hypophosphorous acid solution into a flask. Then slowly and with stirring add about 40 grams of iodine crystals to the hypophosphorous acid solution. The red color produced by the iodine will be sucked up as the hypophosphorous acid reacts with it to make hydroiodic acid. As the solution gets hot, cool it down in ice water. After about half an hour, a fairly clear to yellowish solution should result. If it is still red with iodine, the hypophosphorous acid solution you made wasn't strong enough and needs to be evaporated down some more. One could check for strength of the hypophosphorous acid by first reacting 3.5 ml of the solution with 4 grams of iodine, before moving on to the full batch given here.

Once the hydroiodic acid solution has cooled down, about 20 grams of ephedrine HCl or pseudoephedrine hydrochloride is added to the reaction mixture. A condenser is then fitted to the flask, and over the course of about an hour the mixture is slowly heated up to boiling using an oil bath to warm the flask. The solution should stay clear to yellowish colored as it warms up and cooks.

Now gently boil the solution for about 6 hours. The reaction goes faster when using hypophosphorous instead of red phosphorus, so there is no need to go the full 24 hours.

When the reaction is finished, processing the batch is very similar to the red phosphorus example. The reaction mixture is first diluted with an equal volume of water. Since red P wasn't used, there is nothing to filter out at this point. Then slowly add a solution of lye dissolved in water to the reaction, just as in the example using red P. It will make a lot of heat, so take your time and use shaking or strong stirring between adds of lye to the reaction mixture.

Once enough lye solution has been added to raise the pH of the mixture to 13+, again shake the mixture strongly, and allow it to cool. The layer of meth free base should be floating atop the water.

Now add about 150 ml of toluene or xylene to the mixture and shake it hard for a minute or so. Once the layers have settled, the meth dissolved in toluene layer will be floating atop the water. It should be clear to light yellow colored. Drain off the water layer, then add about 150 ml of fresh water to the sep funnel. Shake this mixture to wash the toluene extract containing the meth, and then let the layers settle. The water layer can then be drained off, and the toluene-meth layer poured into a beaker or similar glass cup. After letting this solution shed entrained water for about an hour, it can be poured into a clean beaker and bubbled with dry HCl gas to get the crystals of meth hydrochloride. The crystals are rinsed with some fresh toluene, then spread out on a plate to dry. The yield will be about 15 grams or so of meth.

Two additional methods exist for making hydroiodic acid, and while not as convenient as just adding iodine to red phosphorus, they produce a more pure acid. Further, they don't require red phosphorus in the making of the acid, so the amount of that material needed for production is greatly reduced.

See *Inorganic Syntheses,* Volume 1. The first of these methods involves bubbling H_2S gas into a suspension of iodine in water. The H_2S gas is made by dripping HCl onto iron sulfide and piping the gas into solution just like the HCl gas made in Chapter Five. The excess H_2S gas dissolved in the product acid is then boiled out of solution under a reflux. This method is quite good.

The alternate method directly produces a very pure acid by direct union of H_2 gas with I_2 vapor in the Pt catalyst bed. This is really good if you aren't afraid of putting together some glass tubing.

There are a few variations upon the standard HI/red P recipe that I have just given. Some people like to use the free base of ephedrine or pseudoephedrine rather than the hydrochloride crystals. This gives them a somewhat better yield of product in the reaction. When using the free base rather than hydrochloride crystals, one has to increase the strength of the acid a bit to compensate for the HI which will be consumed by the free base. In the example batch given, increasing the amount of iodine used to 380 grams will cover the amount sucked up by the free base.

Another variation is to do the heating of the ephedrine/HI/red P mixture inside a pressure vessel able to withstand at least 50 pounds of pressure. This is said to greatly speed up the reaction. This variation has never become very popular because such pressure vessels aren't easily found on store shelves. If one should rupture a pressure vessel containing a cooking batch, a most regrettable fire is likely to result.

The HI/red P variation that has really caught on is called the Push Pull method. Our friend ReadyEddie would like to share his Push Pull recipe for making 10 gram batches. It is a lot faster than the standard HI/red P method, and the equipment used is pretty basic. You'll be quite pleased, I'm sure!

Push Pull by ReadyEddie. The push/pull does not take a lot of skill but does take a bit of understanding of what one is dealing with. First of all this method does require some very watched items. Red phosphorous and iodine crystals can bring unwanted attention if ordered from different supply companies. If by chance one can't seem to obtain these items anywhere, there is still hope. These items can be obtained OTC (over-the-counter) at your nearest supermarket. If one can get one's hands on lab grade without getting a one-way ticket to the slammer, it's well worth it.

The striking pads of matchbooks contain a small amount of red phosphorous. It's not pure, so it should be cleaned up before use. Iodine crystals can be converted from iodine tinctures. Tinctures can be found on the shelf in many different supermarkets, and come in a 2% strength in a 30 ml bottle.

Tinctures can also be found at cattle supply houses, in pint and gallon sizes, and come in a 7% strength. Sometimes iodine powder or pellets can be found in cattle supply stores as well. It's called iodine prill and works great just the way it is.

There are many different ways to collect red phosphorous off of matchbooks. An easy way is to just scrape it off with a razor blade. A better way is to cut the strikers off and soak them in acetone until the red phosphorous falls off. It takes a very large amount of strikers to obtain a large enough pile to do anything with, but it can be done and has been done by many.

After all falls off the paper, take the paper out and filter the acetone/red phosphorous through two coffee filters and now rinse with distilled water and let dry. Now mix up a solution of 20% sodium hydroxide. That is 20% grams of sodium hydroxide in 100 ml of distilled water. Place the dry red phosphorous that was collected in a beaker or flask of choice and add the hydroxide solution to it. Heat on low heat for a few hours, then filter through two coffee filters again. When dry, rinse with hot distilled water a few times and let dry. This will give a pretty pure powder that will fire off a push/pull reaction.

To convert 2% iodine tinctures, the following are used:

500 ml 2% iodine tincture
125 ml muriatic acid (hardware store strength)
235 ml hydrogen peroxide (3% topical solution)
862 ml distilled water

To convert 7% iodine tinctures, the following are used:

1 pint of 7% tincture
3 pints of 3% peroxide
2 oz. of muriatic acid
½ cup of distilled water

Pour the tincture in a one gallon milk jug. Add the muriatic acid and mix it all together well. Let sit for a half an hour. Now add the 3% hydrogen peroxide and mix this in real well and let it sit again for another half an hour. Add the distilled water and shake the hell out of it and then let it sit for another half an hour.

There should be an orange layer on top of a dark grey layer. The grey layer is what you want. The grey layer is iodine crystals that have crashed out of solution. Pour off the orange layer, add more fresh distilled water to them and shake again. Let sit for a few. Pour off the orange solution again. Do this a total of three times. After the third time, pour the contents through two coffee filters. Now the crystals need to be wrung dry. Wring dry and put them in another stack of coffee filters and wring them out again. Keep doing this until one gets a nice solid dry ball of crystals.

That's it. One should have nice iodine crystals that will work in the reaction. Store the crystals in a dark-colored jar or bottle.

Warning: When making one's own crystals from tinctures, wear eye protection and chemical resistant gloves. Remember that iodine is poisonous, so be safe and work outdoors.

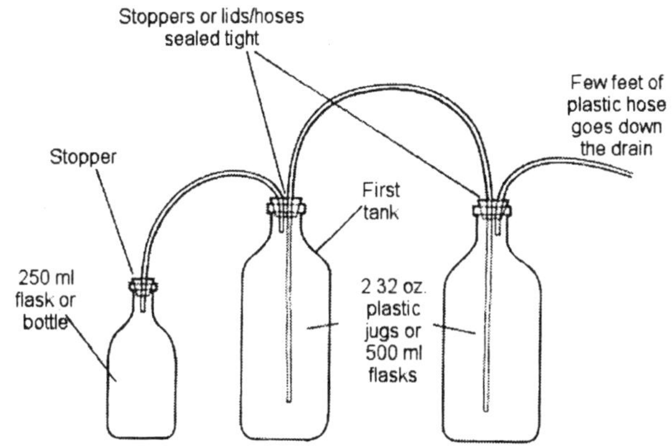

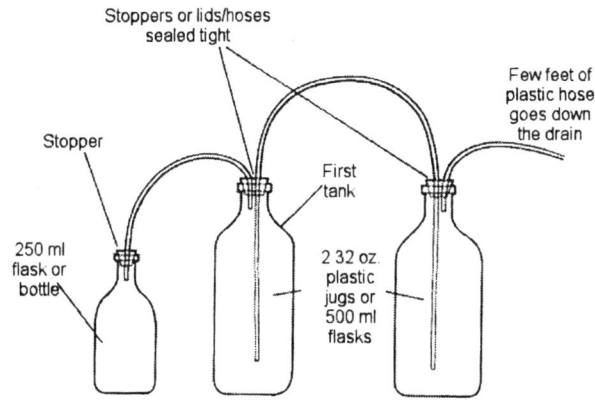

Figure 40
The Push/Pull Set Up

As one can see from Figure 40, it is very easy to construct the apparatus for this reaction. A small clear beer bottle can replace the flask if one is not on hand. This setup can also be made larger or smaller for different size batches.

If, for example, one wants to do a 20 gram batch, use a 500 ml flask or bottle and two 64-oz. plastic Gatorade jugs. Make sure all the hoses are sealed in place because one doesn't want any leaks.

The reaction can be very quick, but other times it may not be. Keep a bucket of dirt or sand around in case it is needed to smother a phosphorous fire. That is a slim possibility, but be on the safe side.

With lab grade red phosphorous and iodine crystals, the ratios by weight are 1 gram pseudo(ephedrine) HCl to 1 gram of iodine crystals to 0.5 gram of red phosphorous. With OTC chemicals, the ratios are 1 gram pseudo(ephedrine) HCl to 1.5 grams of iodine crystals to 0.7 gram of red phosphorous.

So now let's get started with the reaction. First thing to do is set up the push/pull vessel. Fill the first water tank with distilled water ¾ of the way full. The second one leave empty. The drain hose from the second tank is put down the drain past the u trap or in a bucket of cat litter. Weigh out the pseudo(ephedrine) HCl and place inside the flask or bottle, no more than a 10 gram batch in a 250 ml size flask or bottle. Next weigh out the iodine crystals and put them inside with the pseudo(ephedrine) HCl. Mix them together very well and stopper the flask or bottle and place it in the freezer for around 3-4 minutes. Weigh out the red phosphorous. Take the flask from the freezer, the reactants should be dark and may be a dark thick-looking mud. Place the red phosphorous in the flask, mix in good with a glass rod and hook it up to the water tanks. If the drain hose is down the drain, it's time to turn on the water to keep the fumes down.

If using lab grade chemicals, one may have to add a few drops of distilled water to get the reaction going. For OTC chemicals, no water should be needed; just heat the reactants a little and it should fire. Do not flood the reaction with water! Place the flask on a coffee cup warmer or in a hot water bath. When the reaction starts, take it off the heat.

During the first phase of the reaction, the mixture turns to a thick liquid and starts bubbling. The small bubbles will be kind of silvery looking. Some gas may now be pushing into the first water tank and water into the second. The contents inside the flask will rise some. A light yellow-colored mist and white fog inside the flask is normal. Continue to add, on and off, heat to maintain an easy bubbling reaction, at least around 15-20 minutes or until bubbling begins to slow.

In the next phase, increase heat to the flask; slowly raise it up to between 160°-180° F. The reaction should begin to bubble very rapidly. The color of the reaction will change to a dark purple or reddish with a yellow tint. The small reaction bubbles will start turning into big bubbles that collapse into large holes. The reaction should be pushing gas into water tanks, but a lot harder this time. The contents will rise quickly, and may start to smoke. If so, take off the heat and swirl the contents around in the flask to get the smoking to stop. Swirling will also keep the reactants from rising too high. Let react until all seems to be dead inside the flask (no reaction) even with applied heat. A pull may be noted. If not, that's alright too. Sometimes there is no pull with small reactions. Tilt the flask. If reactants slowly flow off the bottom and sides and are no longer stuck, it should be finished. Let cool to room temperature, then add 100 ml of distilled water. Heat in a water bath at 150° F in an open flask for 30 minutes to free the contents. Filter all the red phosphorous out of the mix. A couple of coffee filters also works well for this.

After filtering the red phosphorous out, the color should be clear to pale yellow. If one gets a dark orange or red color that does not filter out, the reaction may not be done or the adulterants in the pills may have messed it up.

If this is the case try putting the red phosphorous back into the water it was filtered out of. Heat it up to about 200° F for two hours to complete it. If this doesn't work, the gak in the pills may have messed the reaction up.

Now it's time to base and extract the meth base with a non-polar solvent such as toluene. Wash the solvent/base layer with distilled water a few times, and dry with dehydrated Epsom salts. Gas the solvent/base layer to get crystals of methamphetamine.

And that is the end of ReadyEddie's Push/Pull recipe. The reader should be aware that the push pull method is much more vulnerable to the yield lowering effects of the new pill ingredients than are the standard hydroiodic acid boil methods given earlier in this chapter. Only very pure pseudoephedrine or ephedrine should be used in this method.

Warning! *Iodine and red P are now very "hot" items. Ordering them is likely to cause the cops to follow right behind!*

Method Five: Zinc Reductions

Check this one out! In *Journal of Organic Chemistry*, Volume 26, page 99, the ester of a benzyl alcohol was reduced to the alkane by using zinc metal and formic acid. Not a bad idea, huh? Simple materials, non-toxic reaction materials and waste, and maybe good yields of meth.

There is a more general reduction method for benzyl alcohols which uses trifluoroacetic acid and sodium borohydride, but trifluoroacetic acid is a potent rat poison and pretty effective people poison as well. I'll be damned if I'm the one who gets blamed for introducing "the rat poison meth recipe." Try this method. If it works, fine. If not, you know how I like those other methods. Stick with them.

In the journal article's procedure, they put 20 grams of an ester of the benzyl alcohol in a flask along with 40 grams of zinc metal, 160 grams (about 130 ml) of 95% formic acid, and 40 ml of water. They refluxed the mixture for 5 hours, then after cooling, the zinc metal left over was filtered out. The filtered reaction mixture was then made strongly alkaline with sodium hydroxide solution, and the product extracted out with toluene.

How would one tinker with this recipe to use ephedrine? First, I would put the 20 grams of ephedrine hydrochloride in the flask. Then I would add the 130 ml of 95% formic acid. I would heat this mixture on a hot water bath for a few hours to make the formic acid ester of the ephedrine. Then I would add the zinc metal and the 40 ml of water, and heat the mixture to reflux for 5 hours.

After cooling, filtering and making strongly alkaline with sodium hydroxide solution, the toluene extract likely contains the formic acid amide of meth. This would be hydrolyzed by boiling with hydrochloric acid, just like in Chapter Five. The product is then recovered from the hydrochloric acid just as in that chapter, distilled, and bubbled with HCl gas to get crystals of meth hydrochloride.

Would you like another recipe? I like to use this one for about 10 gram batches. It's fast and pretty good. The guys who filed for US Patent 6,399,828 lifted their application from my work, and refined the procedure.

First, make about ½ gram of palladium black by adding sodium borohydride to a well-stirred solution containing one gram of PdCl2 in water. See *Journal of the American Chemical Society*, Volume 84, pages 1493-95. I suck up the settled black catalyst with a pipette and squirt it into a filtering flask. Then take a balloon and put the end of it over the vacuum nipple of the filtering flask. Tie it in place with some string so that it doesn't leak air when one blows air into the filtering flask. Add 50 ml of water to the filtering flask as solvent for the reaction.

Now take the ester reaction mixture made by heating 10 grams of ephedrine hydrochloride with about 60 ml of glacial acetic acid and a little perchloric or nitric acid and isolate the ester from the reaction mixture by slowly dripping in naphtha with stirring. After crystallizing in a freezer overnight, filter out the crystals and let them dry on a plate. Then toss them into the filtering flask with the palladium black. Put a magnetic stir bar into the filtering flask, too.

Begin fast magnetic stirring, then toss a piece of sodium borohydride about the size of a split pea into the flask, and quickly stopper the flask to hold in the hydrogen generated. When the fizzing stops, add another piece. Continue this until the balloon stays inflated with hydrogen.

Now filter out the palladium black for reuse. Put the filtered batch in a sep funnel, and add lye solution with shaking until the mixture is strongly alkaline. Extract out the meth with toluene, separate off the toluene layer, and bubble it with dry HCl gas to get about 10 grams of a very nice meth. You'll like this recipe! If you should have trouble getting the $PdCl_2$ to dissolve in the water, add a few drops of HCl. That will make it dissolve.

One can get closer to the Patent's procedure by using Pd on carbon rather than Pd black. To do that, simply add activated carbon powder to the $PdCl_2$ solution mentioned 4 paragraphs ago. In this recipe example, it would be ½ gram $PdCl_2$ and 2.5 grams activated charcoal powder. When using this variation, it's hard to tell when all the $PdCl_2$ has been reduced to Pd metal by the borohydride, so you have to let the carbon settle in the solution to see if the brown color of the $PdCl_2$ has all been removed from the water. Catalyst formation is complete when the water solution is clear after the Pd on carbon has settled.

Another pretty useful procedure for converting ephedrine or pseudoephedrine to meth is to convert the ephedrine to chloroephedrine using Lucas Reagent. This reagent has the advantage of being much more easily obtained than thionyl chloride or PCl_5, so that scientific supply houses can be left out of the supply loop. One can then pretty easily hydrogenate the chloroephedrine to meth.

The method was originally introduced by WizardX, so due credit should be given to him. The cooking procedures can be found in *Vogel's Textbook of Practical Organic Chemistry*, 5th edition, page 555.

Lucas Reagent is a mixture of concentrated 35% lab grade hydrochloric acid, and a chloride salt. Zinc chloride is the preferred chloride for this mixture, but calcium chloride and nickel chloride work as well.

To make chloroephedrine from ephedrine or pseudoephedrine, put 25 ml of lab strength hydrochloric acid into a round bottom flask, then add 8 grams of anhydrous calcium chloride. Calcium chloride is commonly sold at gas stations and hardware stores as ice melter during the winter months. As is, it usually has roughly 20% water contained within the pellets. This can be removed by grinding the pellets and putting them into a glass-baking dish. Bake the powder in an oven at 350° to 400° F for a few hours to drive off the water, then scoop the dried powder into a dry glass container with a lid for storage. The powder will suck up water from the air, so work quickly.

When the anhydrous calcium chloride has been added to the flask, then add 10 grams of ephedrine HCl or pseudoephedrine HCl. Attach a reflux condenser to the flask, and heat the mixture in a boiling water bath for about 10 hours.

Once the mixture has cooled down, it's time to get the product, chloroephedrine. Mix up a concentrated solution of sodium carbonate in water. The most convenient source for sodium carbonate is Arm and Hammer washing soda found in the detergent section of the grocery store. Lye shouldn't be used for free basing chloroephedrine because it can cause aziridine to be formed. Slowly add the sodium carbonate solution to the reaction mixture. It will fizz a lot as it neutralizes the hydrochloric acid.

When enough sodium carbonate solution has been added so that the fizzing upon addition of more solution stops, it's time to shake the solution hard for a couple of minutes. Stop from time to time to vent off any additional CO_2 which may be produced. Then stop and let the solution settle.

The chloroephedrine should be floating atop the water as an oily layer. Check the pH of the water layer with pH paper. It should read around 12. Now extract the chloroephedrine by adding 50 to 100 ml of toluene or xylene, and shaking. The chloroephedrine free base will go into the toluene or xylene. Allow the layers to settle, then drain off the water. Now add about 50 ml of clean water, and shake to wash the toluene solution. Allow it to settle, then drain off the water layer.

The toluene layer containing the chloroephedrine should be poured into a clean beaker and allowed to sit for a few hours to shed entrained water. Then it is poured into another clean beaker, and dry HCl gas is bubbled through it to get crystals of chlorephedrine HCl. They are rinsed with some clean toluene, then spread out on a plate to dry. The yield will be about 75%, and the hydrogenation procedures for reducing chloroephedrine to meth can then be followed.

A more direct one pot method shows some promise for this reaction pathway. That method is to replace the calcium chloride with nickel chloride. Then after heating the reaction mixture to make chloroephedrine, zinc dust could be added to the reaction mixture to produce nickel catalyst and hydrogen for the reaction. I have not yet had the opportunity to personally check out this interesting possibility. Expect to see more information on this in the upcoming second ed of *Advanced Techniques*.

Chapter Sixteen Methcathinone: Kitchen Improvised Crank

One designer variant upon the amphetamine molecule which gained popularity and publicity a few years ago is methcathinone, commonly called "cat." This substance is remarkably similar to the active ingredient found in the leaves of the khat tree which the loyal drug warriors on the network news blame for turning peace-loving Somalis into murderous psychopaths. The active ingredient in the khat leaves is cathinone, which has the same structural relationship to methcathinone that amphetamine has to methamphetamine. It is made by oxidizing ephedrine, while meth can be made by reducing ephedrine.

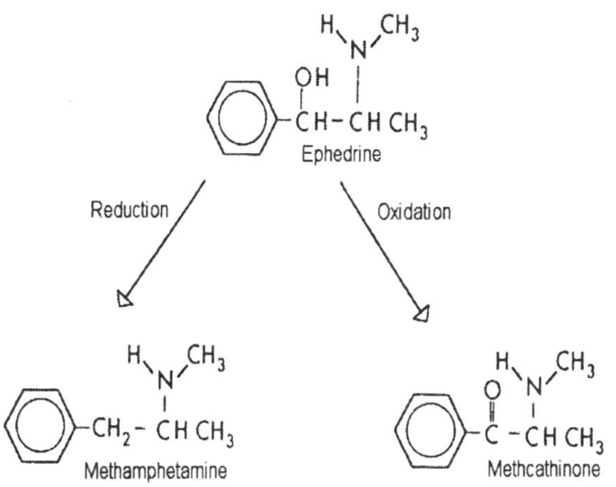

The high produced by methcathinone is in many ways similar to the one produced by methamphetamine. For something so easily made and purified, it is actually quite enjoyable. The main differences between the meth high and the methcathinone high are length of action and body feel. With methcathinone, one can expect to still get to sleep about eight hours after a large dose. On the down side, it definitely gives me the impression that the substance raises the blood pressure quite markedly. This drug may not be safe for people with weak hearts or blood vessels. Be warned!

Chronic use of methcathinone causes a person to become very stinky, as the foul metabolic breakdown products of cat come out of one's pores. Be double warned!

Cat is best made using chrome in the +6 oxidation state as the oxidizer. I recall seeing an article in the narcoswine's *Journal of Forensic Science* bragging about how they worked out a method for making it using permanganate, but that method gives an impure product in low yields. Any of the common hexavalent chrome salts can be used as the oxidizer in this reaction. This list includes chrome trioxide (CrO_3), sodium or potassium chromate (Na_2CrO_4), and sodium or potassium dichromate ($Na_2Cr_2O_7$). All of these chemicals are very common. Chrome trioxide is used in great quantities in chrome plating. The chromates are used in tanning and leather making.

For preparation of this substance, see US Patent 2,802,865. Formerly, back in the days when ephedrine pills were 30-40% active ingredients by weight, and the fillers were mostly water-insoluble, one could just extract these pills with water and directly perform the oxidation in the water extract. Now, with the garbage pills containing guaifenesin and a water-soluble fiber filler, the extraction-free basing-extraction-crystallization-isolation procedure given in this book absolutely has to be used. Using the old procedure, one gets a mess that looks like a milkshake at the end of the reaction.

Pseudoephedrine can also be used to make cat. The pills containing this starting material must be extracted according to the directions given in Chapter Fifteen, and then converted to racemic ephedrine (called racephedrine) by heating with HCl solution as also described in that chapter. This will yield dl or racemic cat, which is almost as potent as cat made from ephedrine.

Note in the patent that for each molecule of ephedrine or racephedrine in the reaction mixture, there are .66 atoms of Cr^{+6} in the solution. As a result, the amount of Cr^{+6} substance used in this reaction will vary with the compound used. For example, in the one-tenth mole-size batch given here, 20 grams of ephedrine or racephedrine hydrochloride will react with:

10 grams of $Na_2Cr_2O_7 \bullet 2H2O$
10 grams of $NaCrO_4$
22.8 grams of $NaCrO_4 \bullet 1OH_2O$
12.9 grams of $KCrO_4$
10 grams of $K_2Cr_2O_7$
6.6 grams of CrO_3

There are two main precautions to be adhered to in doing this reaction. The first one is the need to keep the temperature of the reaction mixture below 100° F. It is better to keep it well below that. To keep the reaction temperature down, the glass container in which the reaction is done should be packed in ice. I have also heard that very fast stirring will so speed up the reaction that the ice bath fails to keep the temperature down. This is only a problem with large-size batches about one mole in size or larger. In these big batches, a favorite agitation technique was to put the reaction mixture contained in a glass jug surrounded by ice in a cooler into the trunk of a car and spend a few hours driving on rough back roads to stir the mix.

The other main precaution is to add the Cr^{+6} solution slowly to the ephedrine or racephedrine with stirring. It is best to do the addition dropwise, but with larger batches, this is just not practical. In any case, use some common sense as to the rate of add for the chrome.

To do the reaction, 20 grams of ephedrine or racephedrine hydrochloride is dissolved in 50 ml of water, then 5 ml of concentrated sulfuric acid is slowly added to it with stirring. The beaker containing this mixture should then be nestled in ice to cool down. Then, in another beaker, mix 45 ml of water with 7.5 ml of sulfuric acid and the amount of Cr^{+6} compound listed above.

Begin stirring the ephedrine solution, and then dropwise add the Cr^{+6} solution to it. The addition should take about ½ hour. The chrome solution will be clear orange-red going into the mix, but it soon darkens to a blackish-red. The stirring should be continued for a reaction time of four hours. Shortening this reaction time gives poor yields and incomplete oxidation. Exceeding this reaction time causes destruction of the product, and again poor yields. The preferred solution temperature during the four hour reaction time is 80-90° F. The amount of cooling required will depend on the batch size.

When the reaction time is over, a 20% solution of lye in water should be added to the reaction mixture with stirring until it is strongly alkaline to litmus. Now pour the mixture into a sep funnel, and shake vigorously for a couple of minutes to ensure complete conversion of the cat to free base. The chrome will come out of solution as a greenish sludge. Extract this mixture with two 50 ml portions of toluene. The extracts should have a mild yellowish tint. The pooled toluene should then be washed once with 100 ml of water, then the toluene layer should be poured into a beaker to sit for about an hour. This will allow entrained water to settle and stick to the glass. Now pour the toluene solution into a fresh, clean beaker and bubble dry HCl gas through it, as described in Chapter Five, to get crystalline cat hydrochloride to precipitate out of solution. Filter this out as also described there, and spread the crystals out on a plate to dry. The yield of white to maybe slightly yellow-tinted crystals is a little over 10 grams.

The yield from this reaction is quite variable, ranging from 10% up to 80%. One often gets unreacted ephedrine instead of cat. Also, at times, it's been reported to me, a very dangerous byproduct can be formed. This byproduct causes one's blood to gel if it is injected. The nature of this byproduct is unknown to me. If there's a possibility that the cat is going to be injected, then one should re-crystallize the crude yellow-tinted cat from acetone.

So that is the standard cat recipe. How can it be improved? Let's consider the time at which this patent was filed, the mid-50s. Just after this time, it was discovered that using acetone as the solvent for Cr^{+6} oxidations gave much better results. It prevents over-oxidation of the product, which is the real problem with the original recipe. Pseudoephedrine is particularly vulnerable to the over-oxidation, so by using acetone solvent, pseudoephedrine could probably be used in the reaction.

In this acetone variation, I would put the 20 grams of ephedrine or pseudoephedrine in a beaker with 50-100 ml of hardware store acetone. Then I would mix 25 ml of water with 10 ml of sulfuric acid, and then mix the prescribed amount of Cr^{+6} compound into this acid-water solution. After cooling, the Cr^{+6} solution should be dropwise added to the acetone-ephedrine hydrochloride mixture with stirring.

After the reaction is complete in a few hours, I would pour the reaction mixture into about ten volumes of water. Then I would make the water solution strongly alkaline by adding lye water solution with shaking. Then I would extract out the cat with toluene, and then bubble this extract with dry HCl gas to get crystals of cat hydrochloride. The yield should be greatly improved, and the reaction should be more reliable.

If one really wants to get elegant, we could use the preferred oxidizer for benzyl alcohols, activated manganese dioxide. See *Journal of Organic Chemistry*, Volume 26, pages 2973-75 (1961).

To make activated MnO_2, one dissolves 1100 grams of manganese sulfate tetrahydrate or 833 grams of manganese sulfate monohydrate in 1500 ml of water. 1170 ml of a 40% solution of sodium hydroxide in water is also prepared. Finally, one makes a solution of 960 grams of potassium permanganate in 6 liters of water. The permanganate solution is heated to 80-90° C, and then the manganese sulfate solution and the sodium hydroxide solution are simultaneously added to the permanganate solution over a period of about one hour, while maintaining the temperature of the permanganate solution at 80-90° C. Strong stirring of the permanganate solution is done during the addition. Manganese dioxide precipitates out of solution as a fine brown solid. Continue stirring for an additional hour.

The hot solution is then filtered to collect the manganese dioxide, and the filter cake in the Buchner funnel is washed with water until the purple color of permanganate no longer comes out with the water rinse. The solid (manganese dioxide) is then baked in an oven at 110-120° C for about four hours to dry it. It is then ground finely, and returned to the oven for another bake at the same temperature. This is activated manganese dioxide.

Then a toluene or Coleman camper fuel extract containing about 20 grams of ephedrine or pseudoephedrine free base is diluted to a volume of about 500 ml with more of the same solvent. Then .4 mole of activated manganese oxide (35 grams) is added. The mixture is stirred, and preferably heated to reflux for five hours. A Dean-Stark trap to remove water will increase yields.

Then the cooled reaction mixture is filtered to remove the activated MnO_2 and the MnO formed by the oxidation. Then dry HCl is passed through the solution to collect the product, cat hydrochloride in about 80% yield. Your Uncle thinks this variation is more bother than it is worth, especially considering that permanganate is a List II chemical, and not easily available at the hardware store.

Chapter Seventeen
Brewing Your Own Ephedrine

I love to guzzle beer. Not that mass-produced swill, but real beer turned out in small batches by microbrewers and homebrewers. Beer that has some body, flavor, and a real kick! Homebrewing is just a joy, as lots of people have found out. Stores selling supplies to the homebrewer have sprung up in every backwater town. They are just all over the place, and newspapers catering to the homebrewer or fans of microbrews can be picked up for free at the local liquor store. The ads in these newspapers are predominately for mail-order brewing supplies at discount prices.

What a fortunate coincidence that the industrial process for making ephedrine is just a fermentation process using brewer's yeast. This process is much cheaper than extracting ephedrine from Ma Huang, and yields l-ephedrine as the product. Other chemical processes give product mixtures that consist of d and l ephedrine and pseudoephedrine. If one wishes to scale up production beyond that which can be sustained by scrounging pills and extracting them, this fermentation is a very viable alternative.

This process uses benzaldehyde as the starting material, so essentially one could consider this method as an alternative to the Knoevenagel reaction back in Chapter Nine. The fermentation action of the brewer's yeast takes the place of that List I chemical nitroethane.

Benzaldehyde is easily available, in spite of the fact that it too is a List I chemical. Oil of bitter almonds can be used as is, once the HCN it contains is removed by applying a vacuum to the oil. On a larger scale, the electric oxidation of the toluene procedure given in Chapter 9 would give all the benzaldehyde that could ever be desired.

The fermentation action of brewer's or baker's yeast converts benzaldehyde to l-1-phenylpropanol-1-one-2 in a yield corresponding to about 70% by weight of the benzaldehyde added to the fermentation mixture. This phenylacetone derivative is then reductively alkylated with methylamine by any one of several procedures to give l-ephedrine.

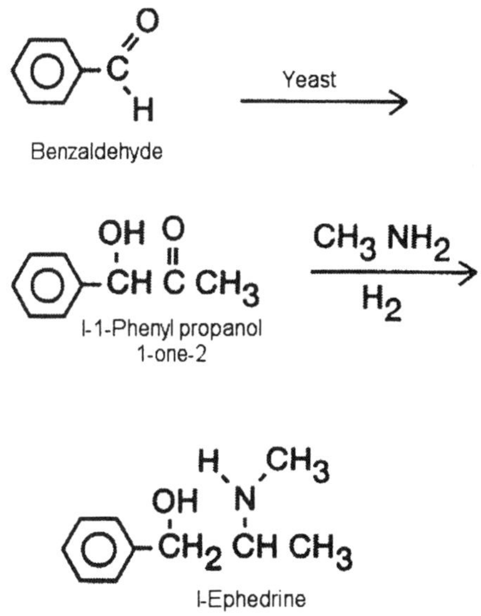

One would think that the reductive alkylation of that phenylacetone derivative would yield d,l-ephedrine, and then that reduction of that d,l-ephedrine would then give d,l-meth, that same racemic meth that results from reductive alkylation of phenylacetone. (Your Uncle prefers the buzz produced by the racemate over the harsher, more nerve jangling buzz produced by d-meth.) Apparently, this isn't the case. The references for this process claim that solely l-ephedrine is produced, and then reduction of this l-ephedrine, which is identical to natural ephedrine, yields that potent but harsh d-meth.

To start with this project, one would first want to read a home beer brewing book, such as *Better Beer and How to Brew It,* since the processes are so similar and much of the same equipment and materials will be used. I have this book, and it is good. This is all you need to sound like a real brewer when you head down to the Brew Shop in your town to pick up supplies.

As with regular beer brewing, one starts with a brew vat, five-gallon plastic pails work just fine for this purpose. They should be cleaned, then rinsed with bleach diluted with several volumes of water to disinfect the surfaces, then rinsed some more with clean water to remove the bleach residue.

Next fill the pails with tap water until they are half to $^{2}/_{3}$ full. We are now ready to brew. See *Drug Trade News*, Volume 16, Number 16, page 27 (1941), (I love that reference) and *Wallerstein Labs. Commun.*, Volume 4, Number 13, page 213 (1941). Also see *Chemical Abstracts,* Volume 17, page 1484, and *Biochemische Zeitschrift*, Volume 115, page 282 (1921), and Volume 128, page 610 (1922). These articles will give you some historical perspective on the process. Then go to *Biotechnology and Bioengineering*, Volume 34, pages 933-41 (1989) and *World Journal of Microbiology and Biotechnology*, Volume 16, pages 499-506 (2000) for more contemporary techniques.

Into the pail the brew mixture is made up by adding molasses to clean warm water. Add roughly 31 grams of molasses for each quart of water in the pail, and fill the pail no more than ¾ full of water because there will be frothing and foaming when the yeast starts to grow. Then live baker's yeast is added. The best yeast to use are the cakes of refrigerated yeast found in the grocery store rather than the freeze dried packets, although both would work. It takes a lot of yeast to do the chemical transformation, so stir in the package of yeast, and let the yeast grow at about 80° F, like one would when making bread.

When the brew mixture in the pail has been fermenting for about eight hours at warm room temperature, it's time to add the benzaldehyde. Start with about 4 ml of benzaldehyde for each quart of water in the pail. Once the benzaldehyde has been added, bubble some air through the culture using an aquarium pump. Then add about ½ ml of acetone per quart of water. Acetone is found in the hardware store's paint section, and a bit of it in the mixture increases the yield of product. Also add about ¼ gram of Epsom salts for each quart of water. The magnesium in Epsom salts aids the conversion of the benzaldehyde. Allow the yeast to work for about four hours, then add an additional 4 ml of benzaldehyde for each quart of water in the pail. Adding all the benzaldehyde at once would tend to poison the growing yeast and ruin the yield. Then continue the aeration with the aquarium pump for at least another eight hours as the yeast completes the conversion of the benzaldehyde.

During the course of the fermentation, an enzyme called carboligase (pyruvate decarboxylase) produced by the yeast converts the benzaldehyde to phenylpropanol-1-one-2. It is believed that the enzyme links acetaldehyde or acetic acid made by the fermenting yeast with the benzaldehyde to give the product. In any case, in less than a day, one gets a yield of product amounting to 70% of the amount of benzaldehyde used.

When the fermentation is completed after about 12 hours or so, it's time to recover the phenylpropanol-1-one-2 from the brew mixture. The yeast in the mixture is a problem. With regular beer brewing, the yeast just settles to the bottom of the fermenter when the fermentation is complete. Siphoning is then done to remove the clear beer from the settled yeast. If you have days to let the yeast settle, that may be an option. The industrial process uses centrifugation of the fermentation mixture to force the yeast to the bottom. I'm sure that works well for them, because once the centrifuge is installed, no materials need be purchased from then on to settle the yeast. The centrifuge pays for itself.

Brewers approach unsettled yeast in two ways. On a small scale, they will add a material called "finings" to the brew mixture, and it settles the yeast. On a larger scale, they will filter the brew. Yeast is some gummy stuff. It will plug a filter paper in no time flat, so in addition to the filter paper, they use filter aid.

Filter aid is stuff like Celite (diatoms), powdered cellulose, or even a bed of sand to catch those gummy yeast particles before they get to the filter paper and plug it up.

Once the yeast has been removed from the brew mixture, the phenylpropanol-1-one-2 can be extracted out of the solution. The original references used ether to do the extraction. I would suggest substituting hardware store toluene. Several extractions with a few hundred ml portions of toluene should be enough to completely remove the product from a 5 gallon pail fermenter.

Next the combined extracts should be distilled to remove the toluene. Once the toluene is mostly all gone, the residue should be fractionally distilled under a vacuum. The product, also called phenylacetylcarbinol, distills over the range of 100° C to 150° C under a vacuum of 14 torr. A really good aspirator using nice cold water will pull a vacuum this strong. Weaker vacuums will result in higher boiling ranges. The yield of distilled product amounts to around 70% of the amount of benzaldehyde added to the fermentation mixture.

An alternative to vacuum distillation is to isolate the phenylpropanol-1-one-2 by means of the bisulfite addition product. Take a volume of water roughly equal to the toluene extracts. Dissolve about 10 to 20% by weight of sodium bisulfite into the water, then cool it down to around 10° C. Sodium bisulfite is commonly sold at brewing supply shops. Now pour the toluene extracts into a sep funnel or other glass container, and then add the sodium bisulfite solution. Shake the two of them together for a few minutes, then let the layers settle. The product will be in the water layer on the bottom, so drain it off and save it. The toluene layer can be thrown away.

To recover the product, prepare a 10% by weight solution of sodium bicarbonate in water. Arm and Hammer bicarb is the most convenient source of sodium bicarbonate. With stirring, drip in the bicarb solution to the bisulfite solution containing the product until no more bubbles of CO_2 are given off. The bisulfite addition product of the phenylpropanol-1-one-2 has just been broken, and the product can be extracted with solvent. Toluene or ether starting fluid are suitable extracting solvents. With ether, simply extract, separate the ether layer, and allow the ether to evaporate away. With toluene, a source of vacuum to aid the evaporation of the solvent would be helpful.

Now the phenylpropanol-1-one-2 can be reductively alkylated to give l-ephedrine. Any one of several methods can be used, just as in the case of reductively alkylating phenylacetone to meth. Method number one has to be catalytic hydrogenation using platinum catalyst.

In the example taken from US Patent 1,956,950, the chemists place 300 ml of the distilled phenylpropanol-1-one-2 in the hydrogenation bomb along with one gram of platinum catalyst, and 85 grams of 33% methylamine solution. They state that it's advantageous to add some ether to the hydrogenation solution. How much is some, they don't say. They then hydrogenate the solution in the usual manner, with up to 3 atmospheres of hydrogen pressure, and magnetic stirring of the contents of the hydrogenation bomb.

When absorption of hydrogen stops in two or three hours, the platinum catalyst is filtered out. Then the ether hydrogenation mixture is shaken with a volume or two of 10% HCl solution to pull the ephedrine out of the ether and into the acid water, forming the HCl salt of ephedrine. The ether layer is separated off with a sep funnel, then the dilute acid is boiled away. The residue is diluted with a little alcohol, and then a lot more ether. Passing dry HCl through this mixture then gives crystals of pure ephedrine hydrochloride. Their yield was around 110 grams.

My commentary on this hydrogenation? That yield is awfully low. Using phenylacetone as a guide, one should be expecting a yield around 300 grams of ephedrine. What's up? Check out the amount of methylamine used. There are about two moles of the phenylacetone derivative, but they don't even use one mole of methylamine. It should be the other way around, an excess of methylamine. Perhaps this is how they only get l-ephedrine from the phenylacetone derivative. In any case, I'd much rather have 300 grams of racephedrine than 110 grams of l-ephedrine. My thoughts are that one would be better served just going to Chapter Eleven, and just plug in this phenylacetone derivative for the regular phenylacetone. That means two or three moles of methylamine for each mole of phenylacetone, alcohol as solvent, and a bit more platinum catalyst in the mixture.

In the patent, they give another reductive alkylation example. They use amalgamated aluminum as the reducer, just like in Method Three in Chapter Twelve. They take 120 grams of the undistilled fermentation product containing the 1-phenylpropanol-1-one-2, and drip it over the course of two hours into a solution of 10 grams of methylamine in 500 ml of ether in the presence of 20 grams of activated aluminum amalgam. Simultaneously, they drip into the mixture 20 to 30 ml of water. Stirring of the mixture is required.

The vigorous reaction that sets in is moderated by periodic cooling. When the reaction is complete after a few hours, they filter the mixture to remove the aluminum. Then they shake the ether solution with 10% HCl solution to draw the ephedrine into the water. The ether layer is separated, then the dilute acid boiled off. The residue is thinned with a little alcohol, then dissolved in a lot more ether. Bubbling with dry HCl gives 25 to 45 grams of 1-ephedrine hydrochloride crystals.

My commentary on this procedure is identical to the last one. So little methylamine used! I haven't tried this, but I would be surprised to say the least if more methylamine didn't greatly increase the yield of product. I would also think that any one of the activated aluminum procedures given in Chapter Twelve could be used, just by plugging in this phenylacetone derivative for the regular phenylacetone. Also the use of ether is to be avoided when possible. One could also use one of the reduction methods from Chapter Twelve, which make use of sodium cyanoborohydride or sodium borohydride to reduce a mixture of methylamine plus 1-phenylpropanol-1-one-2 to ephedrine. Of the two choices, sodium borohydride would be best because it is easily available and produces good yields of product. See *The Journal of Chemical Technology and Biotechnology* Volume 77, pages 137 to 140 (2002) for a sample recipe using sodium borohydride to do this reduction. Note that they zapped the reaction mixture in a microwave oven to kick start the reduction.

You don't like that recipe? Check out this one taken from *Chemical Abstracts,* Volume 47, column 3347. Twenty grams of N-methyl-d,l-alanine and 50 grams of benzaldehyde are placed in a flask and heated on an oil bath at 150-160° C until the mixture stops fizzing off carbon dioxide.

The mixture is then cooled and mixed with a few hundred ml of toluene. Whatever doesn't dissolve in the toluene is thrown away. The product, which is a mixture of ephedrine and pseudoephedrine, is then extracted out of the toluene by shaking the toluene with about an equal volume of 10% HCl. The toluene can be distilled to recover unused benzaldehyde, if there is any in it.

The dilute hydrochloric acid solution which contains the products should be boiled down to concentrate it. The steam will also carry off some byproducts, so vent this steam outside.

Once the dilute acid has boiled down to a volume of 50-100 ml, allow it to cool. Then add a little activated carbon, and stir it around for a while. Then filter it out. This will decolorize the solution.

Add lye pellets a little bit at a time with stirring until the water solution is strongly alkaline. Extract the alkaline water a few times with toluene. The combined toluene extracts should next be bubbled with dry HCl gas to give a crystalline product amounting to about 12 grams. The product will be about 8 grams of d,l-pseudoephedrine, and 4 grams of d,l-ephedrine. It will yield racemic meth upon reduction.

Take note that recovering ephedrine from water solutions is a bit different than recovering meth. That's because ephedrine free base dissolves well in water, while meth doesn't. So for recovery of the ephedrine we take the dilute acid solution of the ephedrine and boil it down, just like in the pill extraction procedure using water. Once it is concentrated, then it is made alkaline with lye, and the ephedrine extracted out. In this way you get good recovery of the ephedrine. Use too much water, and it's difficult to extract it all out.

This recipe is pretty easy to scale up to larger size batches, but it suffers from a really serious flaw. N-methyl alanine is just about impossible to find, and if one could find it, the cost charged for it is astronomical. It's also not so easy to make from the common amino acid alanine.

The way around these problems is to substitute alanine for the N-methyl alanine in the example batch just given. The product obtained then would be phenylpropanolamine (PPA). Reduction of that PPA by any of the methods given in this chapter would then give one Benzedrine if d,l-alanine was used, or Dexedrine if food grade l-alanine was used.

The reaction using alanine rather than N-methyl alanine works exactly the same. The yield of product can be increased if one uses a larger amount of alanine than would be called for if one copied the sample batch exactly. One could double the amount of alanine used, keep the other ingredients the same as in the sample batch and get a much higher amount of PPA than was obtained from the sample batch. Alanine is really cheap, so this is a good strategy. Researchers have found that taking alanine daily helps shrink swollen prostate glands. This is a great excuse for anyone getting the amino acid through health food stores.

People who have tried this reaction using alanine have found that it is best to grind the alanine down to a nice fine powder before adding it to the batch. It doesn't dissolve very well. It is also advisable to add the alanine slowly without stirring as the batch warms up because it has a tendency to clump together in the reaction mixture. Clumped up alanine will not be able to react to make PPA. All things considered, this reaction is a good alternative to scrounging for pills that get more difficult to buy and harder to extract every day.

Chapter Eighteen MDA, Ecstasy (XTC), and Other Psychedelic Amphetamines

The psychedelic amphetamines are a fascinating and largely ignored group of drugs. They all have the basic amphetamine carbon skeleton structure, but show effects that are more akin to LSD than to the amphetamines. The LSD-like effect is due to the presence of a variety of "add ons" to the benzene ring of the basic amphetamine structure. Generally, these "add ons" are ether groupings on the 3, 4, or 5 positions on the benzene ring. Because of these "add ons" one can consider these compounds more closely related to mescaline than to amphetamine. Consider the mescaline molecule pictured below.

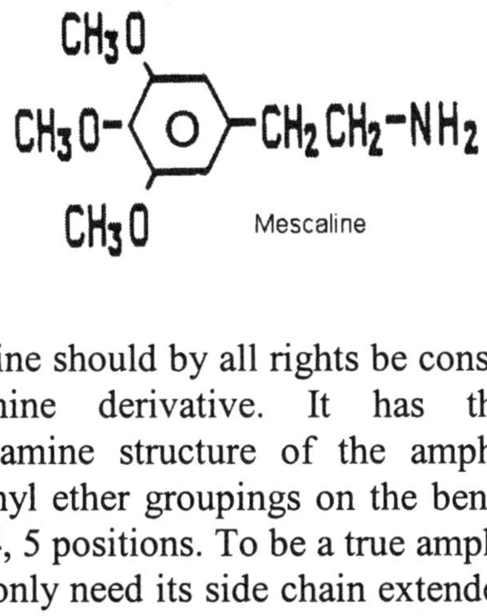

Mescaline

Mescaline should by all rights be considered an amphetamine derivative. It has the basic phenethylamine structure of the amphetamines with methyl ether groupings on the benzene ring at the 3, 4, 5 positions. To be a true amphetamine, it would only need its side chain extended by one carbon, putting the nitrogen atom in the central, isopropyl position. Such a compound does in fact exist. It is called trimethoxyamphetamine, or TMA for short. Its effects are very similar to mescaline in much lower dosage levels than the ½ gram required for pure mescaline. Its chemical cousin, TMA-2 (2,4,5 trimethoxyamphetamine) has similar awe-inspiring characteristics. More on this later.

The most popular and, in my opinion, the best of the psychedelic amphetamines are the members of the MDA family. This family consists of MDA, and its methamphetamine analog, XTC, or Ecstasy, or MDMA. MDA (3,4 methylenedioxyamphetamine) gives by far the best high of this group. Its effects can best be described as being sort of like LSD without the extreme excited state caused by that substance. It was popularly known as "the love drug" because of the calm state of empathy so characteristic of its effect. It could also be a powerful aphrodisiac under the right circumstances.

This substance gradually disappeared during the early 80s due to an effective crimping upon the chemicals needed for its easy manufacture.

This crimping, and the drug laws in effect at the time, gave rise to a bastard offspring of MDA. This substance was XTC, or MDMA, the so-called Ecstasy of the drug trade. This material was a designer variant of MDA, and so was legal. The chemicals needed to make it could be obtained without fear of a bust. It also lacked the best qualities of its parent. While the addition of a methyl group of the nitrogen of the amphetamine molecule accentuates its power and fine effect, the addition of a methyl group to the MDA molecule merely served to make it legal. As fate would have it, the hoopla surrounding the subsequent outlawing of this bastard child served to make it a more desired substance than MDA. This is typical of black-market, prohibition-driven demand.

To understand the various routes which can be followed to make these substances, note the structures of MDA and MDMA shown below:

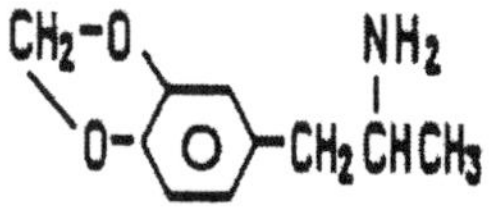
MDA

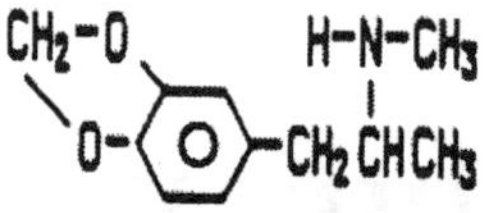
MDMA

To make these substances, and the rest of the psychedelic amphetamines for that matter, the manufacturer has a choice of two starting materials. He can use the appropriately substituted benzaldehyde, which in the case of MDA or MDMA is piperonal (heliotropin), or he can use the correspondingly substituted allylbenzene, which in this case is safrole. These substances are pictured below:

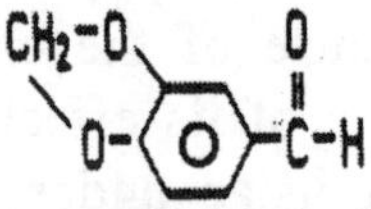
Piperonal

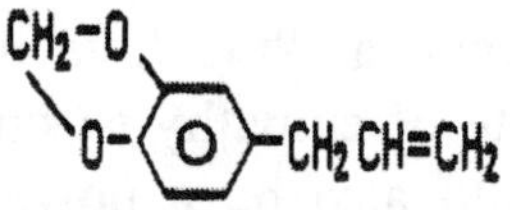
Safrole

Piperonal was the favored starting material for making MDA, as were the other substituted benzaldehydes for making other psychedelic amphetamines. The supply of these raw materials was effectively shut off. Piperonal does find legitimate use in making perfumes, but considerable determination is needed to divert significant amounts of the stuff into clandestine operations.

Once obtained, these substituted benzaldehydes could be converted into amphetamines by an interesting variant of the Knoevenagel reaction as described in Chapter Nine. They could be reacted in a mixture of nitroethane and ammonium acetate to form the appropriately substituted 1-phenyl-2-nitropropene. This nitropropene could then be reduced to the amphetamine by using lithium aluminum hydride. For this recipe, see *PIHKAL* under MDA. The nitroalkene obtained by the reaction of piperonal and nitroethane can also be reduced by the hydrogenation methods given in the Knoevenagal reaction section of this book in a yield of around 50%. Similarly, the electric reduction method given in that section can also be used. Now that both piperonal and nitroethane are List I chemicals, we would have to concede that the narcoswine have won this round, and that this pathway can be considered for all practical purposes to be dead.

This left safrole, and the other substituted allylbenzenes, as starting materials for psychedelic amphetamine manufacture. This route had the advantage of having a raw material source that was nearly impossible to shut down if you are lucky enough to have a grove of sassafras trees nearby. For instance, sassafras oil consists of 80-90% safrole. One merely has to distill the oil under a vacuum to get very pure safrole. Similarly, other psychedelic amphetamines can be made using essential oils that contain the appropriately substituted allylbenzene or propenylbenzene as a major substituent. For instance, calamus oil contains a large proportion of B-asarone, the starting material for TMA-2. Nutmeg contains a mixture of myristicin (potential MMDA) and elemicin (potential TMA). These oils, with the exception of sassafras oil, are all available from herbal supply shops and dealers in the occult. Even without this source, the oils can be easily obtained from the plants.

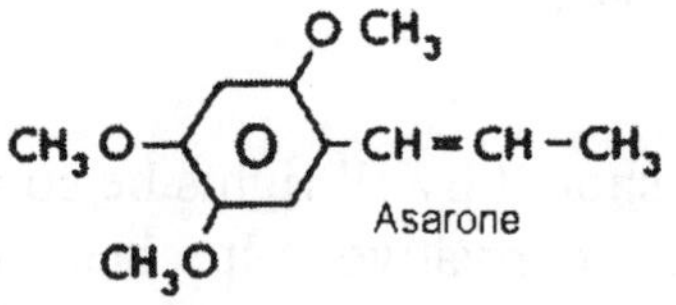
Asarone

Calamus oil is some interesting stuff! Its composition depends upon the country the oil comes from. Luckily, most of the oil on the market comes from India. The vast majority of oil from that country contains about 80% B-asarone, although there are reports (see *Journal of Indian Chemical Society*, Volume 16, page 583, 1939) that some oils from that country contain around 80% allylasarone.

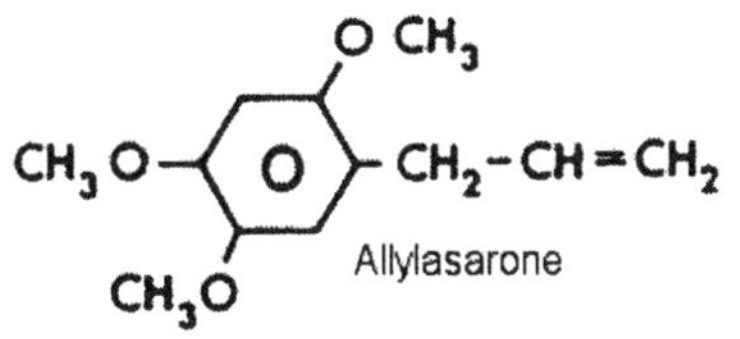

Allylasarone

Other major sources of commercial calamus oil are Japan and Europe. These oils contain lesser and variable amounts of a-asarone. This is the cis-trans isomer of B-asarone. It differs in that a-asarone is a solid at room temperature, and may precipitate out of oils upon cooling in a freezer. It reacts in the same manner as B-asarone. Both can be obtained in a pure form from the oil by fractional vacuum distillation.

On the topic of purifying essential oils, it has been proposed by other underground sources that sassafras oil can be purified by putting it in a freezer, allowing the safrole to solidify, and then filtering out the solid safrole. Let me fill you in on the facts of the matter. Sassafras oil is very stable in a supercooled state. You can put a bottle in a freezer for months, and never see a crystal of solid safrole form. Believe me, I've tried it. To get crystals to form, a seed crystal of solid frozen safrole would have to be added to the supercooled sassafras oil. Where do you get this seed crystal to start with? And at 80-90% pure safrole, the oil will then freeze into a virtual solid block, so what would filter out except the safrole that begins melting during the filtering process? This whole line of pursuit is a waste of time. Moreover, the small amount of impurities are actually beneficial if the HBr route is chosen for production of MDA or MDMA from the sassafras oil.

Starting with essential oils, how does one make the desired amphetamine from them? Let's take the conversion of sassafras oil to MDA or MDMA as the example to illustrate the various processes which can be used. If we go to *PIHKAL*, and read the recipe for MDA, you get the old classical procedure. Safrole obtained from sassafras oil is first converted to isosafrole (a propenylbenzene). This is done by putting safrole into a flask, adding some 10% alcoholic KOH, and then warming the mixture up to 243° C for 3 minutes. This isomerization works just fine so long as absolute alcohol is used, and the alcohol is allowed to distill off. You know that you have gotten isomerization, because the boiling point of safrole is 233° C.

The isosafrole is then mixed with acetone, formic acid and hydrogen peroxide to give the glycol mentioned in Chapter Ten. The reaction mixture is evaporated away under a vacuum, then the residue in the flask is heated with sulfuric acid in alcohol solvent to give the phenylacetone. The phenylacetone is then used to make the amphetamine by any of the methods given in this book.

My opinion on this method? It's a lot of work, the yields are on the low side, and that evaporation of the reaction mixture under the vacuum will destroy your aspirator. Peroxyformic acid is rough on metal. Let's use the more direct approach.

The first problem which confronts the chemist in the process of turning sassafras oil into MDA or MDMA is the need to obtain pure safrole from it. In spite of the fact that crude sassafras oil consists of 80-90% safrole, depending on its source, it is a good bet that the impurities will lower the yield of the desired product. The axiom "garbage in, garbage out" was custom made for organic chemistry reactions. It is simplicity itself to turn crude sassafras oil into pure safrole, and well worth the effort of underground chemists bent on MDA production.

Sassafras oil is an orange-colored liquid with a smell just like licorice. It is a complex mixture of substances which is easily purified by distilling. To obtain pure safrole from sassafras oil, the glassware is set up as shown in Figure 13 in Chapter Three. The distilling flask is filled about $^{2}/_{3}$ full of sassafras oil, along with a few boiling chips, and then vacuum is applied to the system. A little bit of boiling results due to water in the oil, but heat from the buffet range is required to get things moving. Water along with eugenol and related substances distill at the lower temperatures. Then comes the safrole fraction. The safrole fraction is easily spotted because the "oil mixed with water" appearance of the watery fore-run is replaced with a clear, homogenous run of safrole. When the safrole begins distilling, the collecting flask is replaced with a clean new one to receive it. The chemist is mindful that the safrole product is 80-90% of the total volume of the sassafras oil. Under a vacuum, it boils at temperatures similar to phenylacetone and methamphetamine. When all the safrole has distilled, a small residue of dark orange-colored liquid remains in the distilling flask. The distilled safrole is watery in appearance, and smells like licorice.

With a liberal supply of safrole obtained by distilling sassafras oil, work can then commence on converting it into 3,4 methylenedioxyphenylacetone. This is done in exactly the same manner as described in Chapter Ten. Any one of the three Wacker oxidations of the allylbenzene (safrole) to the phenylacetone (m-d-phenylacetone) can be used. When the essential oil contains a propenyl benzene, such as the B-asarone in calamus oil, then the electric cell discussed in Chapter Ten and *Practical LSD Manufacture* should be used to get the phenylacetone in high yield.

With the methylenedioxyphenylacetone obtained in this manner, the chemist proceeds to make it into XTC by one of the methods used to turn phenylacetone into meth. Of all the methods to choose from, the most favored one would have to be reductive alkylation using the bomb and platinum catalyst. The free base is converted into crystalline hydrochloride salt in exactly the same manner as for making meth crystals. It is interesting to note here that XTC crystals will grow in the form of little strings in the ether solution as the HCl gas is bubbled through it. Once filtered and dried, it bears a remarkable resemblance to meth crystals. It generally has a faint odor which reminds one of licorice.

To make MDA from the methylenedioxyphenylacetone, one has three good choices. Choice number one is to use the reductive amination method with a bomb, with Raney nickel catalyst and ammonia. See *Journal of the American Chemical Society*, Volume 70, pages 2811-12 (1948). Also see *Chemical Abstracts* from 1954, column 2097. This gives a yield around 80% if plenty of Raney nickel is used. The drawback to this method is the need for a shaker device for the bomb, and also a heater.

A complete discussion of these two methods can be found in Chapter Twelve. The only difference is that the substituted phenylacetone is used instead of regular phenylacetone, and a substituted amphetamine is produced as a result. One should also see *Advanced Techniques of Clandestine Psychedelic & Amphetamine Manufacture* for a Convenient Tabletop MDA recipe using a Raney nickel cathode to do the hydrogenation, and also for a convenient method of making ammonia-saturated alcohol. MDA distills at about 150° C at aspirator vacuum of 20 torr, and MDMA will distill at around 160° C under the same vacuum. Poorer vacuum will result in higher boiling temperatures.

Another method for converting methylenedioxyphenylacetone to MDA is the Leuckardt reaction. My experience with mixing formamide with phenylacetone to get amphetamine is that using anything other than 99% formamide is a waste of time. You just get that red tar. Two ways have been found around that. These variations use the much more easily available 98% formamide. See *Chemical Abstracts* from 1953, column 11246, and Austrian patent 174,057. In this variation, 40 ml of methylenedioxyphenylacetone is mixed with 110 ml of freshly vacuum-distilled formamide, 2 ml glacial acetic acid, and 20 ml water. This mixture is heated up to about 130° C, at which point bubbling should begin. Then the temperature is slowly raised to keep the bubbling going, as described in Chapter Five, until a temperature of 150° C is reached. This should take at least 5 hours. The yield is 70%, according to the patent.

Processing is then done just as in the case of meth. The formamide is destroyed by boiling with lye solution. In this case, the ammonia gas which is produced is led away in plastic tubing. The formyl amide is then separated, and hydrolyzed by refluxing in a mixture of 60 grams of KOH, 200 ml alcohol, and 50 ml water for an hour. After the reflux, the mixture is made acid with HCl, and the alcohol evaporated away under a vacuum. The residue is then diluted with water, and the free base obtained by making the solution strongly alkaline to litmus by adding lye solution. The free base is then extracted out with some toluene, and distilled.

Most people don't get close to the 70% yield claimed in the patent for this method.

Another choice is to use the European Variation of the Leuckardt reaction, given in Chapter Five. The last I heard from Geert, the heat was closing in on him, but he was going to pass along an XTC recipe that is very popular over there. He says that they do it in an icebox! I haven't heard from him since, and that was nearly 4 years ago. This space is dedicated to him.

The last choice is a very simple, but also very time-consuming (several days!) reaction. Sodium cyanoborohydride in methanol with ammonium acetate and methylenedioxyphenylacetone at pH 6 react to give disappointing yields of MDA. See *PIHKAL* by Dr. Shulgin in the section under MDA, for full cooking instructions.

Reference

Psychedelics Encyclopedia, by Peter Stafford.

The recommended dosage of MDA or XTC is about a tenth of a gram of pure material. TMA-2 is 40 milligrams.

The other good synthetic route of making MDA, MDMA and related psychedelic amphetamines from the substituted allylbenzenes found in essential oils such as sassafras oil is a two-step procedure involving first reacting the substituted allylbenzene (e.g., safrole from sassafras oil) with HBr to make the corresponding phenyl-substituted 2-bromopropane. Then this substance is mixed with an alcohol solution containing excess ammonia or methylamine to yield MDA or MDMA from, for example, safrole. Heating is required to get a good yield of product. Details on this procedure are found in the chapter covering the production of meth or benzedrine from benzene and allyl chloride (Chapter Twenty One). The reason why it is in that chapter is because the final step of heating the 2-bromopropane compound with ammonia or methylamine solution is pretty much identical. Some further commentary on this route not found in that chapter is called for.

The addition of HX (HCl, HBr, HI) to a double bond is a general reaction, meaning most all double bonds, other than those found in benzene rings, will add HX. Of these three acids, HBr adds most easily to double bonds. It is also the only one that will add abnormally, meaning that one can get, besides the 2-bromopropane, the 3-bromopropane also. Exposure to strong light or oxidizing substances promotes the abnormal addition, so this reaction shouldn't be done in full sunlight.

The strength of the HBr used in reaction has a great effect upon the yield and speed of the reaction with safrole. The less free water floating around in the acid, the better it reacts with safrole. So dry HBr gas will react best with safrole, followed closely by 70% HBr, while the ACS reagent 48% HBr is practically useless as is.

Another point to be aware of is cleavage of the methylenedioxy ether by HX. HI is much better at cleaving this ether than is HBr, which is better than HCl. It is because of "ether" cleavage that the temperature during this reaction must not be allowed to rise above the stated limits in the procedures given in this book. If your magnetic stirrer gets warm while working, the batch must be insulated from this source of heat.

An obvious variation upon this procedure which would pop into the head of any thinking chemist reading this tract would center around adding dry HCl to safrole by bubbling dry HCl through a toluene solution of sassafras oil to get the 2-chloropropane, and reacting this substance with ammonia or methylamine like the other phenyl-2-chloropropanes listed in the *Journal of the American Chemical Society* article cited in the meth or benzedrine from benzene and allyl chloride chapter in this book. My observations on this route will be useful if someone is contemplating this procedure.

First of all, dry HCl adds only slowly to safrole at room temperature. A toluene solution of sassafras oil literally reeking with HCl, sealed up and kept at an average temperature of 90° F for three weeks, resulted in only about 10% conversion of the safrole to chlorosafrole. No doubt, some further heat must be applied to the mixture to get reasonably complete conversion of the safrole to chlorosafrole. HCl doesn't cleave ethers very well, so this can be considered safe.

How does this observation jibe with the *Journal of the American Chemical Society* article in which they postulate that when allyl chloride adds to benzene or substituted benzene, the 2-chlorophenylpropanes are the result of HCl adding to the double bond of the allylbenzene? Either the theory was mistaken, or iron chloride is a catalyst for adding HCl to the double bond. I haven't yet checked this out personally, but it's worth a try.

Further, once one has chlorosafrole, what good is it? See the above-cited *Journal of the American Chemical Society* article. You will note that the yields obtained converting similar phenyl-substituted ether 2-chloropropanes is pretty low, down near 10%. That's why bromosafrole is used to make MDA or MDMA. The bromine atom is much more easily replaced with ammonia than is chlorine. It's termed a better leaving group. The iodine atom is a much better leaving group than is the bromine atom, so even better results should be had reacting iodosafrole with ammonia or methylamine. One would expect that lower temperatures could be used, maybe even room temperature. This would avoid all the tar formed as a by-product when heating bromosafrole.

Chlorosafrole can be converted to iodosafrole by refluxing one mole of chlorosafrole with 2 moles of sodium iodide in a saturated solution in acetone for about 15 minutes to ½ hour. After cooling this reaction mixture, the sodium chloride that precipitates out of solution is filtered. Then the acetone is taken off under a vacuum. The resulting residue of iodosafrole and NaI crystals is extracted with toluene to remove the product from the NaI crystals, which can be reused. This toluene extract is shaken with water containing some sodium thiosulfate and a little HCl. This destroys iodine formed by decomposition of the NaI. Snorting iodine really sucks. Exposure to light speeds the decomposition of NaI to iodine, especially in solution. Experimenters using this procedure are invited to write in with their results.

A final word needs to be said about the Ritter reaction. Since safrole and related allylbenzenes from essential oils are all allylbenzenes, one would assume that the Ritter reaction would be directly applicable to them. Such is not the case. See *Chemical Abstracts*, Volume 22, page 86, for an article titled "Cleavage of the Methylenedioxy Group." Here they detail how concentrated sulfuric acid quickly cleaves the methylenedioxy group. As a consequence, brave experimenters wishing to use the Ritter reaction to make MDA must use the substitutes for sulfuric acid which are listed in the *Journal of the American Chemical Society* article cited in Chapter Fourteen. Substitutes include methanesulfonic acid and polyphosphoric acid. Directions for how to make the latter from phosphoric acid and P_2O_5 are to be found in the *Merck Index*. A final caveat for those trying to make chlorosafrole is also to be found in that article. The article states that fuming HCl, heated to 100° to 130° C in a sealed tube, is a potent cleaver of the methylenedioxy group. Heating of safrole with dry HCl must be held well below this level.

Know Your Essential Oils

Sassafras Oil — contains about 80-90% safrole. This is purified by fractional vacuum distillation. Boiling point of safrole is 234° C at normal pressure, about 120° C with an aspirator, and 105° at 6 torr. Yields MDA with ammonia, or MDMA (XTC) with methylamine. Dosage 1/10 gram.

Calamus Oil — that of Indian origin contains 80% B-asarone. Oil from other areas contains much less asarone. Boiling point is 296° C at normal pressure, and 167° C at 12 torr. Yields TMA-2. Dosage is 40 mg.

Indian Dill Seed Oil — contains up to 53% dill apiol (3,4-methylene-dioxy-5,6-dimethoxy-allyl-benzene). Boiling point is 296° C with decomposition at normal pressure. Aspirator vacuum will distill it at about 170° C. Yields DMMDA-2, dosage about 50 mg.

Nutmeg Oil — contains 0-3% safrole, and 0-13% myristicin (3,4-methylene-dioxy-5-methoxy allylbenzene). The boiling point at 15 torr is 150° C. Yield MMDA, dosage 80 mg.

Mace Oil — contains 10% myristicin.

Parsley Seed Oil — contains 0-80% parsley apiol (2-methoxy-3,4-methylene-dioxy-5-methoxy-allylbenzene). Its boiling point is 292° C at normal pressure, and 179° C at 34 torr. It yields DMMDA, dosage about 75 mg. This oil may also contain 10-77% myristicin.

Oil of Bitter Almonds — contains around 95% benzaldehyde. This is a precursor to phenylacetone or amphetamine.

Oil of Cinnamon — contains 80-90% cinnamaldehyde. This can be reduced to allylbenzene with borohydride.

WARNING!! Some wholesale distributors of essential oils are being leaned upon to give up their customer lists. The heat wants to know who is buying sassafras oil, and oil of bitter almonds. They will soon want to know who is getting cinnamon oil, after this book hits the streets. Oils fall under the definition of "mixture" in the chemical division act, and so did not used to be subject to regulation. In the latest version of CFR 21, the DEA has decided that it now has control over mixtures containing List One chemicals. They simply decide by themselves if the List One chemical is "easily obtained" from the mixture. Buying retail is still completely safe, if you are lucky enough to find sassafras oil on any retail shelves. Be warned!

References

PIHKAL, by Dr. Shulgin
The Essential Oils, by Ernest Guenther
Psychedelics Encyclopedia, by Peter Stafford

[illegible]

[illegible] is 60 mg.

[illegible] Oil [illegible] up to 75% [illegible] and [illegible] The boiling point is 270° C. [illegible] decomposition. [illegible] with [illegible] yields [illegible] dosage about 50 mg.

[illegible] Oil — contains [illegible] and [illegible] methylenedioxy [illegible] The boiling point is [illegible] 150° C. Yields MMDA, dosage [illegible] mg.

Mace Oil — contains 10% myristicin.

[illegible] Oil — contains [illegible] methylenedioxy [illegible] methoxy [illegible] boiling point is [illegible] yields DMMDA, dosage about [illegible] This oil may also contain [illegible] myristicin.

Oil of Bitter Almonds — contains [illegible] benzaldehyde [illegible] phenylacetone or amphetamine.

Oil of Cinnamon — contains [illegible] aldehyde. [illegible] can be reduced to [illegible] aldehyde.

[illegible] essential oils [illegible] the [illegible] who [illegible] their oil and [illegible] they [illegible] an [illegible] this book [illegible] Oils [illegible] the [illegible] and [illegible] the latest version of CFR [illegible] DEA has [illegible] that it now has control over

Chapter Nineteen
Ice

At the time of the writing of the second edition, the latest drug craze was the smokable form of methamphetamine called "ice." At the writing of this seventh edition, this material was still popular, with most usage being confined to those with serious drug problems.

I'm not going to endorse or encourage the foolhardy practice of smoking meth. Seeing firsthand what this stuff does to rubber stoppers, corks, and razor blades, I can only imagine what it does to lung tissue. My opinion on this practice is similar to my opinion on injecting the substance. If snorting the hydrochloride salt doesn't get you as wired as you could ever want to get, it is time to give up and find something else to fill your spare time with.

I have never made nor used "ice" as such, but I can tell you how to get smokable forms of meth. Since the godless importers of this stuff have already created a market for it, it's only right that I help American technology catch up.

The regular hydrochloride salt is not ideally suited for smoking, as a lot of the product will get charred during the heating. The free base is quite smokable, but it is a liquid, and as such is not easily sold, as it is unfamiliar. I will cover this matter from two angles: a home technique that works well to base your personal stash for smoking, and a more large-scale procedure for commercial use.

To base your stash and smoke it, mix your stash with an equal amount of bicarb, and then with a dropper, drip a little water onto it with stirring to make a paste. Now take some aluminum foil, and with your finger indent a well into it about an inch deep. Into this well put some of the paste, and heat it from underneath with a lighter. Suck up the smoke with a straw.

For making a crystalline yet volatile derivative of meth similar to crack rocks, one first needs meth free base. All of the production methods in this book yield meth free base. Then to this free base, add dry ice. This will convert the free base to the carbonate, which can be chipped and scraped out of the beaker when the dry ice has evaporated. Use of a solvent during this conversion will be helpful.

Crank rocks similar to crack rocks are pretty simple to make also. These would be just big crystals of pure meth hydrochloride. To get such big rocks, just dissolve the meth hydrochloride into a minimum amount of alcohol. Then let the alcohol evaporate away. As it evaporates away, it will make pretty large crystals of meth hydrochloride.

Chapter Twenty
Calibrating the Vacuum

Before he starts doing the vacuum distillations described in this book, the underground chemist needs to know what kind of vacuum he is able to produce inside his glassware. This is important because the temperature at which a substance distills under vacuum depends directly on how strong the vacuum is. Unless otherwise stated, the distillation temperatures given in this book assume a vacuum of about 20 torr for an aspirator and about 5 torr for a vacuum pump. This chapter describes an easy method by which the chemist finds out just how strong his vacuum is. Once he knows how good his vacuum is, he adjusts the temperatures of his distillations accordingly. The better the vacuum, the lower the temperature at which the substance will distill. He keeps in mind that an aspirator will get a better vacuum in winter because the water flowing through it is colder in that season. The vacuum obtained with a vacuum pump may get poorer over time because solvents from the chemicals he is distilling, such as benzene, may dissolve in the pump's oil. If this happens, he changes the oil.

To begin, the chemist sets up the glassware for fractional distillation as shown in Figure 13 in Chapter Three. He uses a 500 ml round bottom flask for the distilling flask, and a 250 ml flask as the collecting flask. He uses the shorter condenser, and puts 3 boiling chips in the distilling flask along with 200 ml of lukewarm water. He lightly greases all the ground glass joints. (This is always done when distilling, because the silicone grease keeps the pieces from getting stuck together, and seals the joint so that it doesn't leak under the vacuum.)

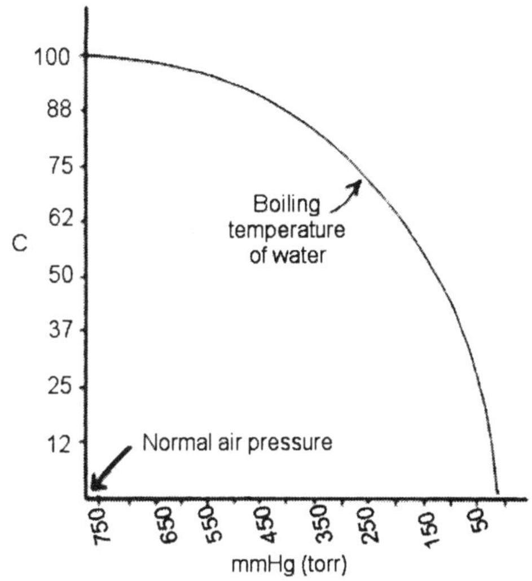

He turns on the vacuum full force and attaches the vacuum hose to the vacuum nipple of the vacuum adapter. The water in the distilling flask should begin boiling immediately. As the water boils away, the temperature shown on the thermometer steadily drops. Finally, the water gets cold enough that it no longer boils. He notes the temperature reading when this happens, or better yet, disconnects the vacuum and takes apart the glassware and takes the temperature of the water in the distilling flask. Using a graph such as the one above, he reads off the vacuum that goes with the boiling temperature.

If his vacuum is bad, the water will not boil. In that case, he checks to make sure that all the joints are tight, and that the stopper in the claisen adapter fractionating column is not leaking. He also makes sure that his vacuum hose is not collapsed. If, after this, the water still doesn't boil, he has to heat the water. He turns on the buffet range at low heat while continuing the vacuum. In a while the water begins boiling. He checks the temperature reading on the thermometer while it is boiling, and notes the temperature. From the graph he reads off the vacuum that goes with that boiling point.

His vacuum should be 50 torr or lower to be able to make methamphetamine. If his vacuum reading is more than 50 torr, he gets a new aspirator or changes the oil in the vacuum pump.

The chemist can use this information to adjust the temperature at which he collects his distilled product. The boiling temperature of phenylacetone is about 105° C at 13 torr, and about 115° C at 20 torr. The boiling temperature of N-methylformamide is about 107° C at 20 torr. The boiling temperature of methamphetamine is about the same as phenylacetone. Phenylacetone and methamphetamine should be collected over a 20-degree range centered on their true boiling points. This makes sure that the chemist gets all of it. The purification scheme he goes through before distilling removes all the impurities with boiling points close to that of his product.

Chapter Twenty One Production From Allyl Chloride and Benzene

At present there are a few viable options left for large-scale manufacture of meth or benzedrine. Phenylacetic acid, benzyl cyanide, and even benzyl chloride are all history. At the time I was writing the Fourth Edition, I had just heard that benzaldehyde and nitroethane had also just been added to the Chemical Diversion reporting list. Allylbenzene is now toast too, although it was never a big item of commerce anyway. This dwindling selection of raw materials is the result of the Chemical Diversion and Trafficking Act of 1988 discussed in Chapter One. Over the years, a continually lengthening list of chemicals has been subject to reporting requirements when sold. These sales reports then go to the DEA, which sorts through the ever-increasing number of reports to try to develop leads. As you can well imagine, the more chemicals listed, the more chaff there is mixed with the wheat, and the less effective this snoopervision scheme is. It is my lifelong aim to get every chemical under the sun put on this reporting list, with the end result being that *none* of them are on a list.

In keeping with this spirit, there are a couple of good methods left out there that are suitable for scaling up to an industrial level of production, while using materials which are not subject to reporting. Sharpen your pencils, and order fast before they are gone. The reporting list has closely mirrored each of my previous editions, and I expect no change in this pattern.

One of these good remaining methods will be presented in this chapter. For the original report on this quite versatile synthetic route, see *Journal of the American Chemical Society*, Volume 68, pages 1009-11 (1946). The route is a Friedel-Krafts alkylation of benzene with allyl chloride to yield 1-phenyl-2-chloropropane. Then this is reacted with methylamine or ammonia to give meth or benzedrine, respectively.

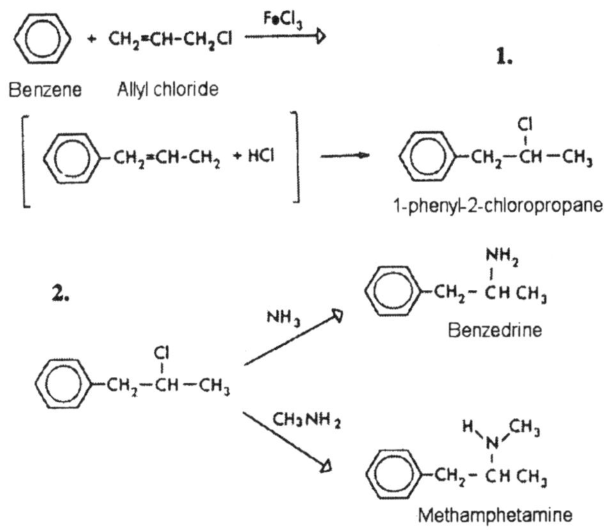

The first reaction is your typical Friedel-Krafts alkylation, with the complication that the HCl produced in the reaction then goes on to add to the double bond according to Markonikov's Rules to yield 1-phenyl-2-chloropropane from the intermediate allylbenzene. The yield from this reaction is unavoidably on the low side because of the tendency of the product to then further go on to react with either benzene to give 1,2,-diphenylpropane, or with more allyl chloride to give multiple-ring substituted products. These can be removed by a fractional distillation, so getting a pure product is no problem, but the yield is going to be only about 35% based upon the allyl chloride used. Unreacted benzene can be recycled back into future batches to cut chemical consumption.

To do the reaction, a suitable glass or stainless-steel reaction vessel is chosen. A 1000 ml round-bottom flask is perfect, but substitutes such as the stainless-steel canister flask depicted in Chapter One will work fine. The size batch given here is right at the upper limit for which magnetic stirring will work. If this procedure is scaled up, only mechanical stirring will work to get the $FeCl_3$ up off the bottom of the flask. Moisture is harmful to this reaction, so be sure that the vessel is dried, and that the reactants are free of water.

Now nestle this vessel into an ice-salt bath, and add 360 ml benzene, then 32 grams anhydrous ferric chloride. When the contents have chilled down to about -20° C (with strong stirring), slowly with stirring add 76 grams (80 ml) of allyl chloride. This addition should take about two hours, and is best done dropwise. After the addition is done, continue stirring for another two hours. The reaction mixture will fume a little HCl, so some ventilation is called for. Don't allow the temperature to climb above -10° C.

Next, pour the reaction mixture into a one-gallon glass jug containing 1 kilo of crushed ice and 100 ml of concentrated hydrochloric acid. Stopper the jug, and shake until the ice has melted. Now separate off the benzene layer floating on top of the water by use of a sep funnel, and wash this benzene layer with some dilute hydrochloric-acid solution, and then with some distilled water.

Next, filter this benzene solution, and dry it over some anhydrous $CaSO_4$. This drying is important because removal of water prior to distillation allows direct recycling of the distilled benzene. If the water was carefully separated off the benzene layer, about 10 grams of $CaSO_4$ in contact with the solution for ½ hour should do the trick.

Now distill this solution through a claisen adapter without glass packing to get a rough separation of the components. The unreacted benzene distills first at a temperature of 80° C or so. When nearly all of the unreacted benzene has distilled, the receiving flask should be changed, and a vacuum applied, slowly at first so as not to cause too vigorous boiling, then at full force. The product, 1-phenyl-2-chloropropane, distills at about 80° C at a vacuum of 10 torr. Tarry gunk remains in the flask, and should be cleaned out with solvent at the end of the distillation.

The crude product should be redistilled through a fractionating column under vacuum to get pure product. The yield is about 50 ml. The recycled benzene should be stored in a sealed bottle until reuse.

This second step of production is also to be found in *Journal of the American Chemical Society*, Volume 68, pages 1009-11 (1946). It is the ammonolyis of the 1-phenyl-2-chloropropane with either methylamine or ammonia to yield meth or benzedrine respectively. This reaction is done in alcohol solution with heating inside a sealed steel pipe. The sealed steel pipe is required because the reaction is done at the temperature above the normal boiling point of the solvent, so a pressure vessel must contain the reactants. The main competing side reactions are further reaction of the product with 1-phenyl-2-chloropropane to give a high molecular weight secondary or tertiary amine. This is suppressed by using a large excess of ammonia or methylamine. Also, the 1-phenyl-2-chloropropane can react with the alcohol solvent to form an ether. It's also possible for the chloropropane to react with water to give the corresponding alcohol. Just plain tar formation is also prevalent. A yield of about 50% and 60% is obtained for benzedrine and meth respectively, based upon the amount of 1-phenyl-2-chloropropane used.

Now, let's vary from the procedure given in the article. In the article they just mixed strong ammonium hydroxide (28% NH_3) into methyl or ethyl alcohol. Then they mixed in the 1-phenyl-2-chloropropane, and heated the mixture inside a pipe. They said that they got the same yield of meth or benzedrine with this watery reaction mixture as they did with a dry one. That may be the case. However, if phenyl substituted starting materials are used, such as bromosafrole, all one will get is tar from this watery reaction mixture. Perhaps this presence of water explains why they got such low yields from the other chloropropanes they made in their experiments. It is your Uncle's opinion that a fairly dry reaction mixture should always be used. The best way to get this dry reaction mixture is to add anhydrous ammonia from a cylinder to the alcohol. First cool down the cylinder and the alcohol in a freezer, then invert the cylinder, crack open the valve (strong ventilation!), and add about 100 ml of liquid ammonia in about 400 ml of alcohol solvent for each 50 ml of 1-phenyl-2-chloropropane or bromosafrole used in the reaction. An alternative method of making fairly dry alcohol solutions of ammonia is given in *Advanced Techniques of Clandestine Psychedelic & Amphetamine Manufacture.*

This same method can be used to get alcohol solutions of methylamine, or one could place some methylamine free base in water solution inside a distilling flask, and apply heat to force the vapors up through the ice cold condenser, and then through some tubing into a chilled and stirred beaker of alcohol. See Chapter Four for a diagram of such an apparatus.

Let's further vary from the procedure given in the article. They used methyl or ethyl alcohol as solvent. One of the side reactions is with the alcohol solvent to form an ether with the phenyl-2-chloropropane or bromosafrole. If one substitutes a secondary alcohol such as isopropyl alcohol, this reaction is less likely to occur. Virtually pure and water free isopropyl alcohol is easily available. It is a product called Isoheet gas line de-icer. Look for it at the gas station or hardware store. This will give better yields than methyl or ethyl alcohol. One would naturally wonder if replacing the alcohol solvent with something inert like toluene wouldn't further improve yields. Maybe; give it a try.

So now we have a solution of 50 ml of the chloro or bromo propane in about 400 ml of solvent just saturated with ammonia or methylamine. Now pour this solution into one or a series of steel pipes. They should be threaded at each end so that the caps may be screwed on at both ends like a pipe bomb. The plumbing section at the hardware store is well stocked with these parts. Screw the cap on tightly when filled.

The reader should be aware that commercial steel pipe and caps are heavily galvanized with zinc. The zinc must be stripped off prior to use in the procedures outlined in this book. Zinc is stripped off by immersing both the pipe and the end cap in 5% hydrochloric acid solution until the violent bubbling slows to a crawl. Then the pipe and caps should be thoroughly rinsed off in clean water and then assembled. A pipe wrench will be required to get the caps on tight enough to prevent leakage while cooking.

This degalvanizing process can lead to some confusion as to when the zinc has been removed. Take for example this conversation I had on the Net:

Posted by piper on March 03, 1998 at 10:04:37:
DEGALVANIZING

What if someone were stripping the zinc from galvanized pipe, and that went good, but the end caps have been going for two days now, and there is STILL zncl2 coming off them. Someone needs those caps soon! Can the process be accelerated? Just up the strength of the HCl solution? Maybe some DC current? Also, Teflon tape melted @ 130° C. What's up with that? Is it bad?

Posted by Uncle Fester on March 06, 1998 at 17:21:27:
In Reply to: Re: DEGALVANIZING-hypothetically!
Posted by metal man on March 06,1998 at 04:59:06:

Metal Man is correct when he states that concentrated nitric will strip plates off steel without attacking the steel. If, however a little bit of water gets in the mixture, the steel will be history. The end caps of pipes have the thickest amount of zinc on the inside. I strip zinc off steel all the time at work. The end caps should have been completely stripped within a couple of hours using 5-7% HCl. That's concentrated HCl diluted 6 to 7 times. Under that zinc is mild steel. It too will fizz in HCl solution, but much slower. Using a scrub pad, one can go into that end cap and scrub off the surface layer of gunk that forms while metals are being stripped. Zinc while it dissolves will be black. Exposed iron or mild steel will be a brighter color. After rinsing, application of some copper sulfate solution will show exposed steel. The old copper displacement reaction, leaving a copper deposit. For a part to take two days to strip in HCl, either such a small amount of HCl solution was used that it has been exhausted by the stripping action, or the dissolution of the underlying iron has been mistaken for more stripping of zinc.

This possible source of confusion can be eliminated by using stainless-steel pipes and caps rather than galvanized steel. These can be used as is, without any treatment.

Now with the reaction mixture inside the pipe, it's time to heat the mixture. For production of meth or benzedrine from 1-phenyl-2-chloropropane, the preferred heating procedure is to heat at 160° C for about 9 hours. For production of MDA or MDMA from bromosafrole, the preferred heating is at about 125° C for 3 to 4 hours. One does this heating by putting a pan with cooking oil on a stove burner, and then immersing the pipes in the cooking oil. The temperature of the oil bath is then held at the desired temperature for the required period of time.

After the cooking period is complete, the pipes are removed from the heating bath, and allowed to cool down. Once they have cooled, they can be opened, and the contents poured into a distilling flask. Most all of the alcoholic ammonia or methylamine should be distilled off. In the case of methylamine, great care should be taken to catch its vapors for reuse. This is done using that apparatus shown in Chapter Four, making a fresh load of methylamine in alcohol for use in the next run. The last portion of alcohol should be removed using a vacuum, down to a volume of about 100 ml.

The residue in this flask should be shaken very vigorously with 10% HCl solution. This converts the amine products into their hydrochlorides, which are water-soluble. The shaking should continue for at least 5-10 minutes to get all of the product extracted out of the gunky tar matrix. Now extract this 10% solution which contains the product with a couple 50 ml portions of toluene. This removes entrained gunk. Finally, make the solution strongly alkaline to litmus with lye. This generates a lot of heat, and should be done cautiously with shaking between adds of lye. When the solution is strongly alkaline, shake vigorously for about 5 minutes more, then check again the pH of the water to make sure it is still quite alkaline. There should be a healthy amphetamine layer floating on top of the water. With the hot water, it will give a strong aroma of amphetamine when sniffed. Cool the solution, and extract with two 50 ml portions of toluene.

The toluene extracts contain the amphetamine. This should be distilled in the usual manner as described in Chapter Five to yield about 25 ml of amphetamine or meth. This is converted to the hydrochloride salt as also described in Chapter Five, to give about an ounce of pure benzedrine or meth. This procedure can be scaled up as desired. In case you were wondering, the boiling point of benzedrine free base is about 15° C lower than meth.

Distilling must be done when using this method, because there is just no other way to remove the higher amines made as byproducts. If one is making MDA by this method, those higher amines produce a scary and paranoid trip. If one can't distill, this method shouldn't be used.

Making Bromosafrole From Safrole, and 1-phenyl-2-bromopropane From Allylbenzene

To use this "pipe bomb" method to make MDA or MDMA from bromosafrole or amphetamine and meth from 1-phenyl-2-bromopropane, one of course first needs the bromo compound. Good luck finding that stuff. Luckily, it's not a very complicated procedure to cook your own. Let's take, for example, making 1-phenyl-2-bromopropane from allylbenzene. Allylbenzene was covered back in Chapter Nine. It's pretty easily made from cinnamon oil, or cinnamaldehyde. This conversion to the bromo compound and then the "pipe bomb" reaction is the alternative route to the Wacker oxidations to phenylacetone. Both routes are quite practical, and can be scaled up at will. Directions for making 1-phenyl-2-bromopropane can be found in the *Journal of Biological Chemistry,* Volume 108, pages 622-23, by H.E. Carter. Reaction with hydrobromic acid gives the bromopropane from allylbenzene:

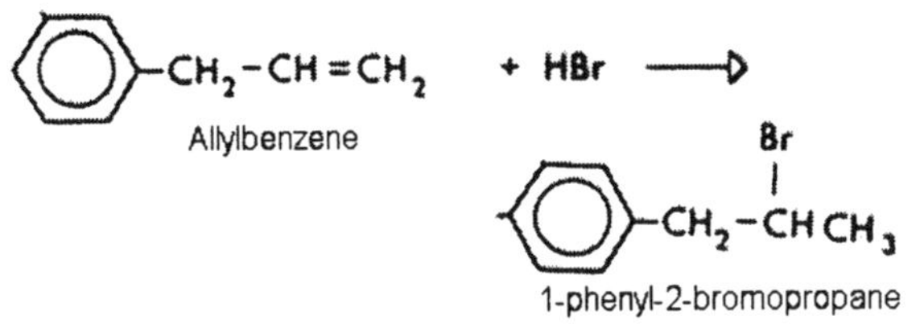

His procedure is to put 200 ml of glacial acetic acid in a bottle along with 200 ml of 48% hydrobromic acid. This mixture is chilled in an ice bath, then 100 ml of allylbenzene is added to the bottle. A stopper is wired in place on the bottle, and the mixture is slowly allowed to come to room temperature with occasional shaking. After 10 to 12 hours, the original two layers merge into a clear red solution. After 24 hours, the contents of the bottle are poured onto crushed ice.

When the ice has melted, the 1-phenyl-2-bromopropane will have formed an oily dark liquid layer separate from the acid-water solution. It may be at the bottom of the beaker if lots of crushed ice was used, or it may be floating on the top if less was used. This crude product should be separated from the acid-water using a sep funnel. Then the acid-water should be extracted with about 100 ml of toluene. This extract should be added to the crude product. The combined extract and crude product are then washed with water, and then with bicarb solution. Fizzing from the bicarb solution will be produced as it neutralizes acid in the crude product, so beware of pressure building up in the sep funnel. Always wear eye protection so that mists of this stuff doesn't end up in your eyes.

Then when all the acid has been neutralized, as shown by lack of fizzing when put in contact with fresh bicarb solution, the toluene-bromopropane solution should be placed in a distilling flask, and fractionally distilled to remove the toluene-water azeotrope, and then the remaining toluene solvent. When the toluene has mostly distilled away, a vacuum should be applied and the 1-phenyl-2-bromopropane distilled. It will boil at a temperature similar to phenylacetone, roughly at 120° C under a good aspirator vacuum of around 20 torr. Less efficient vacuum will result in higher boiling points. The yield is around 235 grams (180 ml).

So that method using 48% HBr apparently works fine with allylbenzene. When using substituted allylbenzenes such as safrole, however, it is quite useless. There is just too much water in the reaction mixture (48% HBr is 52% water), and the HBr in the mixture just refuses to add to the double bond to give bromosafrole. The general method, which works equally well with allylbenzene or the safrole found in sassafras oil, can be found in the *Journal of the American Chemical Society.* The paper dates to 1946, Volume 68, pages 1805-6.

In this general procedure, the chemists mixed 100 ml of allylbenzene with 250 ml of glacial acetic acid. One could simply add 100 ml sassafras oil with 250 ml of acetic acid and get basically the same solution. Then to this solution with rapid stirring, they bubbled a rapid stream of anhydrous HBr gas from a cylinder for a period of about two hours, while keeping the mixture cooled with an ice bath. The fumes of HBr are injurious, so ventilation out a window or working outside is recommended. The bubbling of HBr produced the formation of two layers in the reaction mixture. Then they added 200 ml more glacial acetic acid to make a homogenous solution, and kept the mixture cold overnight.

In the morning, they poured this reaction mixture onto ice, and recovered the bromo compound in exactly the same way as in the first example. This general method is good, but that HBr in a cylinder isn't that easy to come by, and the dangerous fumes from the bubbling are something one would want to avoid. There is a way around that.

Thirty-six percent HBr in glacial acetic acid is commercially available. This makes doing the reaction much simpler. One just mixes one volume of sassafras oil or allylbenzene with two or three volumes of ice cold 36% HBr in acetic acid, and then stirs the mixture with cooling for about a day. Then upon pouring the mixture onto ice, the bromosafrole or 1-phenyl-2-bromopropane is recovered just as in the first example. Bromosafrole smells a lot like phenylacetone, and its boiling point is about the same as m-d-phenylacetone, around 150-160° C under a vacuum of about 20 torr.

So how much water can you have in the acid, and still get bromosafrole from sassafras oil? See *Chem. Abstracts* 1961, column 14350.

Unfortunately, it uses 70% HBr, a quite uncommon reagent. To do this procedure, a beaker containing 100 ml of 70% HBr is chilled to 0° C in an ice bath. Then, with stirring, 50 ml of sassafras oil is added dropwise. The stirring and cooling is continued for an additional 14 hours, then the mixture is poured onto a few hundred grams of crushed ice. This mixture is stirred or shaken until the ice melts, then the product is extracted with ether or toluene. The organic layer should then be separated, and washed with some water, followed by some bicarb solution, to remove traces of acid. The solvent is then evaporated away, either by distillation or under a vacuum to give a virtually 100% yield of the product, bromosafrole. Its aroma is similar to phenylacetone.

Let's suppose that all one can get is the standard ACS reagent 48% HBr in water. How can one get bromosafrole using this stuff? If one looks in the *Journal of the Alabama Academy of Science,* Volume 64, pages 34-48 (1993), one can find a claimed method. It uses the standard ACS reagent HBr, which is 48% strength. In this variation, 50 ml of sassafras oil, 250 ml of 48% HBr, and a magnetic stirring bar at least one inch in length are placed in a 500 ml volumetric flask. The top is stoppered to keep the nasty vapors inside (HBr is bad to breathe), and fast stirring is continued for one week at room temperature. A layer of cardboard between the stirrer and the flask will help keep the stirrer from warming up the solution. Within a few hours, the reaction mixture takes on the color of a cheap burgundy. A homogenous mixture never results. When stirring stops, the product just floats on top of the acid. At the end of the week of stirring, the bromosafrole is isolated just as in the above method. Two hundred and fifty grams of crushed ice are used. It is best if the product is distilled to recover unreacted safrole, and remove the colored matter. Bromosafrole distills at about 160° C with aspirator vacuum. Note here that the Alabama article chemists used their crude bromosafrole without distilling it, and got quite pure XTC as the product. Others and I have found that 48% HBr is quite ineffective in reaction with safrole.

An alternative procedure using 48% HBr gives superior results. 48% HBr can be dehydrated to HBr gas upon contact with phosphorus pentoxide, P_2O_5. Sulfuric acid can't be used as the dehydrator because it breaks the HBr down to bromine gas. To use this variation, a gas bubbler is set up as shown in Figure 20. In the round bottom flask, place a bed of phosphorus pentoxide. Into the sep funnel or dropping funnel, place about 210 ml of 48% HBr. Lead the bent glass tubing into a beaker containing about 160 ml of sassafras oil. It is best to dilute the safrole with a couple of volumes of solvent such as toluene or glacial acetic acid, because this will help to catch and hold the HBr gas as it is bubbled into solution.

Now chill down the sassafras oil solution in ice, and begin dripping 48% HBr onto the bed of phosphorus pentoxide. When it hits the P_2O_5, the water in the 48% HBr reacts with the phosphorus pentoxide to make phosphoric acid, and the HBr puffs off as a gas which escapes down the glass tubing into the sassafras oil solution. All joints and plugs on the gas bubbler must be tight so that the HBr is forced into the sassafras oil solution. It is also a good idea to magnetically stir the sassafras oil solution so that the HBr bubbles are more likely to either react immediately or go into solution, rather than escaping. HBr gas is very foul and dangerous to breathe, so good ventilation must be provided.

As the approximately 210 ml of 48% HBr drips onto the bed of P_2O_5, pay attention to its reaction when it hits the P_2O_5. If the rate of gas generation goes down, the bed of P_2O_5 may need some sloshing around or stirring up with a glass rod, or even the addition of some more P_2O_5.

When the bubbling of HBr is completed, the sassafras oil solution should be poured into a stoppered bottle, and kept cold overnight to complete the reaction to bromosafrole. It is then poured onto crushed ice as in the other procedures, separated, washed, and distilled to yield about 100% bromosafrole.

One might want to consider doing this drying of the 48% HBr the other way around. By cautiously adding P_2O_5 to a stirred mixture of 48% HBr and glacial acid, one should be able to dehydrate it enough to react with safrole. How much would one have to add? I don't know. One mole of P_2O_5 reacts with three moles of water, and the phosphoric acid formed then exerts a further drying action by associating with water molecules. One could only find out by trying.

A possible way of increasing the activity of 48% HBr is to add the common lab chemical potassium bromide, KBr, to the acid. I have no idea if this works, but it would produce a mixture a lot like the Lucas Reagent used in Chapter 15 to make chloroephedrine. It's worth a try if 48% HBr is the only form of hydrobromic acid you can obtain.

Another way of drying 48% HBr is one I came up with a few years ago. I got this procedure to work on the first try, so I considered it a slam dunk gimme. Apparently, other people have had some trouble doing it, so let me give more detail.

To get good yields of bromosafrole from 48% HBr and sassafras oil, mix one part sassafras oil with one part glacial acetic acid and two parts 48% HBr in a nearly full Erlenmeyer flask. Chill this mixture down in ice, then with strong magnetic stirring pass a stream of dry HCl gas into the solution for about an hour. See Chapter 5 for the dry-HCl gas-generator. How much HCl to pass into solution? Well, for a batch using 50 ml of sassafras oil, the amount of dry HCl generated by dripping 75 to 100 ml of sulfuric acid onto a half-full 500 ml flask of salt-hydrochloric-acid paste is about right. A little bit more wouldn't hurt. Good ventilation is required!

A nearly full Erlenmeyer is used to give maximum column depth for the bubbles of HCl to rise up through. The drying is a surface phenomena. An Erlenmeyer is used because the inward sloping walls slow up the rising bubbles. A plug of glass wool stuck in the neck of the flask down into the solution would slow them up some more.

As the dry HCl passes into the solution, it dehydrates the 48% HBr, causing it to react with the safrole. The dehydration and the reaction both generate a good deal of heat, so fresh ice will periodically have to be put into the bath around the reaction flask. The temperature of the reaction shouldn't be allowed to rise above 10-15° C. The reaction mixture first turns green, then blue, then purple, and finally burgundy. When the bubbling with dry HCl is finished, stopper the flask and continue stirring in the cold for two days. Sometime around a day into this stirring, no separation of phases can be seen when stirring ceases.

I would say that the maximum temperature seen by the reaction mixture during this time was roughly 20° C. Keeping things cold in Wisconsin is easy.

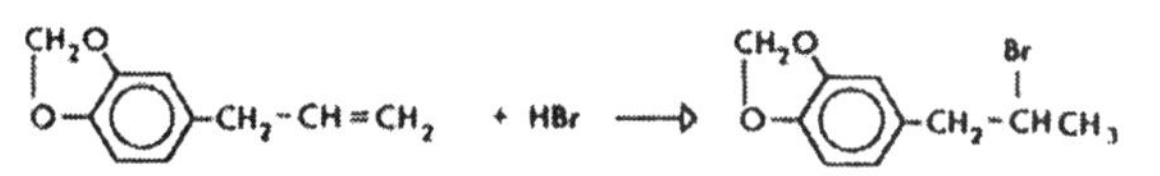

Safrole

3,4-methylenedioxyphenyl-2-bromopropane
a.k.a. - bromosafrole

The amount of dry HCl produced by dripping sulfuric acid onto salt will vary with the exact conditions, so the batch should be checked for reaction before quenching it on ice. It doesn't hurt to add too much dry HCl, within limits, but too little won't dehydrate the acid sufficiently. To check this, after the day of stirring is done, pour some of the reaction mixture into a beaker, then from the beaker, return it to the reaction vessel. This leaves a coating of the reaction mixture on the glass in the beaker. Fill the beaker with water to rinse away the fuming acids, empty it, and sniff inside the beaker for the aroma of the organics clinging to the glass. If it still smells like the candy shop fragrance of sassafras oil, an additional bubbling with dry HCl is going to be required, followed by another day of stirring in the cold. After the first batch or two, it's easy to gauge how much dry HCl one is getting. If the aroma has changed to something more chemical and fruity, yes, just like phenylacetone, sufficient HCl has been added.

When two days of stirring are completed, the batch is poured onto crushed ice, as in the other methods. When the ice has melted, a little bit of toluene is added (a volume about equal to the amount of sassafras oil used), and the water-bromosafrole mixture shaken. Prior to adding toluene, the bromosafrole will likely be on the bottom of the container, but after adding toluene and shaking, it should be floating on top. It's still burgundy-colored. Separate the bromosafrole layer with a sep funnel, and then wash it with about 3 volumes of water. Add bicarb slowly until the fizzing stops. This will knock out the carried-over HBr, HCl and acetic acid. Shake some more, then add a little more bicarb to make sure all the acid has been neutralized.

Separate the toluene-bromosafrole, and place it in a distilling flask. Distill off the toluene at normal pressure, then vacuum-distill the remaining bromosafrole. A vacuum that distills safrole at 110° C will distill bromosafrole at about 140-145° C. Some chlorosafrole distills at about 125° C. It can be used as is, or the chlorosafrole can be converted to iodosafrole according to the directions found in Chapter Eighteen in this book. The yield is about 66-75% conversion to bromosafrole, with the remainder being unconverted safrole and chlorosafrole. Bromosafrole smells a lot like phenylacetone. It may turn pink on standing, and should be stored in a freezer until used.

Last, but certainly not least, check out the Pugsley Bromosafrole Recipe in *Advanced Techniques of Clandestine Psychedelic & Amphetamine Manufacture.* People have been getting very good results using this procedure. Essentially, it involves reacting sulfuric acid with sodium or potassium bromide in ice cold DMSO solvent to give anhydrous HBr solution. Sassafras oil is then added to the reaction mixture to give virtually 100% yields of bromosafrole.

Chapter Twenty Two Phenylacetone From Benzene and Acetone

This procedure makes use of the simple and common solvents, benzene and acetone, and links them together to form phenylacetone. Back when I was cooking phenylacetone, I often fantasized about how this could be done. Little did I know that it had been accomplished by a couple of Russians a few years before my cooking began.

This isn't a procedure to get overly excited about, as the yields are low (36% based upon the manganese III acetate used), and a quite dilute solution is required. This procedure is most suited to someone willing to do large-scale cooking, not the typical basement experimenter.

The interested reader should see *Chemical Abstracts*, Volume 77, column number 151620 (1972), and *Journal of the American Chemical Society*, Volume 93, pages 524 to 527 (1971), and *Bulletin of the Academy of Science of the USSR,* Volume 21, number 7, page 1626 (1972).

In a large glass pot, as for instance one could get from people who sell milk pipeline equipment to dairy farmers, place 20 moles of acetone (900 ml). Hardware store acetone can be used by drying it with $^{1}/_{10}$ volume of calcium chloride, also available at the hardware store as ice melt. Then add 5 moles of benzene (340 ml), and 1000 ml of glacial acetic acid, and one mole of manganese III acetate (268 grams; price about $500). Equip the flask with a reflux condenser.

These ingredients are mixed together, and then heated at 70° C until the brown color of Mn^{+3} disappears (about 2 to 3 hours). Then the contents are poured into a large stainless-steel distillation set up as described in Chapter One of this book, and the acetone, benzene, and part of the acetic acid are distilled off under reduced pressure. The weak vacuum produced by cheapie vacuum pumps, or low-powered aspirators, is about right for this vacuum distillation. By chilling the receiving flask in ice, the unused acetone and benzene can be recovered for reuse.

Then the residue, which consists of phenylacetone and other organic products dissolved in acetic acid along with Mn(II) acetate, should next be diluted with several volumes of water. The product phenylacetone can then be extracted from this watery mixture, with toluene. The extracts should next be washed with dilute sodium hydroxide solution, and then the toluene-phenylacetone solution can be distilled to yield around 25 ml phenylacetone.

To get decent results from this reaction, the amount of water in the reaction mixture should be held to under 1%, and preferably under ½%. Improperly dried acetone is a prime culprit when tracking down sources of water in the reaction mixture. Water introduced from Mn(II) acetate can be removed by distilling off the water-acetic acid azeotrope at 76° C.

The other main problem with this reaction, besides the large dilution used, is the need for Mn(III) acetate. This expensive material isn't something one can find on hardware store shelves. Mn(II) acetate, on the other hand, is a very common industrial chemical, used as a mordant in dying, and as a drier for paints and varnishes. It's also pretty cheap, so long as one doesn't want 99.9% pure chemical.

To get Mn(III) acetate from Mn(II) acetate, we return to a recurring theme in industrial chemistry — the electric generation of Mn(III) from Mn(II). We saw one example of this kind of conversion back in Chapter Nine in the benzaldehyde recipe. For this next one see *Acta Chemica Scandinavica*

B 33 (1979), pages 208-212. At a graphite or platinum anode in a simple, undivided cell, using a cathode much smaller than the anode to minimize reduction of the Mn(III) formed, the chemists produce Mn(III) acetate from Mn(II) in glacial acetic acid solvent.

One mole of Mn(II) acetate is dissolved in one liter of glacial acetic acid. A little bit of sodium lithium fluoroborate (a few grams) is added as current carrier to the solution. One could also try sodium or potassium acetate as current carrier; it may not interfere in this reaction. The fairly large graphite or platinum anode is placed in the solution, along with the smaller cathode. The mixture is warmed, and then with stirring, DC current is made to flow through the cell. One should apply 4 milliamps of DC current for each square cm of anode surface facing the cathode. This is a one-electron oxidation, and one can count on getting around 66% efficiency in the oxidation. So one should pass about 1.5 faradays of current. One faraday is 96,500 amp seconds, so if for example one is passing one amp through the solution, the electrolysis to Mn(III) will take 40 hours. At four amps, it will take 10 hours, and so on.

At the end of the electrolysis, one has the Mn(III) acetate solution in acetic acid. Then to this solution, one can add the benzene and acetone with stirring, and react as usual. It's a lot of work to get 30 ml of phenylacetone, but those chemicals certainly are low-profile, cheap, and easily available.

For another example of electric generation of Mn(III), see US Patent 4,560,775. One can also use permanganate in glacial acetic acid to oxidize the Mn(II) acetate to Mn(III). For an example of this procedure, see *Journal of the American Chemical Society*, Volume 96, pages 7977-7981 (1974). On a smaller scale, this procedure is preferable to the electric oxidation. The phenylacetone synthesis can be made to work, and it is just plain unstoppable from a policing point of view. It won't be long now, and every chemical under the sun will be on the "watched list."

A kind of related reaction can be found in *Journal of Organic Chemistry*, Volume 54, pages 733-34 (1989). Here phenylacetone is made in 85% yield. Just two problems with this reaction. It runs in a pretty dilute solution, and it uses 2-nitropropene as a reactant. A quick look through my chemical catalogs doesn't turn it up for sale anywhere. I would imagine it isn't too hard to make one's own 2-nitropropene though. The reaction proceeds as follows:

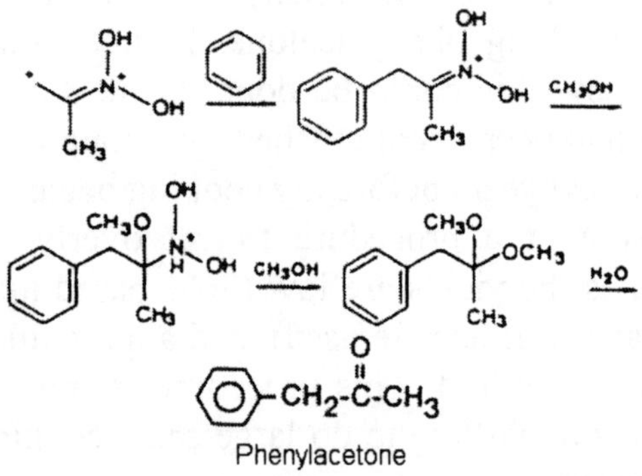

The strong acid trifluoromethanesulfonic acid protonates 2-nitropropene, and this intermediate then links up with benzene. After pouring the reaction mixture into dry methanol, and then adding water, the product, phenylacetone, is formed in 85% yield.

In one of their typical examples, the chemists mix 3 grams of 2-nitropropene in 45 ml of benzene. In another container they have a solution of 31 ml of trifluoromethanesulfonic acid in 50 ml of benzene along with an unspecified amount of the methylene chloride co-solvent. The co-solvent acts as antifreeze for the mixture, so 50 ml of methylene chloride is probably about right. They cool this solution down to -40° C with a dry ice-acetone bath. Then with vigorous stirring, they add the nitropropene in benzene solution to the trifluoromethanesulfonic acid in benzene and methylene chloride solution. They allow this mixture to react for one minute.

After the reaction time, they then pour this reaction mixture into 1000 ml of dry methanol cooled to -78° C (dry ice-acetone bath) with vigorous stirring. This reaction mixture is then allowed to warm to room temperature. The yellow-colored mixture is then diluted with 1500 ml of water.

To recover the product, they start by adding bicarb to this reaction mixture until it is neutralized. This is when the added bicarb no longer causes fizzing. They next add salt until the solution can dissolve no more. Then they extract out the phenylacetone with methylene chloride. One could also use toluene. Distilling this extract then gives them pure phenylacetone, around 5 ml.

As is, this isn't a clandestine-suitable process. It just uses way too much solvent to get such small amounts of product. If one could reduce the amount of benzene used from the thirty fold excess relative to the 2-nitropropene down to around 10 fold, and if then one could also reduce the amount of methanol used, this method would have some promise. Good luck, and happy cooking!

For another somewhat related reaction, I have only the abstract. The research was done in 1959 by Robert Levine. It gives a 34% yield of phenylacetone. To liquid ammonia, one first adds sodamide and acetone. This forms a sodio derivative of acetone. Next, bromobenzene is added, and the mixture allowed to react for 10 minutes. Then the reaction mixture is quenched by adding ammonium chloride. After the ammonia evaporates away, the residue is extracted with toluene. This extract is washed with some dilute hydrochloric acid to remove aniline and diphenylamine formed as byproducts. Then the toluene extract is distilled to get pure phenylacetone. Dibenzyl ketone is formed as a byproduct also.

[illegible]

... color to ... [illegible] ... the ... [illegible]

A modified ... [illegible] ... this ... [illegible] ... by adding ... [illegible] ... until the solution ... [illegible] ... Then ... [illegible] ... with ... activated ... [illegible] ... could ... [illegible] ... give them ... dry ... around 5 ... [illegible]

As ... this ... [illegible] ... much ... to get ... small amounts of product. One could ... the amount of benzene used ... [illegible] ... down to around ... and that one could also reduce the amount of ... used ... would have ... [illegible]

Another ... [illegible] ... the ... of the reaction ... by Robert ... [illegible] ... acetic ... [illegible] ... The ... [illegible] ... is added and the ... allowed to ... for 30 minutes. Then the reaction mixture ... by adding ammonium chloride. After the ammonia evaporates ... was ... extracted with ... [illegible] ... washed with ... chloroform to ... the ... formed as byproduct. ... the ... to get pure ... Otherwise ... formed as a byproduct ... [illegible]

Chapter Twenty Three
Last Resort —
Extracting l-methamphetamine
From Vicks Inhalers

By popular demand, this method of last resort will be covered in this edition. The Vicks Vapor Inhaler is available off the shelf at your local grocery or drug store in the cold- or allergy-remedy section. It contains 50 mg of the free base of the weaker isomer of meth, along with the "Vicks vapors" which are bornyl acetate, camphor, lavender oil, and menthol.

Of the above ingredients, only the meth free base (l-desoxyephedrine) has a basic nitrogen, so separation is possible. To extract and separate the l-meth from the other ingredients, we first disassemble the inhaler to get at the cotton-like wadding that contains ingredients. This wadding should be immediately soaked in 10 ml of the 10% hydrochloric acid. The hardware-store brands of hydrochloric acid are about 20% — 30% strength, so dilute accordingly. Using surgical gloves, squish up this wadding repeatedly to get the HCl into contact with the meth free base and convert it to the hydrochloride, which is water soluble. After a good thorough squishing, pour the hydrochloric acid into a sep funnel. If solids are floating around, filter the solution. Now add another 10 ml of plain water to the wadding, squish it around again to rinse out more product, and pour this too into the sep funnel.

Now extract out the entrained vapors with a couple of 20 ml portions of toluene. Throw away the toluene, and keep the hydrochloric acid solution. Now make this hydrochloric acid solution strongly basic to pH papers by adding some lye or lye solution, with strong shaking between additions of lye.

The meth has now been free based, and is freed of most of the Vicks vapors. Extract out the meth free base with about 20 ml of toluene. Separate off the toluene, and bubble dry HCl gas through it as described in Chapter Five. The crystals of 1-meth hydrochloride should be spread out to dry after filtering, and their aroma noted, once they are free of toluene. If they still smell like the Vicks vapors, one should first try drying them under a vacuum for an hour or so. If this still doesn't render them odor-free, they can be recrystallized by first dissolving them in a minimum amount of alcohol (91% isopropyl from the drug store shelves), and then adding toluene with shaking until about 10 volumes of toluene have been added. After some standing in the cold to get complete precipitation, the crystals can be filtered out. At this point the smell of Vicks vapors should be gone.

I have heard an unconfirmed report from a correspondent named Tammy that new versions of this inhaler don't respond to HCl extraction so well. The wonders of polymer science. If this is the case, the first extraction should be with 91% isopropyl alcohol. After two extractions with isopropyl alcohol, add a couple of drops of HCl and then this extract should be evaporated under a vacuum, or barring this, just mix with 20 ml of toluene.

Now extract this toluene solution with two 20 ml portions of 10% hydrochloric acid. From here, proceed as with the 10% hydrochloric acid solution.

Chapter Twenty Four
Keeping Out of Trouble

Making methamphetamine, it should be remembered, could be a dangerous activity. But, in addition to any dangers inherent in the activity, underground chemists making methamphetamine face dangers of another sort. The source of these other dangers are the agents of the various law enforcement agencies. This chapter will discuss some of the dangers and how underground chemists avoid them.

How then does the underground chemist minimize his risks? The first and most important thing is to use hit-and-run tactics. He makes a lot of product at a time, and then closes up shop. It is much safer to spend a week or so on steady work and make a supply of product that will last for a while than to keep setting up and supplying a lab every few weeks to make smaller amounts. This cuts the chemist's exposure to a minimum. Secondly, all the chemicals to make methamphetamine are only brought together when the chemist is ready to begin production. Having all the chemicals together could result in a conspiracy charge. For example, having phenylacetic acid, acetic anhydride and pyridine together could result in a charge of conspiracy to manufacture phenylacetone, if the knuckleheads at the state crime lab are aware of this method of making phenylacetone. To avoid this, phenylacetic acid and methylamine are kept at one location, and the other chemicals and glassware at another. After the chemist is done making his supply of methamphetamine, he washes all the glassware in hot, soapy water, rinses them a couple of times with hot water and then with rubbing alcohol. He lets the glassware drip dry, and then bakes the glassware in the oven at 400° F for an hour or so. This removes all traces of product from his glassware. The empty glass jugs of chemicals are rinsed out with water and the labels scraped off. Then they are broken and the pieces taken to a far away dumpster.

A very important precaution for the underground chemist is to keep his mouth shut. While his friends may mean him no harm, they would tell their friends and eventually the wrong ears would hear about it. The streets are crawling with snitches who keep themselves out of jail by reporting what they hear. Without a snitch, police agencies are incapable of detecting a cockroach crawling across a loaf of bread.

The people to whom the chemist sells his products have no business knowing where it comes from. In fact, he is constantly on guard against his customers, because they are his main source of danger. If one of them should foul up, he may very well try to set up the chemist to get out of his own problems. This is the way that Johnny Law makes his busts, so the underground chemist is on guard. If one of his customers has a newfound buddy who wants to buy from him, he starts babbling crazy nonsense or claims ignorance. He decides how to deal with them later.

As long as the chemist does not deal with strangers, the only way that the narcs can get at him is to have one of his customers make what is called a "controlled" buy on him. This is when they send his customer in to make a purchase from him while they wait and watch outside.

The underground chemist protects himself by only making deliveries to his customer's home. He never does business out of his own home, or at bars, parks, parking lots or any other place suggested by his customers. He knows his dealers well, and knows their schedules. His dealer does not know exactly when he will be showing up with the next shipment; he just shows up unannounced and makes the delivery. A street-legal dirt bike is a good delivery vehicle. If the narcs try to jump the chemist at his customer's home, he takes off cross-country, leaving a cloud of methamphetamine powder behind him. He can melt the baggie on his tail pipe. If the narcs eventually catch him, he says they looked like a sleazy gang of hit men. He never lets his customers talk him into meeting at a bar, park or other public place where Johnny Law can watch and make a controlled buy. He ignores excuses such as not wanting a roommate to know about the shipment.

If the underground chemist must store significant quantities of methamphetamine in his home, there is a good way to keep it undetected. He dissolves the uncut material in 190-proof grain alcohol. He uses uncut methamphetamine because alcohol dissolves it better than cut. Alcohol dissolves a surprisingly large amount of methamphetamine. He records the exact amount dissolved per hundred ml of alcohol. He pours the alcohol into a dark whiskey bottle and adds it to his liquor collection. It smells just like any other booze. It will go undetected in a search.

When he is ready to sell undissolved methamphetamine, he measures out the required amount of alcohol and pours it into a filtering flask along with a couple of boiling chips. He stoppers the flask and attaches the vacuum hose to the vacuum nipple. He boils off the alcohol under a vacuum. He can heat the flask with hot water to speed the process, but does not use any stronger heat. In a little while, the alcohol is gone, leaving the crystals in the flask. He scrapes them out and chops them up. He can now add the cut to the crystals. The filtering flask can be rinsed clean with hot water.

The maker of methamphetamine, like the user, may be subjected to urine testing, and so he is aware of the following information. A single dose of methamphetamine can be detected in the urine for three days after taking it. When repeated doses are taken over an extended period of time, it builds up in the cerebral spinal fluid, lymph, and other noncirculating bodily fluids. As a result, it is detectable for considerably longer than three days.

The three-day period mentioned above assumes a normal fluid intake. Since the kidneys are the main way the body has to get rid of methamphetamine, the process can be considerably sped up by putting the kidneys on overtime. I can think of no more enjoyable way to do this than to drink a lot of beer over a period of a few days. This process can be sped up even more by increasing the efficiency of the kidneys. This is done by drinking a lot of cranberry juice. This increases the acidity of the urine, and shifts the partition coefficient in the nephrone of the kidney in favor of excreting the methamphetamine more rapidly.

A last-ditch method to avoid detection is to get some Snowy Bleach or other similar powdered bleach, put it on the fingertip and under the fingernail, and rinse it off into the urine sample. The bleach attacks the methamphetamine by oxidation, converting it to a harmless set of fragments. This technique works better with THC and other more easily oxidized drugs, but it works satisfactorily with methamphetamine if the urine is dilute. In order to avoid wasting the limited oxidizing power of the powdered bleach on the other normally occurring compounds in the urine, a lot of water is drunk before giving the urine sample. And, in order to keep the concentration of bleach high enough to ensure the destruction of the methamphetamine, as small a sample as possible is given.

Remember: *The information in this book is intended for informational and research purposes only! If you do anything illegal and get caught, you will have to be prepared to face the consequences! This is a letter that I recently received from one of my readers:*

Dear Uncle Fester:

I am the subject of the first methcathinone (Cat) case in the Fifth Circuit (Texas, Louisiana, Mississippi, Alabama, etc.). I was one of many whom a dude named Bill Killion "fingered" to avoid prosecution in Wichita, Kansas. You may know him; his correct name is William Killion.

I still await sentencing sometime in early January, 1996. And many of the investigations for my defense involve research of ephedrine, white-cross and methcathinone, which I thought might interest you. Since I didn't discover your book, ***Secrets of Methamphetamine Manufacture***, until after my arrest, I learned of cat through its American patent (2,802,865). There is a German patent, and one other foreign patent with R+ and S- methcathinone types. I wish I knew how to obtain those patents, because it may be argued that all of these types are not covered by statute, and the yields from the ephedrine may be helpful in establishing "drug-quantity" guidelines for my sentencing.

Currently, under federal laws, once an individual has been found guilty of manufacturing any amount of any drug, he may then be sentenced according to any amounts that can be established by a light standard of proof. The sentencing in all federal drug offenses (except simple possession) is governed by the drug weights involved. In my case, where no drug seizure was made, the drug quantity involved is estimated based on the lab seized or the established amount of precursor seized.

In my case, Killion fingered me as having ordered mini-thins from two distribution companies (T&M and Olympus), and stated that I was in possession of a methcathinone lab. The first thing the Kansas DEA did was call and subpoena the companies' records, which showed that 39,000 mini-thins had been shipped to my parents' and sister's addresses (both living next door to me). Then the DEA simply awaited the next order and shipment.

I was found guilty, based largely on the DEA's surveillance of me picking up 3,000 mini-thins at the post office. Soon, after agents followed me to my sister's home, my wife was observed going to buy a can of Red Devil lye and delivering it to me. A warrant was sought, and seven hours later the DEA rushed in to find me in a detached garage on my sister's property, where two five-gallon paint buckets containing toluene paint thinner were discovered, along with some Batman drinking glasses (one with a trace amount of chrome salt), a jug of HCl, a container of water mixed with sodium hydroxide, and one broken-up, duct-taped 3,000 ml flask.

My point is that on a bust where these white-cross mini-thins are involved, the distribution company's records (and any other records that the law can obtain) are used to establish the amount of precursor that had been involved in the laboratory manufacture, and from this the total quantity of drugs involved is determined, which in turn sets the length of the sentence to be imposed. The same thing has been done with methamphetamine precursors for years, but the yields of cat from ephedrine have not yet been established.

In early 1993, two cases of cat manufacture arose in Marquette, Michigan. One is published: *US vs Baker*, 852 F. Supp. 609 (W.D. Mic 1994). Affirmed on appeal in the Sixth Circuit, this case established that 50% of the weight of the ephedrine pills will reasonably equal the weight of the cat which could be produced from it. This amount was established by a government-employed chemist who testified at the trials in these cases. The prosecutor in my case is using *Baker* to calculate my sentence. I do not feel that 50% is unreasonable, unless I could find proof that it is. I am now arguing that insufficient proof exists that the 39,000 white-crosses were actually dropped into a cooking pot! They were ordered COD over a four-month period in 3,000-lot batches. And white-crosses have been sold for many years themselves as a "speed" on the black markets. The FDA, which is now blasting against these pills, is a good source for this information.

I wrote the FDA for information about white-cross mini-thins being illegally abused and sold as speed and look-alike-speed. The FDA sent me a copy of their proposed laws or regulations on ephedrine under published law, from the *Federal Register*, part 310 & 341, Volume 70, October, 1994 or 1995, I believe (I sent it to my lawyer, so I'm going by memory). It's a section on bronchial medications which are sold over-the-counter (OTC). In the FDA's proposed regulations, they mean to designate all ephedrine, norephedrine, and racephedrine products as prescription drugs. I do not believe that the DEA will succeed in its proposal, but they also propose (and have moved forward considerably) to remove all single-ingredient ephedrine products from OTC sales. They focus on the white-cross thins (which companies are selling now as stimulants) being sold as bronchial-aid products. Well, it's obvious that the distributors of white-crosses that sold them as stimulants before they were outlawed by the FDA have now relabeled them as asthma medicine, and continued their sales. Stimulants were prescription drugs at the time of those former sales, and that's why white-crosses were commonly sold as speed on the black markets years ago. There have been official discussions about outlawing OTC sales of the white-cross bronchial products, as recently as November 14, 1995. Also noticeable is that the FDA has mentioned how the DEA has moved in to dictate distribution-company regulations and notification rules on these single-ingredient ephedrine sales until they can persuade the FDA to outlaw the OTC sales. The DEA recognizes the manufacturing implications, and claims that the FDA is hampering their attempts to control precursor chemicals.

I have also noticed that the feds have added benzaldehyde and nitroethane to their list of "hot" chemicals, under 21 USC 801 in 1995. However, I do recognize that the grocery-store "extract" oil of bitter almond has distillation potential for quantities of benzaldehyde. I am not aware of where the nitroethane is commonly sold. But I like this recipe the best of any in ***Secrets of Methamphetamine Manufacture***, because it is so easy to get the ingredients. I ran across your book when my lawyer gave it to me as part of the government's evidence against me. They never used it at my trial, though.

Secrets of Methamphetamine Manufacture was found in Killion's abandoned car after I had run him off from my Texas home because he was getting too crazy on drugs. He went a short distance towards Kansas before his car conked out, and then he abandoned it, leaving a copy of your book and others to be found by the DEA. When Killion got back to Kansas, he started manufacturing cat again in that state, but then got popped and started fingering a lot of folks. I had to settle for a crappy court-appointed attorney who lost the case. If the 39,000 white-crosses are sufficiently proven to be involved, my sentence will be from 10 to 12½ years without parole. I've been kept in solitary for my protection, due to blindness in one eye (medical), for over a year now, while fighting this case. Please, if you think of something that might help me, let me know. I'd be greatly obliged and most appreciative.

I thought you might be interested in taking a look at the *Baker* case and the *Federal Register*, part 310, at your local law library. They are very interesting. I will send you a copy of them if you like. I'd send them now, but my attorney is currently reviewing them.

Sincerely, a fan and friend,

Mr. X

P.S.: I plan to leave the USA and do my own thing with chemistry in a better country, after I win my appeal and get out!

Posted by Uncle Fester on February 24, 1998 at 22:35:20: A Story for Hive Bee
You asked for a story, so...

I'll tell you a bedtime story, and I'll make it a scary one because I know that the scary ones make you hot.

Once upon a time there was a happy little cooker. He loved to cook, and between consuming the product himself and some dealing, he produced a few pounds of product over the years.

Our happy little cooker thought he was pretty safe because he never kept much of the goodies he cooked around his place, and he always cleaned up his glassware when finished cooking.

But our happy little cooker had been making some tragic mistakes. Our cooker loved his meth, and so did his friends. He ordered lots of ephedrine, and then pseudoephedrine pills, from various mail-order companies. He paid for them with his credit card. This credit card had his real name!

He also bought chemicals with this credit card. Things like iodine and red P. Also a couple of quarts of sassafras oil, $PdCl_2$, ammonium chloride, and formaldehyde.

One day while checking over reports of mail-order pill purchases, the evil narcoswine became interested in our happy little cooker. They decided to check him out. Just the purchase of large amounts of pills is enough to get a search warrant.

In the happy little cooker's house, they found about a gram of stash, and just a little trace of meth on some glassware. Our happy little cooker was now not so happy, but he thought he'd get off lightly because he only had a gram around the house.

Oh, he was so wrong! Using his credit card records, the evil federales were able to reconstruct all of our shell shocked cooker's past activity. That includes the X he made last year!

Our now very unhappy little cooker was portrayed as a menace to society at trial and lost. In the federal system, sentence is based upon how much they can show you made. A very flimsy standard of proof suffices. Using those credit card records, they were able to give our tragic little cooker 15 years in the Big House. He doesn't like it there, but at least now he has gotten rid of that American Express Card.

The end.

Posted by flaskjockey on March 06, 1998 at 21:10:55:

In reply to: A story for Hive Bee, posted by Uncle Fester on February 24, 1998 at 22:35:20:

The state system doesn't care how much or how complete, which is scarier. California has an amusing way of convicting you, called "Intent to Manufacture." All they need now is extracted ephedrine (any amount or trace) OR any reducing agent, even if the agent doesn't reduce ephedrine. These police state bastards can get away with anything. The only solution is to shoot them.

The Telltale Trashcan

by Uncle Fester

Once upon a time in the sleepy, little town of Dullsville, there lived some people who liked to go fast and stay up late. These fast people kind of stood out from the other folks in that sleepy little town, but they thought they were safe because they were a tight-knit group and not in business in a big way. Yes, my friends, these fast people were meth cookers.

Tight-knit and generally careful as they were, these fast people were not safe at all because they were sloppy in one small area of their lives which was about to become very important to them. They were sloppy in what they put out in their trash.

Over time, word of these fast people who stood out from the rest of Dullsville made it to the attention of the dreaded narco swine. A very fast and easy investigative technique for them to use against our heroes was to simply grab the trash which they put out before the garbage man could pick it up.

Oh, the things they found in that trash! Emptied bottles of ephedrine pills, emptied packages of Sudafed, and coffee filters which had been cut to size to fit various filters were all just laying in their trash to be found. A quick walk around the outside of their house revealed the presence of a propane cylinder. The brass fitting on that cylinder had turned green from exposure to the anhydrous ammonia it contained.

It took no time at all for the dreaded narco swine to obtain a search warrant for the fast people's house. They found a little bit of stash meth, and some items which the narco swine described as "a meth lab capable of churning out multi-pound batches of meth, and blow the neighborhood up to boot."

After the narco swine took off their moon suits for the TV cameras, our heroes were led off to jail. There they are now going even slower than the rest of Dullsville. With plenty of time on their hands, they now realize that they should have been more careful with their trash. It's too late for them, but not for you.

The End

PENNYWISE AND POUND FOOLISH
by Uncle Fester

There once was a man named Jack who had the normally virtuous trait of thriftiness. When Jack was able to save a few cents on a small purchase, he felt that he had won a victory and held his head up high as he walked out the store. A penny saved is a penny earned, after all. Jack shopped sales, and Jack cut coupons, but above all Jack loved his "store card."

Jack's fondness for his collection of "store cards" would have done him little harm except for one little detail... Jack liked to cook some crank once in a while, too! He picked up his packages of Sudafed as part of his weekly shopping runs and just tossed them in with all the other stuff he was buying. Jack thought that this would cause less suspicion towards him than if he just showed up at various stores and simply loaded up at the cold medicine counter. On this point Jack was right, but his penury was setting a trap for him because at each checkout he flashed his "store card" to get his few penny discounts.

One day, Jack's girlfriend had all of his stinginess that she could take. The final straw was the anniversary gift he brought home from the rummage sale. She stormed out of the house, and vowed to make Jack's life a living hell.

Jack's former girlfriend knew all about Jack's meth hobby, and she ran to the police. After she told her tale, the police did some checking to gather evidence for a search and bust. When they checked Jack's garbage, it was simply pristine, as he rarely threw anything out. Watching and tailing Jack yielded them nothing either. He wasn't a "dealer" and lived a frugal and spartan life.

Finally the narco swine stumbled upon Jack's Achille's Heel... his addiction to "store cards." You see, whenever that little card is flashed at the checkout, the entire purchase is recorded in the files of the central scrutinizers with your name attached. Once there, it could linger for centuries...

When Jack's "store card" records were checked, what a gold mine they stumbled into! Box after box of "cold medicines," bottles of ephedrine pills at the gas station, along with cans of solvents at the hardware store, acids and drain openers at other stores. It was a gold mine for them, and the shaft for Jack!

Now Jack is whiling away some time at a place where he doesn't have to pay for his meals or housing. Thanks to those "store card" records, they were able to string together years of cooking and claim that he made outrageous amounts of go-go powder. They were proud to give him more time than his cellmate Bubba who stabbed three guys because he didn't like the way they had their hats screwed on. Welcome to the War on Drugs.

The End

Chapter Twenty Five
Legitimate Uses of Some Chemicals

Acetic Anhydride — commonly used in the chemical industry, especially for making dyes. ✶✶✶

Benzene and ether — common solvents, but they are sometimes used for free basing coke. ✶✶

Formic acid — used for taxidermy and tanning leather. ✶

Hydrochloric acid and sulfuric acid — the two most common mineral acids, with too many uses to list. When buying them, underground chemists say they want them for electroplating. ✶

You can pick up these two at the hardware store. Hydrochloric will often be labeled as muriatic acid, and will generally be 30% HCl. This is good enough for most uses. I found an industrial grade concentrated sulfuric at my local hardware store in the plumbing section. It was a product called Liquid Fire by Amazing Products. It sells for around $6 a pint, and is used as drain opener. This is plenty good enough for dripping on salt to make HCl gas.

Methylamine — used in photography, as an additive to racing fuels, and as an ingredient in rocket fuel and tanning solutions. ✶✶✶✶

Phenylacetic acid — used in perfume to produce the smell of honey, and added to the nutrient broth of penicillin mold to increase the yield of penicillin. ✶✶✶✶

Platinum and Raney nickel — catalysts used in all hydrogenations. ✶✶

Pyridine — a common but expensive solvent and reagent. ✶✶

All other List I chemicals ✶✶✶✶
All other List II chemicals ✶✶✶

One should obtain toluene and acetone at the hardware store in the paint section. There they carry zero stars, so long as you don't buy so much at one place that you get them wondering.

✶ Least suspicious to purchase
✶✶✶✶ Most suspicious to purchase

Chapter Twenty Six
Web Sites

Learning about clandestine chemistry is a lot of fun, as you well know after finishing this book. It's also a lot of fun to follow along with conversations and "posts" that people put on the Internet. There are a few web sites exclusively devoted to such conversations and postings. They are great places to drop in on and spend hours keeping up with the latest news or reading about clandestine processes and equipment.

Before you eagerly dive into the Internet, let me give you a few necessary caveats. To start with, go ahead and read, but keep your mouth shut. If you are doing any cooking, you certainly don't want to draw attention to yourself by posting questions or mentioning any results you are getting. These bulletin boards require registration before you are allowed to post. The heat trolls on these web sites, either by asking questions that giving an answer to would involve you in their "conspiracy," or by checking out the e-mail addresses of posters who seem like they are actively cooking. By keeping quiet and just reading, no one will know you are there.

Caveat number two is a corollary to the first one. Since no one who is actively cooking would be stupid enough to wave around their e-mail address for everyone to see, the people doing the posting will fall into two classes: retired cookers who still love to talk about the excitement of their glory days and relate war stories, and simple students of the field, ranging from rank amateurs to the fairly advanced. The problem is to discern the two classes, and separate the wheat from the chaff in the huge amount of material posted on these boards.

Making this differentiation is a difficult task for the beginner. The second class tends to be skilled at sophistry, and in the process they give birth to misconceptions that carry onward in time with the tenacity of urban legends. Let me give you some help in this sifting process; a good source posting on the net will cite references for the readers to follow up on rather than just making claims. This is the same style your Uncle uses in this book and all my other ones. It is the only legitimate style to use when making posts as well. Suspect sophistry and underlying agendas when references aren't mentioned during chemical discussions.

Now that you have been sufficiently warned, let me pass along two web sites where one can read about the world of clandestine chemistry at great length. Site number one is a site I like to visit and pass along cooking tips when I'm not busy writing books, raising my kids, working my day job, or running a publishing company. It can be found at www.wetdreams.ws. The site is commonly referred to as The Zonez.

Another site with a long history and a really extensive archive which can be accessed through their search engine is called The Hive. It can be found at www.the-hive.ws. Several boards can be found at both sites dealing with various aspects of clandestine chemistry, and the latest news in the field. Enjoy, and keep your mouth shut if you are cooking!